Touching and Being Touched

Touching and Being Touched

Kinesthesia and Empathy in Dance and Movement

Edited by
Gabriele Brandstetter, Gerko Egert and
Sabine Zubarik

DE GRUYTER

ISBN 978-3-11-029186-5
e-ISBN 978-3-11-029204-6

Library of Congress Cataloging-in-Publication Data
A CIP catalog record for this book has been applied for at the Library of Congress.

Bibliographic information published by the Deutsche Nationalbibliothek
The Deutsche Nationalbibliothek lists this publication in the Deutsche Nationalbibliografie; detailed bibliographic data are available in the Internet at http://dnb.dnb.de.

Cover image by Jean-Michel Cima: Herses (une lente introduction),
Choreography: Boris Charmatz, Le Quartz, Brest, France, 1997
Printing: Hubert & Co. GmbH & Co. KG, Göttingen
♾ Printed on acid-free paper
Printed in Germany

www.degruyter.com

Table of Contents

Introduction

Touching and Being Touched

Motion, Emotion, and Modes of Contact

"Rühren und Berühren," "Touching and being touched" – this complex semantic field will be the thematic focus of the collection of essays presented in this volume.[1] As a form of both haptic and affective contact, the act of touching is not merely a physical gesture or a pragmatic action. Touch constitutes a complex form of connection in which the relationship between the one touching and the one being touched is almost never clearly defined.

The essays collected in this volume discuss acts of touching and being touched from a twofold perspective: as the movement and touching of the body and as a form of being moved and touched emotionally. What kinds of relationships do we establish with one another through touching? How can spectators be touched by a dance or music performance or when watching a film? In what way are these forms and practices of touch related to the evocation of feeling? And how can we describe the contact and communication between dancers, or between performers and their audience?

Touch is a fundamental element of dance. The (time) forms and contact zones of physical touch are a means of expression of both the dancers' self-reference and their interaction. Liberties and limits, options and taboos of touch give us insight into the *aesthesis* of different forms of dance, their dynamics and communicative structure, as well as into the production and regulation of affects. However the configurations of touch in the interaction of touching and being touched are as different and varied as the practices of dance themselves. They reflect historical and cultural differences in the understanding of how touch happens, which feelings it produces and conveys, and which esthetic concepts and communicative rules it expresses. An essential form of touch in dance is the seizing of the partner's hands.[2] The form in which this happens reveals a lot about the contact rules of proximity and distance in a given society, the reverential gesture of seizing the partner's hands and bowing in courtly dance, or the different ways of forming a circle or chain by seizing each other's hands – as they can be observed in folk as well as in stage dances. Different rules of touch apply to artistic dance for example 19th-century classical ballet, where

1 The project "Rühren und Berühren – *movere* im Tanz" (Affecting and Touching – *movere* in Dance) was directed by Prof. Dr. Gabriele Brandstetter at Freie Universität Berlin (2008–2012), within the scope of the cluster of excellence *Languages of Emotion*.

2 Cf. Nordera, Martina, "Prendre par la main," in: Marie Glon/Isabelle Launay (ed.), *Histoires de gestes. Ouvrage collectif sous la direction de Marie Glon et Isabelle Launay*, Paris 2012, 165–180.

the touch between the dancers forms part of the esthetics of the performance: ease, the illusion of weightlessness, and the floating elegance of movement constitute the ideals of ballet. In order to achieve these effects, specific techniques of movement are employed which control the body's balance and orientation. With relation to touch, this entails for example minimal floor contact, the ballerina's toe dance. The contact of male and female dancers in the *pas de deux* is dominated by forms of touch that convey the impression of light, floating movements when holding, supporting, or lifting the partner. An interesting tension between the haptic interaction of the dancers and the emotional being touched of the spectators may derive from the esthetic principle of dissimulation: the relatively mechanical forms of contact in the *pas de deux* conceal their technical nature in the masterly game of seemingly "light," fleeting and affective gestures of touch, thus conveying a "feeling" that appeals to the spectators' empathy.

Not only courtly dance and ballet, but also the different practices of *social dance*, culturally and socially varied, use numerous gestures of touch. Argentine tango in particular cannot be conceived of without touch – it begins with an embrace that everything that follows depends on. Due to the strong improvisational character of tango, touch has to be understood as a nonverbal form of transmission and communication of the leading dancer and the partner who follows. Through the different forms of touch, the partners communicate with each other, kinesthetically aligning all parameters of their movement: dynamics, direction, intensity, orientation of the body axes, and floor contact.

As these seem to be mostly prosaic, physical techniques, we may ask ourselves why Argentine tango in particular is considered to be a highly emotional dance and the quintessential expression of passion and erotic desire. At which point and how exactly do emotions come into play? Are the emotions that spectators perceive even relevant to the dancers, or are the dancers confronted with completely different feelings, less "encompassing" and "psycho-logical," and concerning instead the "smaller" physical realities?[3] It seems that a pleasant embrace will inevitably produce positive feelings such as harmony, sympathy, and joy, as these are conveyed automatically with the touch itself. Feelings of oppression, anger, self-doubt, or power can develop when touch is more of a hindrance than a help to the dancers' communication, i.e., when it is not well-coordinated

3 The above-mentioned research project "Berühren und Rühren: *movere* im Tanz" (cf. n. 1) dealt with these matters, looking at various aspects of the problem. The researchers conducted individual and group interviews as well as questionnaire surveys, and organized experimental workshops that they observed and participated in, in order to analyze different states of mind (e.g., grief), states of transmission and perception (e.g., flow), and situations of touch (e.g., closeness, embrace).

with the partner or the movement of the couple as a whole. Judgments such as "too close" or "too tight" are generated subjectively, connotations of an emotional state of mind that cannot be separated from the specific physical situations of the dance partners.[4] Beyond the actual relationship of the couple, the importance of touching and being touched in tango is also relevant in the social setting of the dance, the *milonga*. The admittance, execution, and display of touch are socially encoded and regulated and undergo specific socio-cultural differentiation, depending on the given setting and context.[5]

One dance practice, whose name already points to touch, is Contact Improvisation. Developed in the 1970s by the Judson Dance Theater, it draws on the concepts of postmodern dance. The communication between two or more bodies derives from the various possibilities of contact. Touch is generated predominantly through the transmission – giving and taking – of weight. Thus "touch" does not mainly refer to an active form of touching as a haptic experience. It rather results from a play with complex transformations of balance, using gravity, momentum, energy, and inaction. These factors determine the duration, quality, and contact zones of touch. Several contributions to this volume discuss the connection between feeling and a "politics of empathy" in Contact Improvisation,[6] as well as the relation between "kinesthetic awareness," touch, and the principle of "sharing" established in this dance practice.[7]

In contemporary dance performances, we can also observe different forms of touching and being touched: they arise, among others, from the (affective) relationship of the dancers, from medial constellations, or from the specific use of language and movement.[8]

Contemporary choreographers keep drawing on elements of Contact Improvisation. Sasha Waltz, Meg Stuart, Boris Charmatz, and others came in

---

4 Cf. Sabine Zubarik's essay "'Touch Me If You Can': The Practice of Close Embrace as a Facilitator of Kinesthetic Empathy in Argentine Tango" in this volume, 275–291.

5 Cf. Sonia Abadi's essay "Dancing Tango: The Weave of an Emotional, Corporal and Social Network" in this volume, 309–318.

6 Cf. Ann Cooper Albright's essay "Feeling In and Out: Contact Improvisation and the Politics of Empathy" in this volume, 263–273.

7 Cf. Gabriele Brandstetter's essay "'Listening': Kinesthetic Awareness in Contemporary Dance" in this volume, 163–179.

8 For these forms of touch cf. Gerko Egert's essay "Movements of Touch in MAYBE FOREVER" in this volume, 63–82; for the medial constellations cf. the essays "*The Fault Lines* of Touching" by Krassimira Kruschkova and "Cold Burn (Teion Yakedo): On Touch in Contemporary Japanese Performance" by Katherine Mezur in this volume, 105–111 and 83–104. For the relation between language and movement cf. Erin Manning and Brian Massumi's essay "Just Like That: William Forsythe – Between Movement and Language" in this volume, 35–62.

contact with this dance form in the early days of their professional training, and they adopt its principles and ideas for their own stage work today. In Meg Stuart's pieces, for example, the individual movements of the dancers, which often appear to be isolated and scattered, are frequently interrupted by touch for example in the interaction of passive and active bodies in her early "Disfigure Study" (1991), in the beginning of "No One is Watching" (1995) that strongly reminds of Contact Improvisation, or in the extreme contrast of separation and proximity in "VIOLET" (2011). These instances of touch are almost never characterized by intimate communication. Stuart frequently incorporates elements of everyday relations into touch that (quite differently from Contact Improvisation) are dominated by miscommunication, inequality, a-symmetry, and a-synchronicity.[9] Thus she critically challenges the utopian dream of an egalitarian community, free of power and violence, desire and emotions, as it is developed in the performances of Contact Improvisation. Torn between the wish for contact and the knowledge of a necessary failure of touch, the characters on stage are neither isolated individuals nor members of a specific community. And although these relationships are strongly characterized by affective dynamics and forms of desire, they do not reveal any distinct, individual, definable emotions. Just like the dancers, the spectators are touched, yet in a different way.

Both the different examples mentioned above and the essays collected in this volume illustrate that touch is linked with emotions in multiple ways. Gestures performed by the dancers, such as blows, embraces, and caresses, are often interpreted by the spectators as being the expression or cause of an inner movement, and they convey something of the one touching to the one who is being touched.

But can the messages always be clearly deciphered, and can forms of touch be "read" as signs that express emotions? May we assume that touch delivers messages that (merely) need to be decoded and understood? Or does the execution of touch initiate and produce these feelings in the first place? Can the connection between emotion and touch even be understood temporally, as a se-

9 "Most Contact dances are about mutual exchange and democratic goals. In real life I saw that relationships are not equal, that people are very rarely in sync with each other and if they are it is only for precious moments. For the most part there are power plays, manipulation, and demands or expectations that are granted or denied. There is always someone weaker and more vulnerable in an exchange. I wanted to bring these ideas in. To do this, it was necessary to set up partnering situations where the roles were not fluid but fixed and each action had to have consequences." (Meg Stuart in: Stuart, Meg: *Damaged Goods: Are We Here Yet?*, ed. Jeroen Peeters, Dijon 2010, 52).

quence of cause and effect, or is the concurrence of both the specific feature that has such a strong impact in dance?

Two key concepts characterize the different forms of and relations between touching and being touched in dance: kinesthesia and empathy. The results of our research and the different contributions to this volume (deriving from interdisciplinary contexts of dance, arts, philosophy, and psychology) suggest that *touch* – the sense that connects "sensing" and the emotional "feeling"/being affected – is closely related to processes of kinesthetic perception and empathy.

In recent times, dance studies have drawn on the findings of neuroscientific research on kinesthesia. In this context, however, kinesthesia is not merely understood as the sensual perception of the position and movement of the body in relation to its surroundings. Scholars are more interested in questions relating to the "Kinesthetic Communication in Dance."[10] How do we communicate, and how do we touch through kinesthetic processes? This does not only happen in a relation of movement that affects the sensory (self-)perception for example in a slight or more vigorous touch. The same is true of distant relations of movement, in those imaginative processes of transfer that Dee Reynolds calls "kinesthetic imagination."[11] In which way does "touch" happen between "kinaesthetic and proprioceptive bodies"[12] and the spectators of dance performances, and could it be described as emotional or kinesthetic responses?[13] How can we describe these processes with the concepts of philosophical phenomenology?[14] And in which ways does the "kinesthetic" not only refer to the "energetic" principle and the quality of touch, but also to formations of cultural modes of kinesthesia? Part of the experience of touch – in its two respects – is the complete concentration on feeling the movement for example by blocking out the visual dimension: closing one's eyes and engaging in "blind movement." In the practice of dance, this element of kinesthesia is used for purposes of sensitization:

---

10 Smyth, Mary M., "Kinesthetic Communication in Dance," in: *Dance Research Journal* 16.2 (1984): 19–82.

11 Reynolds, Dee, *Rhythmic Subjects: Uses of Energy in the Dances of Mary Wigman, Martha Graham and Merce Cunningham*, Hampshire 2007. For the relation between kinesthesia and empathy, contagion and affect, cf. Reynolds' essay "Empathy, Contagion and Affect: The Role of Kinesthesia in Watching Dance" in this volume, 211–231.

12 Cf. Mark Paterson's essay "On 'Inner Touch' and the Moving Body: Aisthêsis, Kinaesthesis, and Aesthetics" in this volume, 115–131.

13 Cf. Corinne Jola's essay "Do You Feel the Same Way Too?" in this volume, 181–210.

14 Cf. Hilge Landweer's essay "Choreographies With and Without a Choreographer: Intuitive and Intentional Corporeal Interactions" in this volume, 133–161.

"Moving with eyes closed."[15] What does this mean for the transmission in movement, in terms of empathy?[16] In a history of this "way of feeling," this corresponds to an "inner sense" ("innerer Sinn"), "an inner touch by which we perceive ourselves." In this context, Daniel Heller-Roazen quotes Aetius (*Placita Philosophorum* 4.8.7),[17] analyzing the relation between "coenesthesis"/kinesthesia and the conveyance of this basic feeling as a "community spirit." Touching and being touched do not describe clearly distinguishable, definable forms of sensual perception, but a "sphere of possibility."[18]

Neuroscientists conducting research on empathy, on the other hand, provide us with clear definitions. Tania Singer writes:

> There is empathy if: (i) one is in an affective state; (ii) this state is isomorphic to another person's affective state; (iii) this state is elicited by the observation or imagination of another person's affective state; (iv) one knows that the other person is the source of one's own affective state.[19]

Neuroscientific experiments on empathy (a "social feeling") specify those cognitive and emotional components that derive from the assumption that empathy is a form of "emotional contagion." Yet how can we describe empathy if it does not entail the (empathic) feeling of affects for example joy or fear, as they are revealed in the facial expressions of the other,[20] but the *movements* of (kinesthetically affected) touching? Based on Theodor Lipps' theory of empathy, and further developed in modern-day philosophical studies, the concept of empathy has become a decisive factor of esthetic appeal in film studies.[21]

How can we understand the movement between dancers, their touching (and distance), and the emotional effects on the spectator as "Choreographing Empathy"?[22] Isn't the dual principle that links touch, kinesthesia, and empathy – the

15 This was the title of a workshop organized by choreographer DD Dorvillier at the "Impulstanz" festival, Vienna 2011.
16 Cf. Kornilia Chatzimasoura's essay "Lost and Found in Interpretation: Senses and Emotions in Contexts of Argentine Tango" in this volume, 293–308.
17 Quoted in Heller-Roazen, Daniel, *The Inner Touch: Archeology of a Sensation*, New York 2007, 11.
18 Cf. Niklaus Largier's essay "Figure, Plasticity, Affect" in this volume, 23–34.
19 de Vignemont, Frédérique/Singer, Tania, "The Empathic Brian: How, When and Why," in: *Trends in Cognitive Science* 10.10 (2006): 435–441, 435.
20 Cf. Ekman, Paul, *Emotions revealed: Understanding Faces and Feelings*, London 2003.
21 Cf. Curtis, Robin/Koch, Gertrud, *Einfühlung: Zu Geschichte und Gegenwart eines ästhetischen Prinzips*, Munich 2009. Cf. Robin Curtis' essay "Is the Movement of the Filmic Image a Sign of Vitality?" in this volume, 249–262.
22 Cf. Foster, Susan L.: *Choreographing Empathy: Kinesthesia in Performance*, London 2011.

movement "towards"[23] – open and at the same time redirected by a third factor,[24] marking cultural and social differences: communication and its observation? At this intersection, touch, tactile gestures, and affective perceptions,[25] the forms of touching and being touched between art and (kinesthetic) experience differ from each other and connect at the same time.[26] The responsivity – being touched in touching – also comprises non-touching (*noli me tangere*) and touching before the actual (tactile) touch.

The following essays discuss a range of questions relating to touch, kinesthesia, and empathy, as well as their specific entanglement. How can a form of empathy be understood that cannot be reduced to the visual representation and mirroring of the other, but also operates in kinesthetic modes? And how does touch configure itself in dance and art, while also, immanently, relating to forms of being touched (of the other dancers, performers, and actors, but also the spectators)?

We would like to thank all those who have supported and contributed to the publication of this volume. Firstly, we would like to express our gratitude to all those who helped organize the conference documented in this volume: Angela Alves, Maia Traine, and Inka Paul for their support in the different stages of planning and realization, Bettina Sluzalek and Hannes Frey of *Radialsystem V* for their cooperation and hospitality, as well as Nina Diezemann of the research cluster *Languages of Emotion* who was in charge of public relations matters and organized the evening events. As to the editorial work on this volume, we particularly wish to thank Sabine Frost for the processing and formatting of the texts, Christine Henschel for her thorough copy-editing work, and the translators for their indispensable contributions. A huge thank you finally goes to the cluster of excellence *Languages of Emotion* at Freie Universität Berlin and to the *Deutsche Forschungsgemeinschaft* for their financial support of both the conference and the publication of this volume.

Gabriele Brandstetter, Gerko Egert, and Sabine Zubarik
Translated by Christine Henschel

---

23 Cf. Manning, Erin, *Politics of Touch: Sense, Movement, Sovereignty*, Minneapolis, 2007.
24 Cf. Breithaupt, Fritz, *Kulturen der Empathie*, Frankfurt/M. 2009.
25 Cf. Jean-Luc Nancy's opening contribution to this volume, "Rühren, Berühren, Aufruhr: Stirring, Stirring up, Uprising," 13–21.
26 Cf. Marie-Luise Angerer's essay "Affective Modulations in Politics, Theory and Art" in this volume, 235–248.

References

Breithaupt, Fritz, *Kulturen der Empathie*, Frankfurt/M. 2009.

Curtis, Robin/Koch, Gertrud, *Einfühlung: Zu Geschichte und Gegenwart eines ästhetischen Prinzips*, Munich 2009.

de Vignemont, Frédérique/Singer, Tania, "The Empathic Brian: How, When and Why," in: *Trends in Cognitive Science* 10.10 (2006): 435–441.

Ekman, Paul, *Emotions revealed: Understanding Faces and Feelings*, London 2003.

Foster, Susan L., *Choreographing Empathy: Kinesthesia in Performance*, London 2011.

Heller-Roazen, Daniel, *Der innere Sinn: Archäologie eines Gefühls*, Frankfurt/M. 2012.

Manning, Erin, *Politics of Touch: Sense, Movement, Sovereignty*, Minneapolis 2007.

Nordera, Martina, "Prendre par la main," in: Marie Glon/Isabelle Launay (ed.), *Histoires de gestes: Ouvrage collectif sous la direction de Marie Glon et Isabelle Launay*, Paris 2012, 165–180.

Reynolds, Dee, *Rhythmic Subjects: Uses of Energy in the Dances of Mary Wigman, Martha Graham and Merce Cunningham*, Hampshire 2007.

Smyth, Mary M., "Kinesthetic Communication in Dance," in: *Dance Research Journal* 16.2 (1984): 19–82.

Stuart, Meg, *Damaged Goods: Are We Here Yet?*, ed. Jeroen Peeters, Dijon 2010.

I Touch

Jean-Luc Nancy

Rühren, Berühren, Aufruhr

Stirring, Stirring up, Uprising

Rühren, Berühren, Aufruhr: German makes it possible to gather three notions in the semantic family of *ruhr*, which can be matched in French with *le bouger, l'agiter, le toucher* and *le soulèvement* [and in English with "moving", "agitating", "touching" and "uproar" or "uprising"]; and each of these terms can be understood with its own array of possible values. "Moving" and "agitating" convey some physical as well as moral senses, as do "touching" and "uprising". The latter term, for its part, gives its moral value a socio-political orientation.

This semantic family pertains to movement, which is neither a local movement (a displacement, *Bewegung* in German) nor a movement of transformation (metamorphosis, *Verwandlung* in German; for example, generation and corruption, increase and decrease), but instead a movement one may most suitably call "emotion", a term modalizing "motion", the closest transcription of the Latin *motus*, from the verb *movere*, which also gave us "motility" and "emoting".

"Touching", *toucher* in French, seems rather foreign to the semantics of movement, while in German it manifestly belongs to it. "Touching", "tact" or "contact" seem to involve an order that is more static than dynamic. Granted that one has to move in order to touch and one has to "come into contact", as we say, but touching itself appears to designate a state rather than a movement, and contact brings to mind firmly joining something rather than engaging in a mobile process.

And yet French is also well acquainted with the mobile, motorized and dynamic value of touch: it crops up when we speak of a person or a work that "touches" us; when we allude to the special "touch" of a pianist, the "touch" of a painter or the "touch" of divine grace.

Touching shakes up and sets in motion. As soon as I move my body closer to another body (even an inert body made of wood, stone or metal), I displace the other (be it infinitesimally), and the other sets me apart, holding me up in a way. Touching acts and reacts at the same time. Touching attracts and rejects. Touching propels and repels – impulsion and repulsion, rhythm of the outside and the inside, of ingestion and rejection, of the clean and the unclean.

Touching begins when two bodies distance themselves and set themselves apart. Children come out of the belly and in turn become the belly, which can swallow and spit something out again. They take the mother's breast or finger into their mouth. The first touch is a suckling. And yes, the child sucks and inhales the milk that nurtures him. But suction is more than that: it latches the mouth on to the body of the other. It establishes or re-establishes a contact thanks to which the roles are reversed: in turn, the child who was contained becomes the container of the body that contained him. But he does not enclose it; on the contrary, he holds it before him at the same time. The movement of sucking lips keeps redoing the alternation of proximity and distance, penetration and escape, which presided over the downward drop from the belly to the way out of the body of this new body, which is ready at last to separate.

By separating, the new body conquers this new possibility whose sketchy design is all it knew: the possibility of intercourse and contact. The sketch was mostly auditory, and hearing itself was diffracted according to the whole prism of the small body immersed in the resonating liquid with which the other body had been enfolding him. The sounds of this body, its heart and its guts, and the sounds of the outside world would touch his ears, his closed eyes, his nostrils and his whole infused skin at the same time. Yet "touching" would be saying too much. Each possible sensation was still diluted in a dim way, a permanent, quasi-permeable exchange between the outside and the inside as well as between the various entryways of the body. Touching would be saying too much and yet it is there already: it is the first *rühren*, the first flow and flotation, and something mindless swings with it, which has not yet got to be born.

At birth he will separate. But he will remain this thing, this he or she floating within an element or world in which everything relates to everything, everything strains towards everything and pulls away from everything – but this time according to the numerous scansions of all the insides/outsides of separate bodies.

Only a separate body on its own is able to touch. On its own, it can also entirely separate its touch from its other senses, which is to say, constitute as an autonomous sense that nonetheless traverses all the senses, differentiating itself within them while distinguishing itself as a kind of common reason. Reason or passion, impulsion, motion.

Where he was immersion, an all-encompassing floating and enfolding, in the relative lack of distinction between his outside and his inside, tending to confusion in the common sway of two bodies, sucking his own finger, there he detaches himself, and once outside, finds himself on his own before this outside. That is to say, he is no longer inside the inside, within this immanence. He transcends, in the most proper sense of the word: going beyond being in itself.

His mobility leaves the suspension behind, the almost complete absence of gravity and the viscous indifference of directions. His mobility becomes true movement as other bodies are further off. Far from seeking a return into immanence and immersion, on the contrary his gestures assert his distinction, a separation that is neither a privation nor an amputation of anything at all. The separation is the opening of the intercourse. The intercourse is not seeking to restore a lack of distinction: it celebrates the distinction; it announces a meeting, which precisely is contact.

In truth, contact begins when the child begins to occupy all the space in which he had been floating. He comes to touch the walls, and his movement becomes that of the slow reversal which puts him in a position to exit, to let himself be pushed from the inside and sucked out by the outside – which is to say, decidedly this time, to embrace the order of an inside/outside. Touching the limits of the vase and the belly, he becomes himself the same as another wall as well as a wave ready to insinuate itself and glide between the labia that are going to spread apart for him. This sliding gives its final form to the passage from flotation to friction, from immanence to transcendence, and by opening the vulva, it also opens up all the other breaches that his separation is going to stir up, through which contact will become properly possible, itself both breach and adherence, intimate extimity.

Contact does not cancel the separation – on the contrary. All the varieties of (metaphysical or psychological) logic that posit the primordial attraction of a supposedly lost unity and the need to accept a forced separation (sectioning, sexuation, the plurality of senses, guises and aspects) belong to the logic of a kind of monotheism or morbid mono-ideism. They are patho-logic without being a logic of *pathos* or *dunamis tou pathein*, which is the ability potentially to receive and the capacity to be affected. And affection is first of all passion and the movement of passion, a passion whose very nature is "to touch": to be touched, touching in its turn, self-touching one another with the touch coming from outside, from the one who touches me and the one through which I am touching.

To be affected does not mean that an erstwhile subject comes to receive an affection, in a given circumstance. How could one receive without being capable of doing so? But this capacity itself has to be capacity in the most proper sense: the power to receive. Being capable of receiving implies that one is already receiving and is affectable. To be affected requires that one has been affectable and that one has always already been this way. That is why there has always already existed an outside and always already an opening towards it – always already an opening straining towards the outside, a desire for the outside such that it can only have been preceded by the outside – otherwise it could not have a desire for it. The subject is neither prior nor exterior in relation to the outside; it is (if we choose to speak of the subject, that is) much rather, *sujet au dehors*, as we can put it in French [and in English: "subject to the outside"]: that is, prone or subject to the other, subject to the other's touch. In this vase, the amnion, where the homunculus is bathing, it is this touch of the outside that is set off as a flotation turning into a friction.

When the vase lets its contents pour out, water spills and the little one emerges, drenched. His entire body (whole and detached for the first time) bears the humid imprint that becomes his skin, melting into the outline of his skin but making this skin always able to receive the outside, and to be bathed and lulled, rocked by the swells of the outside.

Thus touching, at first and always, is this rocking, this flotation and this friction, which suckling then repeats, reigniting and playing up again the desire to feel oneself touched and touching, the desire to probe and feel oneself in contact with the outside. More even than "in" contact: being contact oneself. My whole being is contact. My whole being is touched/touching. Which is to say, also open to the outside, open with all its orifices – my ears, eyes, mouth and nostrils, not to mention all the channels of ingestion and digestion, like those of my humours, sweats and sexual juices. As for the skin, it sets out to extend an envelope around these openings, these entries-and-exits, which locates and specifies them while at the same time developing for itself this ability to be affected and to have a desire for this. Each sense specializes affection in a specific setup (seeing, hearing, smelling and tasting), but the skin never fails to connect these schemes while it avoids blurring them. The skin that envelops is itself nothing but the development of the entire circumscription of the body (of its entire detachment), brought into play and generally exposed. French allows the wordplay *ex-peausition.* In German one could make up *Aus-sein/ Haut-sein* (while in English, one could say *exhibi-skin*).

But most importantly in any language, the exhibition or exposition, the *Ausstellen* that the body is, as well as its *Ausdehnen* (Freud wrote, "Psyche ist ausgedehnt", "Psyche is extended"), does not consist in spreading things out fixedly as on to the picture railing of an art gallery. On the contrary, we can only understand this exposition as a permanent movement, an undulation, an unfurling and refolding, an air of continual change in contact with all the other bodies – that is, in contact with all that comes near and all that is approached.

Since Aristotle, we have known that the identity of the sensible and the sentient in feeling or sensing (which is thus a being-felt or being-sensed as well), similar to the identity of the thinkable and thought in the act of thinking, implies at the focus of the sensation (in vision, hearing, olfaction, gustation and contact) a kind of compenetration of both, in the act and as this act. The act of sensation, which is to say, *energeia* according to Aristotle's concept of act, constitutes actual effectiveness, the event being produced by the sensation. The sentient soul is itself sensible and for this reason feels itself as it feels. And nowhere is this clearer (nowhere is it more sensed) than in the sense of touch: not the eye, nor the ear, nose or mouth, feel themselves feeling with the intensity and precision that the skin feels. Images, sounds and tastes remain distinct in some way from the sentient organ, even if they fully occupy it. This may be true for touching as well in the event that I represent the touched substance to myself (by thinking, "this fabric is rough" or "this skin feels fresh"). But, although it is in truth impossible to determine these things, one could say that representation is less immediate when we touch. With the other senses, representation comes on more quickly, though in different ways, depending on each instance (an image is concurrent with its vision; a melody and a timbre with their audition, but a little less so; a flavour is still less concurrent with the sense of taste; and a smell is even more removed from the sense of smell, which brings it into the order of the sense of touch).[1]

Now, one can only understand the identity of touching and touched as the identity of a movement, a motion and an emotion – precisely because it is not the identity of a representation and the thing that it represents. The fresh skin that I mentioned is not that at first (a "fresh skin") in the action of my hand as it touches it. But it "is" my gesture; it is my hand and my hand comes across because my hand is its contact or its caress (in reality, except for a medical touch, there is not any contact with skin that is exempt from a potential caress). Motion and emotion – they themselves a single thing – envelop the act, the sensitive *energeia*. And this *energeia* is nothing but the effectiveness of the contact, which is the effectiveness of a coming-towards and a reception-of-something – a double, exchangeable quality: I come towards the skin that welcomes and receives me; my skin welcomes the coming that the welcoming reception of the other is for itself. The coming-towards-one-another meets them at a point of quasi-confusion. Nor is this very point itself a motionless dot: it only is the

---

1 Within the scope of this conference, I cannot possibly linger here. But we should refine the differential analysis of the senses. All the senses partake of touch in that they all carry the possibility that the sentient is identical with the sensed. But everyone modulates this identity as he or she pleases, and the difference in modulations is part and parcel of sensibility, which cannot be one and general. If it were so, it would only have an abstract "sensible", a concept of the sensible. But in each setup, it brings to the fore both a (visual, auditory and so on) sensibility *and* the plurality of sensibilities, that is, the fact that they refer the ones to the others in a differentiated and inexhaustible manner. One will thus be able to go through all of them according to the model of the sense of touch, *and* to differentiate all of them by referring one to one or another among them. Now, to keep this short, taste and smell have a different involvement with the inside/outside relation: for them, absorption and assimilation occur in a very particular mode. Moreover, taste mostly has to do with a sensible that has some (solid or liquid) consistency, and olfaction mostly a fleeting, gaseous or airborne sensible. In each instance the relation differs from the extension and movement proper to the sense of touch. Each time they are special *touches*, whose full meaning varies from one body to the other: so and so "has a nose", as we say in French, whereas someone else "has an ear", and so on. This "having" is a way of touching/being touched.

image of a "point"; its reality is motion and emotion, something that budged, traction and attraction, and at the same time, continual variation and fluctuation. At the same time it is a vibration, a palpitation from one to the other, a swaying to and fro, and for this reason it is an "identity" that does not identify although it gathers the one and the other and lets their presences partake in a shared coming.

Such is the *rühren* of touching. The liquid movement of a rhythm, a swell, an undertow of ex-istence, which is "being outside" because the "outside" is the inflection, the curve and the scansion of this flotation and friction ruling the way in which my body is steeped among all the bodies and my skin alongside all the other skins.

The movement of touching, therefore, is not what another term designates: *tasten* in German, *tâter* in French (which also has *palper*) – which might seem more appropriate. Indeed, *tâter* [feeling, handling, patting down, examining by touch] designates a cognitive behaviour, not an affective one. One extends feelers in this way [on *tâte*] in order to recognize or appreciate a surface or a consistency, or to assess a density or a flexibility. But that is not the way we stroke or caress. Touching is stroking; it is essentially a caress, that is, the desire and the pleasure to come as close as one can to a skin – be it human, animal, textile or mineral, and so on – and to engage this proximity (namely, this superlative and most extreme approach) to play off two skins grappling with each other.

This play takes up again the rhythm that is essentially and originally the game of the inside/outside – perhaps the only game there is, if all playing consists in taking and leaving an area, in opening breaches, filling and voiding places, boxes and schedules. Touching is a movement in that it is rhythmic and not in that it might consist in the process or the steps of an exploration. "Approaching", here, does not amount to coming into some area, and "contact" does not amount to establishing an exchange (of signs or signals, information, objects or services). "Approaching" rates as the superlative movement of proximity, never cancelled out in an identity since what is "closest" needs to remain at a distance, an infinitesimal distance, so as to be what it is. "Contact" amounts to a shake-up of sensitivity (also superlative and extreme), that is, of the very thing that makes up the capacity to receive and be touched. (*Rühren* can also have the meaning of playing an instrument. In French, one used to say *toucher le piano* [to touch the piano]: it always has to do with waking, shaking and animating.)

This play and this rhythm of the tactile are the *Rühren* [stirring] of a desire – perhaps desire itself: indeed, is there even a desire that has not a desire to touch, if touching gives the pleasure of desire itself, the pleasure of desire straining towards the proximity of intercourse, since the intercourse is nothing but the sharing of an inside and an outside brought into play.

The first and once most widespread meaning of *ruhr* has been the sense of pleasure in love and sex. The rhythmic movement and overflow, gushing forth not only juices but whole bodies spilling against and into one another, and one setting itself off from the other only to take it up and move in again together in succeeding waves that they become thanks to one another – this movement does not belong to any process of action or cognition (and let's not mention here the finality of generation – which opens up another body; since this pleasure is without finality, or only has an end in the ending that suspends it on to itself in the overflow, exhausting it and opening it beyond itself).

It is understandable that the most widespread of taboos corresponds to the sense of touch. Freud noticed this, as can any anthropologist and ethnologist. In our own cultures, we know the importance of this taboo very well: while it no longer has an ostensibly sacred component, it watches with jealous care over all the conditions, permissions and modalities of bodies coming into contact. We do know exactly up to which point we have permission to touch as little as the hand of the other, not to mention the rest of the body; and up to which point and how permissible it is to embrace and to kiss, to hug and to stroke.

The knowledge we have of the degree to which touching involves one's being is solid – and consequently of the way in which being is strictly impossible to dissociate from intercourse. There is not – absolutely not – "being" followed by a relation. There is "to be", the verb whose act and transitivity are formed in connection(s) and only in this manner. Descartes' "I am" does not contravene this necessity, no more than Kant's "I" (or Fichte's, or Husserl's), or Heidegger's *jemein*. Each "I" is – and is nothing but – the act of its intercourse with the world, stretched out towards it, towards what one calls the "other" and whose otherness reveals itself in the touch or as touch.

Now, the touch (which did not just accidentally give its name to a mode of divine intervention in the soul), as motion and emotion of the other, consists both of the tip of a contact and of the reception or accepting of its pressure and its reach. It grazes and pricks, punctures or seizes, indiscernibly and in a vibration where it immediately withdraws. The touch itself is its own trace already, which is to say, it wipes itself away as a mark or the point of its imprint while propagating its effects of motion and emotion.

Saint John of the Cross speaks of "touches and impressions of the divine union towards which I am directing the soul", and he specifies:

> the activity of the understanding can very easily disturb and destroy this delicate knowledge, which is a sweet supernatural intelligence, which no natural faculty can reach or

> comprehend otherwise than by the way of recipient, and never by that of agent. No effort, therefore, should be made, lest the understanding should fashion something of itself...[2]

An understanding that is not "active" is a passive understanding, the taste of a flavour, the feeling of a touch. Mystics have no monopoly on these metaphors – if they are metaphors. The "touch" of a painter, the "touch" of a pianist (and the keys of a piano, known as *touches* in French, and why not a computer touch pad), an added "touch" of something (a touch of fantasy, a touch of melancholy), to a decor or a text, as well as an erotic "touch",[3] share the same quality, both on the dot and vibratory.

Now, it is never a matter of metaphors. It is always a matter of sensible reality, thus material and vibratory. When the soul quivers, it really is quivering, just as one may speak of water about to boil. What we commonly call the "soul" is in fact nothing other than the waking and welcoming – both mixed – of motion/emotion. The soul is the touched body – vibrating, receptive and responsive. Its response is the sharing out of the touch, its rise towards it. The body rises up, as the German word *Aufruhr* suggests, designating, as I pointed out, a socio-political uprising. Indeed, there is some insurrection (and sometimes some erection) in the motions of touching. A body rises up against its own enclosure, against being locked up within itself, and against its own entropy. It rebels against its death. It may not be impossible that the very touch of death triggers a last surrection, heart-rending and abandoned at the same time.

Whether it is about the coming of another (him or her), or the absolute alteration of death, it is the body that opens up and extends outside. It is its pure act: just as Aristotle's prime mover is pure *energeia* in which there is no remaining "potency" (*dunamis*), that is, nothing to expect, nothing that could come from the outside, likewise, when I am *touched*, I have nothing to expect: the touch is all act, in its mobile, vibratory and sudden action. And as for Aristotle's god, this act is accompanied by its own excess, which is its pleasure, the climax that is the flower or spark of the act – sun or darkness, always an abyss, and towards it, always the stirring *ruhr* of *berühren* gushes forth or spills.

Translated by Christine Irizarry

---

2 Saint John of the Cross, *The Ascent of Mount Carmel*, London 1906, 237 f. and 278.

3 "Touche moi", "tu te touches", in an absolute sense, are erotic words in French (and in English: "touch me", "you are touching yourself").

Reference

Saint John of the Cross, *The Ascent of Mount Carmel*, London 1906.

This text was first published in: *Scores 1: Touché* (2011). We would like to thank Tanzquartier Wien and the editors of *Score*, Walter Heun, Krassimira Kruschkova, Sandra Noeth, and Martin Obermayr, for their kind permission to reproduce the text.

Niklaus Largier

Figure, Plasticity, Affect

Dance is the production of figures in space and time. It is an art of figuration that affects and touches both the performer and the perceiver. As I will argue here, the 18th-century philosopher, writer, and theologian Johann Gottfried Herder is maybe the thinker who can help us best to understand this production of figures and their effects in touch and being touched, the topic of this essay on dance. Herder is most invested in a very specific understanding of touch. An exquisite writer of essays, sometimes confused and confusing, as often lucid and surprising in his observations, he is an author who brings to our attention that touch is not just one sense among the five senses, but – as I will show here – the very sensation that stands at the origin of all sensuous perception and affect. What Herder analyzes is the intimate connection between touch and sculpture and, to speak in broader terms, between touch and figure or, even more broadly, touch and the art of figuration with its sensual, emotional, and cognitive implications.

In a short sketch "Von der Bildhauerkunst fürs Gefühl" ("On the art of sculpture in view of touch"), written in 1769, Herder observes:

> When our dance will change, when it isn't anymore the play of little feet and the forced bowing of chests and hips that we call grace, then the body will be able to speak. Now everything is forced into one thing: little feet and their movements, the shielded hips, the forced grace – it is all there is, and it is all gothic.
>
> [Wenn unser Tanz nicht mehr *ein Spiel kleiner Füsse* und ein *gezwungnes Beugen der Brust und Taille*, die man Grazie nennt, seyn wird: so wird der Körper besser reden. Jetzt hängt alles zusammen – *kleine Füsse* und ihr kleines Spiel: *gepanzerte Taille* und *gezwungne Grazie:* das ist Hauptstück, der ganze Tanz also Gothisch!][1]

What "gothic" ("gothisch") means here is an obscure, confused, and contrived form of expression.[2] Instead of this "forced" form of articulation, Herder is searching for a way in which the body starts to speak and, maybe better, turn into a figure of evocation, interaction, expression, perception, and experience. Herder raises the very same issue and question with regard to sculpture when he writes at the beginning of the same essay:

---

1 Herder, Johann Gottfried, „<Entwurf (1769)> Von der Bildhauerkunst fürs Gefühl," in: Helmut Pfotenhauer/Peter Sprengel (ed.), *Bibliothek der Kunstliteratur*, vol. 3: *Klassik und Klassizismus*, Frankfurt/M. 1995, 95–107, 101 (my translation, N.L.).

2 Cf. Grimm, Jacob and Wilhelm, *Wörterbuch der deutschen Sprache*, vol. 8, Leipzig 1854–1961, 1002f.

> A statue has to live; her flesh has to come alive, her face has to speak. We have to believe that we are touching her, that she becomes warm under our hands. We have to see her standing before us and to feel that she is speaking to us.
>
> [Eine Statue muß *leben:* ihr Fleisch muß sich beleben: ihr Gesicht und Mine sprechen. Wir müssen sie anzutasten glauben und fühlen, daß sie sich unter unsern Händen erwärmt. Wir müssen sie vor uns stehen sehen, und fühlen, daß sie zu uns spricht.][3]

Herder continues:

> Thus the eye moves into the tips of our fingers and we forget the cold surface and the comparison to painting. We don't see anymore. Instead we feel tender skin, a round knee, the pleasant cheek, the beautiful breast and hip, in short, the beauty of the body.
>
> [Das Auge tritt in die Spitzen der Finger: wir vergessen die kalte Oberfläche, als obs Malerei wäre: wir sehen nicht, wir fühlen die zarte Haut, das runde Knie, die sanfte Wange, die schöne Brust, die weiche Hüfte – den schönen Umriß des Körpers.][4]

But how, then, should we imagine the body in expression, the dance that isn't caught up in the "gothic" small movements of the "little feet" and the "forced grace," the dance that takes shape in sculptural figures in order to produce an experience that is unique and at the same time the basis for all other experience, including emotions, sensation, and cognition? How do we have to think of the plasticity of the figure and the figure in motion as the very basis of this experience? And how do we have to understand this very experience in terms of touch and being touched – an experience, so to speak, below the horizon of meaning and allegorical image? I will try to follow the track of Herder's thought a bit in order to understand these very questions and Herder's answers to them.

I

When Herder thinks about touch and being touched he does so in connection with what he calls the "ground of the soul."[5] An instance where he addresses the issue can be found in his reflections and elaborations on aesthetics, a body of texts that is possibly more relevant for his anthropology than for what we would call his aesthetics nowadays. For Herder the "ground of the soul" is

---

3 Herder, „Von der Bildhauerkunst," 95 (my translation, N.L.).

4 Ibid. (my translation, N.L.).

5 I am following here my own more comprehensive analysis in: Largier, Niklaus, "The Plasticity of the Soul: Mystical Darkness, Touch, and Aesthetic Experience," in: *Modern Language Notes* 125 (2010): 536–551.

a realm of darkness and possibility from which all perception emerges. It is, however, not just a negative counterpart of our discursive and intellectual activity and knowledge. It constitutes a complex intersection of sensuality, affect, perception, feeling, and imagination, and it thus forms the basis for what in his view is an alternative and comprehensive way of cognition where the dualism of soul and body, spirit and matter is being replaced by a playful intersection of sensual and affective experience.

Herder takes the notion of a "ground of the soul" from a long mystical tradition. While the medieval and early modern Christian mystics point to the open receptivity of the ground of the soul, to its aspects of sheer virtuality, to its function as a pure mirror of god and creatures, and to its unity with god, Herder turns it into the place of perception of the whole cosmos ("des ganzen Weltalls") in the embodied soul. When Herder mentions that all mankind, that each human being is born as a "dunkel fühlende Auster" ("a darkly feeling and perceiving oyster") he emphasizes the naturalistic aspect of his notion of a ground of the soul even more strongly. He understands it as a primary plane of perception and as a horizon of emergence for all possible worlds of experience in the relation between the ground of the soul and the figures in front of it that make this realm of endless experiential possibilities take shape.[6] So, when one reads "ground of the soul," one should not think of an esoteric 'ground,' not even of a soul, but rather of a horizon, an empty and dark stage full of unknown possibilities where all kinds of objects, men and monsters, things, affects, terror, and joy can emerge in an interaction between figures that affect each other and the viewer. It is not just a place or a defined way of perception, instead we might want to speak, as I did before, of a plane of perception, a horizon of possibilities that emerges together with the figures we encounter and that takes shape in correspondence with these figures.

This notion of the soul resonates with elements of contemporary philosophy of nature and aspects of French sensualism Herder is drawing on. However, he never entirely covers up the mystical provenience of the trope of the ground of the soul. Thus, he writes in his *Metakritik der sogenannten Transzendental-Ästhetik*, that "nothingness in nothingness" reveals itself "in the dark ground of the soul of the mystics" ("das Nichts im Nichts [...] offenbaret sich im dunklen Seelengrund der Mystiker") in order to become the "only possible condition of the revelation of the sensual as well as the comprehensible world" ("um die einzigmögliche Bedingung aller Offenbarung des sinnlichen sowohl als verständlichen

6 Herder, Johann Gottfried, *Werke in zehn Bänden*, vol. 8, ed. Hans Dietrich Irmscher, Frankfurt/M. 1991, 363.

Weltalls zu werden").[7] In other words, the ground of the soul is not only the natural and generic ground of our becoming and of all our abilities that Herder grasps in the image of the oyster, it is also the "dark ground" of all our feelings, and it is "the dark ground of the soul of the mystics," the nothingness and the abyss that we find as the point of origin in his radical theory of perception, sensation, and emotion. In other words, the "ground of the soul" is the very possibility of being touched by a multiplicity of figures that arouse it.

It is on this basis that Herder figures and reconfigures the sense of touch in a new way, namely as a correlation between the "darkness" of the ground of the soul and its very expression in "Gefühl," that is, in the way the soul takes shape in perception. This emphasis on touch in Herder's thought has been analyzed time and again.[8] In my focus on the notion of the ground of the soul I am adding one aspect that has been largely ignored, namely that in rehabilitating touch Herder rehabilitates a quasi-empirical notion of mystical "darkness" and of an experimental art of figuration that corresponds to it. To put it more concisely: with the concept of a "dark ground of the soul," Herder introduces both a notion of an empty plane of possible perception and of the effect of figures on it. This is the way in which he conceptualizes aesthetic experience – and it is a way of understanding aesthetic experience that is most productive when we think about the intersection of touch and being touched in the encounter with dance as an art of figuration. Dance, we might say, is an exemplary art of figuration that plays and experiments with figures and their effects in the realm of endless forms of possible perception. This play is – on the level of the experimental deployment of figures – devoid of all meaning, devoid of allegorical sense. It is nothing else than the production of figures and effects in the tactile realm of sensual, emotional, and cognitive perception.

One of the most beautiful passages that let us understand what Herder means in his turn towards touch and being touched can be found in his attempt to portray and characterize what happens to a viewer who observes a sculpture. The example he chooses is – not surprisingly – the art historian Winckelmann, "der tiefsinnige Betrachter am Vatikanischen Apollo" ("the profound observer of

---

7 Ibid.

8 For a discussion of the significance of touch in the 18th century, see Mülder-Bach, Inka, *Im Zeichen Pygmalions: Das Modell der Statue und die Entdeckung der 'Darstellung' im 18. Jahrhundert*, Munich 1998; Zeuch, Ulrike, *Umkehr der Sinneshierarchie: Herder und die Aufwertung des Tastsinns seit der frühen Neuzeit*, Tübingen 2000; Braungart, Georg, *Leibhafter Sinn: Der andere Diskurs der Moderne*, Tübingen 1995; Binczek, Natalie, *Kontakt: Der Tastsinn in Texten der Aufklärung*, Tübingen 2007.

the Vatican Apollo"). In the following passage Herder describes how Winckelmann observes the sculpture of Apollo. "Didn't Winckelmann," Herder asks,

> have to destroy the very properties of the object of his contemplation, i.e., all the properties and qualities that make our visual perception, namely color, surface, angle of view? And didn't he have to give his eyes seemingly a new sense, namely feeling ['Gefühl' means 'feeling' and 'touch', N.L.]? And wasn't the sense that he used [namely the eye, N.L.] a reduction and replacement of a more originary sense? A sense that was open to the proper or essential effect of art? Now let's suppose he [the viewer] reaches this sense? His engaged viewing, his obvious touching and feeling transport the beautiful in its form and shape into the imagination, giving it to the imagination in an embodied way [...] and thus the effect happens: the beautiful body is perceived as a body [...] the imagination becomes active and engaged, and the imagination speaks as if she felt and touched: speaks of pleasant fullness, of admirable curves, of beautiful roundness, of soft aspects, of the moving marble that is brought to life by the touching hand. These are all feelings! Why feelings? And why feelings that are *not* metaphors? They are experiences. The eye that started to collect them wasn't eye anymore when it collected; the eye turned into hand [...] perception into immediate touch. The imagination speaks feelings and touch.
>
> [hatte er [der 'tiefsinnige Betrachter am Vatikanischen Apollo'] also nicht nöthig, eben die Eigenschaft seines Gegenstandes zu zerstören, die das Wesen der Augenvorstellung ist, Fläche, Farbe, Winkel des Anscheins? und musste er sich nicht mit dem Auge gleichsam einen neuen Sinn geben, das Gefühl? und war der Sinn, den er anwandte, nicht also eine Verkürzung, die Stellvertretung eines ursprünglichern Sinnes? für den die eigentliche Würkung der Kunst möglich war? – Nun setzt, er erreiche diesen. Sein vielverändertes Umherschauen, oder sein sichtliches Umhertasten gebe seiner Einbildungskraft das ganze Schöne in Form und Bildung gleichsam einverleibet über: die Täuschung ist geschehen: der schöne Körper, als Körper wird empfunden – sehet! nun empöret sich die Phantasie, und spricht – als ob sie tastete und fühlte: spricht von sanfter Fülle, von prächtiger Wölbung, von schöner Rotundität, von weicher Erhebung, von dem sich regenden, unter der fühlenden Hand belebten Marmor. Lauter Gefühle! warum lauter Gefühle? und warum Gefühle, die keine blosse Metaphern sind? Sie sind Erfahrungen. Das Auge, das sie sammelte, war sammlend nicht Auge mehr; es ward Hand: der Sonnenstral Stab in die Ferne, das Anschauen unmittelbare Betastung: die Phantasie spricht lauter Gefühle!][9]

What Herder describes in this difficult and exciting passage – where he follows Winckelmann's gaze in order to deconstruct it – is the transformation of the eye into an organ of touch. As Herder points out, this happens literally, not metaphorically. He doesn't describe ways in which the eye, as we sometimes say, touches a surface or a body, using the notion of touch in order to express the intensity of seeing. Instead, he describes an event, an encounter with plastic figures and figuration below the threshold of meaning and signification. He de-

9 Herder, Johann Gottfried, *Sämtliche Werke*, ed. Bernhard Suphan, Berlin 1877–1913, reprint Hildesheim 1967/68, vol. 8, 126 (my translation, N.L.).

scribes a transformation that happens while we see and observe, a transformation in which the eye loses its specific character when the viewer perceives the effects of the figure or sculpture. It loses the determinate character of looking at form, surface, color, shape, and perspective. In short, it is not looking at an object or a representation anymore. In this transformation the eye turns into touch and the visual field is 'reduced,' so to speak, to the tactile, which is the name of the very ground of possibility of all sensation and feeling. This ground takes shape in aesthetic experience, that is, in the experience of the effects of the figure. Touch here means simultaneously to touch and to be touched. And touch refers to nothing else than the fact that figures, sculptures, moving figures make sensations and emotions emerge from a dark and open plane of perception. It is touch that moves along the figures we encounter and that in moving along gets absorbed in the very perception. In other words, touch is the very possibility of getting absorbed into the effects of the figures that art deploys. This makes the experience of touch and being touched one of an experiential replication of the figures in the soul and of a communication between the soul and the object. In doing so, it also turns into an engagement of the imagination and the affects that immediately arise from touch. In this process, the visual, characterized by its tendency to make the experience into an experience of an object, disappears and turns into touch, the possibility to experience itself.

The very paradigm of visuality, often seen as the highest faculty among the senses, is here presented as "Verkürzung," as an abbreviation and representation, as a reduced version of sense experience, which replaces a more fundamental and a more comprehensive sense and sensation ("Stellvertretung [...] eines ursprünglicheren Sinnes"). We see this more comprehensive form of sensation appear where the language of the viewer – in the quoted example Winckelmann observing a statue, analyzed by Herder in a wonderful phenomenological way – betrays that he has already moved on, that he has left the visual behind, that in his experience we are forced to move on from the visual to the realm of touch and feeling, touching and being touched. It is important to take Herder's argument seriously that we are not dealing with metaphorical language here. He does not present us with an allegorical or metaphorical explanation of the act of viewing and the ways in which emotions and affects are involved in it and in which they are seduced by figures. What happens instead is to be described as a kind of transfiguration of seeing into touch whereby the ground of the soul, better, the plane of possible perception, is being aroused in the experience of figures and where something new takes shape in this experience – a new object, a new soul, and a new sphere of experience.

To be more concrete: the eye sees determined, limited, defined shapes, forms, and colors. Touch however appears as a moment of liberation from this determination, as a reminiscence of something decidedly more open (maybe comparable to "das Offene" in Rilke's *Duineser Elegien*),[10] and as the level of experience where object and subject, figure and viewer merge in a slow temporal and necessarily spatial process. Touch is the name for this temporal and spatial involvement and abandonment where all our possibilities of perception are formed and informed by the figures that emerge before our eyes.

II

In reading closely the passages where Herder discusses this transfiguration, among them the passage quoted above, we might want to speak of an "eroticizing processuality" ("eine erotisierende Prozessualität")[11] since touch implies an erotic in its spatio-temporal unfolding. In the moments of perception that Herder describes, in the reduction of the visual to touch, the experience moves the ground to take shape through the animation of senses, emotions, and imagination. This happens in a temporal process where the plane of perception forms the open stage of such an experience in a close and erotically arousing encounter with the things that come to us in space. Dance, more generally figures and figures in motion produce scenes of reciprocity, thus the notion of eroticism, where – as Herder writes – "the 'warm, creating hand,' the organ of artistic creation, turns into the organ of the very reception of the work of art" ("wo 'die warme, schaffende Hand', das Organ der künstlerischen Produktion, zum Organ seiner Rezeption [wird]").[12] In other words: the plane of perception, its receptivity and openness, is informed by the effects of the very plasticity of the figures and takes shape in a 'tactile' exchange with it. Or, to put it differently, the 'tactile,' touching and being touched, is the name for an experience that deconstructs the objectivity of vision and its plane of reification in order to allow the figures to fully deploy their effects upon the soul, to act upon the soul, and to unfold its potential of experience in a phenomenology of rhetorical, figural effects. Through this process the soul is increasingly absorbed in the very experience of the effects that take place in this exchange between the viewer and the figures in motion – to the point that the life of the soul cannot be distinguished anymore from the liveliness that the figures deploy.

---

10 Rilke, Rainer Maria, *Werke: Kommentierte Ausgabe in vier Bänden*, vol. II: *Gedichte 1912 bis 1926*, ed. Manfred Engel/Ulrich Fülleborn, Frankfurt/M. 1996, 224.

11 Mülder-Bach, *Im Zeichen Pygmalions*, 73.

12 Ibid.

What happens is neither in the figures alone nor in the soul alone, it is in the movement that embraces both and that produces the absorption in sensation, affect, and cognition.

I want to emphasize that both aspects, the "eroticizing processuality" and the tactile reciprocity can and should be linked to the notion of the ground of the soul and the empty plane of perception that Herder presupposes. Thus, he writes in another passage about a classical Greek statue:

> A spirit has poured itself into the statue, guided the hand of the artist, held the work of art and made it one. Whoever stood next to the famous statue of the hermaphrodite (to use a most difficult example) and who didn't feel how Bacchic dream and hermaphroditism reign in every trembling and curvature of the body, in everything he touches and doesn't touch – to whomever, who didn't feel the torture of sweet thoughts and voluptuousness like a fire that inflames his body, who didn't perceive involuntarily the resonances and consonances of this play of chords, neither my words nor any words can explain this. It is the very nature of a sculpture that she speaks to us as an act since she is all mankind and living body. It is the character of a sculpture that she captivates and pierces our being, and that she arouses the chords of human compassion.
>
> [Ein Geist hat sich über die Statue ergossen, hielt die Hand des Künstlers, dass auch das Werk hielt, und Eins ward. Wer (um so gleich ein Schwerstes anzuführen), wer je am berühmten Hermaphroditen stand und nicht fühlte, wie in jeder Schwingung und Biegung des Körpers, in allem, wo er berührt und nicht berühret, Bacchischer Traum und Hermaphroditismus herrschet, wie er auf einer Folter süsser Gedanken und Wollust schwebt, die ihm, wie ein gelindes Feuer, durch seinen ganzen Körper dringet – wer dies nicht fühlte und in sich gleichsam unwillkürlich den Nach- oder Mitklang desselben Saitenspiels wahrnahm; dem können meine nicht und keine Worte es erklären. Eben das ist das so ungemein Sichere und Veste bei einer Bildsäule, dass, weil sie Mensch und ganz durchlebter Körper ist, sie als Tat zu uns spricht, uns festhält und durchdringend unser Wesen, das ganze Saitenspiel Menschlicher Mitempfindung wecket.][13]

It is this moment of being touched that makes all the elements of perception emerge – the erotic, the desire, the imagination, the emotions, and the fundamentally relational nature of the soul. It does so beyond words, in darkness and silence below the horizon of meaning. What happens is a "Saitenspiel" ("play of chords"), an arousal of the soul and of all its potential to resonate through figures in space and time, letting it take shape in an experience that we call aesthetic. This form of experience is, as the quote clearly illustrates, not a form of cognition unified through an act of rational intervention but through the "spirit" that dwells in this darkness and that emerges from it in the very moments of aesthetic experience. This "spirit" doesn't transcend the ex-

---

13 Herder, Johann Gottfried, *Werke*, vol. 2, ed. Wolfgang Pross, Munich 1984, 517.

perience, however, but it forms the plane of immanence that connects the figure and the viewer in the very act of perception.

III

As I have shown, Herder considers the sense of touch to be the highest of all our senses. This has often been understood in terms of a rehabilitation of touch in a historical context that predominantly and quite conventionally puts vision at the top of a hierarchy of the senses. While Herder indeed rehabilitates the sense of touch he actually does more than that. He turns touch not only into a specific sense that is the foundation of all sensation. He also turns it into the basis for a new understanding of figuration and figures in movement. Touch is the very sphere of possibility for the temporal deployment of all effects of sensation and aesthetic experience in the soul. It is the sphere of an experimental use of figures and figuration before or below the moment when they turn into allegories.

In aesthetic experience, one might summarize Herder's position, all sensation turns into touch – and touch encompasses virtually all sensation. Or, to put it differently, touch is the aspect of sensation and of the life of the soul that constitutes the realm of possibility, even more, the possibility to become everything in sensation, to take shape in sensation, and to experience what I am calling the plasticity of the soul in its relations with itself, things, works of art, and the world. Touch is the place where the soul takes shape in an experience that is made possible by the very openness, darkness, and potentiality that has its foundation in the "ground of the soul." Touch is, to put it in the words of the title of this text, the name of both, the plasticity of figures out there and of the plasticity of the soul, of its very ability to experience aesthetically, and of the possibility to dwell in pleasure or terror under the influence of sensation. In other words, touch is intimately related to the arts of figuration, the erotic plasticity of figuration that appears in sculpture and dance and that produces this very sensation.

Herder's understanding of touch and its relation to the plasticity of sculptures also implies an important distinction between figure and ground – and, as I want to point out, a doubling of that distinction that takes the soul of the viewer as a kind of living mirror of the process of figuration. The plasticity of the sculpture, the plasticity of the figure distinguishes it from the ground and from the surroundings. It is only in this distinctiveness that the figure attracts our attention, and what Herder describes in his phenomenology of aesthetic experience is at first nothing else than moments of attraction. These moments emerge in the very encounter with the plasticity of the figure and in the resonance they provoke.

Let me return to Herder's short remarks on dance that I quoted at the beginning of my presentation. There, he discusses dance not in terms of a formally defined social practice, e. g., the baroque courtly dance. Instead, for him, dance can be, can become that very art of figuration, an exemplary cultivation of the realm of possibilities that we call soul: affective possibilities, sensual possibilities, and cognitive possibilities. It can be the very art of figuration that is free of allegory and metaphor in order to be nothing else than a play with figures and their effects in and on our empty plane of perception. Dance is thus a realm of figuration before meaning emerges, before we identify pictures, images, and allegories. It is a realm of figural play that Herder identifies with touch, or better, that he identifies with a transition from the visual to the touch – a transition that escapes both intention and meaning, and a transition that time and again opens a field of emergence.

Thus, figures in motion elicit possibilities and they produce – to use Deleuzian terminology – perception-events, sensation-events, affect-events in time and space. Depending on the specific situation they can indeed represent something, they can acquire meaning, they can give birth to allegorical meanings. Strictly speaking, however, as plastic figures they don't represent anything and they don't have meaning. They are just what they are, figures in movement on a background of darkness, a screen of possibilities and interactions, a screen or plane of touch and being touched. Thus, they don't represent events but they constitute events. In other words, figures in movement, bodies in movement are not media and they are not embodied forms of meaning and ideas – although this can indeed be the case as well. For our purpose, however, it is important to be aware of this difference. They are places, spatial moments and movements, configurations, transfigurations, and disfigurations where touch unfolds. And ever again we become aware of the fact that the visual, that our gaze turns into touch, into sensation, emotion, cognition. In other words, the visual, which is there without doubt, loses its dominant aspect, we don't look at the dancer and his or her figurations as an object – but we are getting caught up in moments of absorption that happen before our eyes but express themselves on a plane of perception where touch and being touched converge in perceptions, sensations, affects.

IV

I want to conclude this essay with a short coda. The notion of the "ground of the soul" returns – with explicit references to Meister Eckhart – innumerable times in the writings and interviews of John Cage. In his Darmstadt lectures on "Composition as Process" (1958), Cage constructs an intimate relationship between the "function of the performer," his acts of "giving form," and "the Ground of Meister Eckhart."[14] In bringing together the acts of "giving form" with a specific notion of "indeterminacy," Cage also views the ground of the soul as a horizon of possibility, a realm of darkness, "from which all impermanencies flow and to which they return."[15] He does so in a way that is experimental, transposing the relation between figuration and the ground of the soul into a realm of aesthetic creation and experience where "the outcome" can never be "foreseen." In turning away, Cage writes, "from himself and his ego-sense of separation from other beings and things," both the performer and the listener or viewer "faces the Ground of Meister Eckhart" and "does what is to be done, not splitting his mind in two, not separating it from his body, which is kept ready for direct and instantaneous contact with his instrument."[16] What results from this cannot be fixed in time. It is the product of a correlation between an art of figuration and a horizon of possibility where an experimental play explores forms of cognition, perception, and affect. According to Cage, this experimental play is to be found also in dance, namely in the relation between "clarity of rhythmic structure" and "grace," which form "a duality":

> Together they have a relation like that of body and soul. Clarity is cold, mathematical, inhuman, but basic and earthy. Grace is warm, incalculable, human, opposed to clarity, and like the air. Grace is not here used to mean prettiness; it is used to mean the play with and against the clarity of the rhythmic structure. The two are always together in the best works of the time arts, endlessly, and life-givingly, opposed to each other.[17]

In this quote Cage reflects, from an entirely different point of view, on the question of grace that Herder raised two hundred years earlier. Like Herder, Cage makes use of the trope of the ground of the soul in order to think about dance and grace. Beyond "forced grace" and "prettiness," dance turns into the art of figuration that evokes – "live-givingly" for Cage and as a body that "speaks" for Herder – not specific meanings but a realm of possibilities of sen-

14 Cage, John, *Silence*, Hanover 1961, 35.
15 Ibid., 39.
16 Ibid.
17 Ibid., 92.

sation, affect, and cognition. For both the *chiffre* that points to this realm is the "ground of the soul," and for both it is dance that produces the experimental figures that make these possibilities emerge from the darkness of the ground in events of perception, of touch and being touched.

References

Binczek, Natalie, Kontakt: Der Tastsinn in Texten der Aufklärung, Tübingen 2007.

Braungart, Georg, *Leibhafter Sinn: Der andere Diskurs der Moderne*, Tübingen 1995.

Cage, John, *Silence*, Hanover 1961.

Grimm, Jacob and Wilhelm, *Wörterbuch der deutschen Sprache*, Leipzig 1854–1961.

Herder, Johann Gottfried, "<Entwurf (1769)> Von der Bildhauerkunst fürs Gefühl," in: Helmut Pfotenhauer/Peter Sprengel (ed.), *Bibliothek der Kunstliteratur*, vol. 3: *Klassik und Klassizismus*, Frankfurt/M. 1995, 95–107.

Herder, Johann Gottfried, *Sämtliche Werke*, ed. Bernhard Suphan, Berlin 1877–1913, reprint Hildesheim 1967/68.

Herder, Johann Gottfried, *Werke in zehn Bänden*, ed. Hans Dietrich Irmscher, Frankfurt/M. 1991.

Herder, Johann Gottfried, *Werke*, ed. Wolfgang Pross, Munich 1984.

Largier, Niklaus, "The Plasticity of the Soul: Mystical Darkness, Touch, and Aesthetic Experience," in: *Modern Language Notes* 125 (2010): 536–551.

Mülder-Bach, Inka, *Im Zeichen Pygmalions: Das Modell der Statue und die Entdeckung der 'Darstellung' im 18. Jahrhundert*, Munich 1998.

Rilke, Rainer Maria, *Werke: Kommentierte Ausgabe in vier Bänden*, vol. II: *Gedichte 1912 bis 1926*, ed. Manfred Engel/Ulrich Fülleborn, Frankfurt/M. 1996.

Zeuch, Ulrike, *Umkehr der Sinneshierarchie: Herder und die Aufwertung des Tastsinns seit der frühen Neuzeit*, Tübingen 2000.

Erin Manning and Brian Massumi

Just Like That

William Forsythe – Between Movement and Language

In 2001, The Forsythe Company created *Woolf Phrase*,[1] a piece conceived from Virginia Woolf's novel *Mrs Dalloway*, generated, as Forsythe says, by "moving around the rhythm of Woolf's language".[2] From its conception, the piece is about movement and language coming together and in relay. Its project: to make a movement phrase of the movement of her phraseology.

The piece begins and ends in a recurrent circling that asks us to encounter how language opens itself to movement, and how movement co-composes with this opening. Refrains return again and again: "just like that," "somehow," "all of a sudden," "PAAF!" as the piece circles the twisting of bodies around the mewling of a seagull, the barking of dogs, the buzzing of bees, the surging of waves. These refrains haunt the piece, and echo beyond it. When we would leave the rehearsals, they would follow us. We would find ourselves re-performing what we had just participated in as spectators. At any moment, we would catch ourselves parsing our actions and comments with a *just like that. PAAF!* We would sound to a hand gesture recurring from the dance, punctuating the passage from this moment to the next.

What is it, in a work of dance, that gives it this capacity to linger, to reactivate itself, contagiously, beyond its own duration? What is it, in the techniques mobilized by the Forsythe Company, that is capable of agitating language to the limit so that it begins to move, movement reciprocating by taking on the inflections of language?

Here, we take up the refrain again, hoping to restage the contagion in writing.

---

1 The descriptions in this essay are based on video documentation of the September 29, 2001 presentation at the Frankfurt Opera House performed by Prue Lang and Richard Siegal; the authors' experience of several of the performances of the 2011 revival run of the piece (November 18–20, 24–27 and December 2–4, 2011, Bockenheimer Depot, Frankfurt); and the authors' studio observations of the rehearsals leading up to the 2011 performance, which included the creation of a new piece, *Whole in the Head*. In the 2011 run, *Woolf Phrase* was performed by three pairs of dancers on different nights, Cyril Baldy/Esther Balfe, Cyril Baldy/Prue Lang, and Roberta Mosca/Tilman O'Donnell. The piece originally premiered on March 15, 2001, at the Frankfurt Opera House.

2 Sulcas, Roslyn, "Woolf Phrase and Bees of the Invisible", in: *New York Times*, http://www.nytimes.com/2001/12/09/arts/dance-using-forms-ingrained-in-ballet-to-help-the-body-move-beyond-it.html (24.1.2013).

Somehow, Seashore

Two dancers. She comes from the left, at the back of the stage, dancing a circular movement, her gestures roll, repetitive, a mewling of a seagull emanating from the movement. Movement, rolling, returning, ending where it rebegins, wrapping around a vocalization.

He comes from the right, stands at the microphone and pauses, allowing her movement to settle into its never-ending rebeginning. And then he speaks: ***Somehow ... music.*** The space fills with an undulating sound wall, catching the mewling of the seagull in its rising. ***Somehow***, he words, ***somehow seashore.*** At words' end, a movement cuts in. He gestures to the left, in the direction of the mewling, musically enwrapped. ***Somehow, somehow, collectively beginning, collecting, falling. Somehow suspended in mid-air. Just like that.*** He gestures again. ***Just like this.*** He falls, tumbling into the recesses of the rising soundspace.[3]

Somehow, music. Forsythe Company composer Thom Willems has created a sound surround. As Forsythe says, Willems gives the performers an "acoustic space to dance."[4] The undulating of the music is the space of the movement. The piece takes place in this rhythmic space, more so than on the stage. The stage is just the launching pad. Everything that will be seen, heard, sounded, spoken, gestured, and evoked, will collect in the rhythmic milieu hovering above, wrapping around.[5] Everything will belong to it, everything will collect in it, swaddled together in its undulations. For everything has rhythm, and rhythms have a way of coming together. They resonate and accrue. Twists and turns and turns of phrase, inflections of movement and of language, will assemble and accrete, to compose this piece. Just like that.

From the first moment, a space of mutual envelopment is set in place. It would be more precise to say that a milieu of mutual envelopment is activated. It takes, it is set, set alight. The set itself, the literal stage space and the things it holds – bodies, costumes, microphones – are but the kindling. The activation is between registers. It is set acoustically as well as linguistically, as speakingly as movingly, as much in sound effect as in the vocalization of words, in gesture as in intonation. What is activated envelops the difference between these registers,

---

3 In what follows, words spoken in the performance of *Woolf Phrase* are indicated in bolded italics. All of these phrases are from Virginia Woolf's *Mrs Dalloway*, with the exception of the apocryphal anecdote about Virginia Woolf and the Dalai Lama (discussed below).

4 Forsythe, William, *Interview with John Tusa. Balletco Magazine*, 2003, http://www.ballet.co.uk/magazines/yr_03/feb03/interview_bbc_forsythe.htm (24.1.2013).

5 On the rhythm and milieu, see "Of the Refrain," in: Gilles Deleuze/Félix Guattari, *A Thousand Plateaus*, Minneapolis 1987, 310–350.

as well as between the events occurring on each, folding into rhythm. ***Waves collect. Overbalance. And fall***, he words.

At words' end, suspended by the rising sound surround in the mid-air of the dance's rhythmic milieu, he falls. What is composed, at this moment in the piece, is the tension between the rising and the falling, language and gesture: their differential coming together. Just like that is not a comparison between the contrastive aspects. It is the difference staged, directly sensed. Just like this – feeling, like at no other moment. *Like* just this moment's felt intensity.

The piece will be made of these differentials, these tensions, these contrastive intensities of the moment, on every register, as what happens on one also rhythmically accrues to the others. The composition will jump from gesture to wording, and back again, forever circling, swaddling feeling. Movement will cut into the wording, speaking will interrupt the moving. At times they will coincide. Cut in, cut out, fold around and together. Rising, falling, resonating, just like this, just like that, disjunct, rolling in difference together.

> "The sparrows fluttering, rising, and falling in jagged fountains were part of the pattern; the white and blue, barred with black branches. Sounds made harmonies ... the spaces between them were as significant as the sounds."[6]

Woolf's in the Middle

"What is of interest," Gilles Deleuze always maintained, "is the milieu, what happens in the middle. It is not by chance that the greatest velocity is in the middle."[7] The middle: between rising and falling. It takes extreme speed, perhaps infinite speed, to pass between rising and falling. It takes extreme speed, at least at the velocity of thought, to pass between language and gesture.

What happens in the middle is that the either–or is held fast together in passing contrast. It is the holding together that is felt, in excess of one or the other. The in-excess of the one or the other is not a both–and. The either–or is taken as such into the passing. The differential is sustained. This is what Deleuze and Guattari term a "disjunctive synthesis."[8]

---

6 Woolf, Virginia, *Mrs Dalloway*, New York 1992, 24. The first part of this phrase is spoken repeatedly by both dancers in the 2011 revival of *Woolf Phrase*.

7 Deleuze, Gilles, "Un manifeste de moins," in: Gilles Deleuze/Carmelo Bene, *Superpositions*, Paris 1979, 87–131, 95.

8 Deleuze, Gilles/Guattari, Félix, *Anti-Oedipus*, Minneapolis 1983, 12–13.

"The middle," Deleuze says, "is not an average, but an excess. It is through the middle that things grow. This was an idea of Virginia Woolf's."[9] The middle is the extremity of disjunctive synthesis, as a growth medium: the milieu of the performance's accruing to itself. The pattern of the passings-through the middle give texture to the rhythmic space of the piece's occurring. Somehow, collectively beginning, collecting difference. Somehow, suspended in disjunctive synthesis.

The dancers' *somehow* is not the somehow of a practical procedure that is hidden or unapparent. It is transparent – like when we say that a story is "somehow sad." The sadness has been directly felt. Just like that, no question. But all the same, there is more. The "somehow" evokes other sadnesses, unspecified. It invokes them into this one, echoing. This feeling resonates with many another, not fully expressed, in all transparency. It overfills itself with their difference – and the contrast of all sadnesses with happiness. ***And now, the same moment, supposedly without happiness.*** He takes her hand. The gesture passes between sadness and happiness, at the speed of a supposition, at the velocity of thought.

There is always an excess of the feeling in the feeling, by which it overflows, or overfills, its designation. In "just like this," the "this" is too full to be designated just one. ***One separation, one, two. One. Two. One, one-two.*** Disjunctive synthesis.

It's like this. Just, like, sad. Supposedly without happiness. Like, sad. Teenage speech patterns get it: "like" does not designate an identity or resemblance. It marks an affective overflow in speech. This feeling is just sad. More than that, it is just like sad is. It refracts all sadness, and its difference from happiness, in the singular feeling this is. The words "somehow" and "just like that" mark the affective tonality of the moment. They mark affect in language. They linguistically gesture to the feeling tone of the moment. Although marked in language, the feeling tone has risen from the middle, in the passing between language and movement.

And between music and voice. And between movement phrases. Disjunctive synthesis upon disjunctive synthesis, passing each other in the middle, crissing, crossing, rolling together into the rhythm of the moment.

The Any-Point of Movement

"Where does the movement start and stop?" (Forsythe 2011–11–10). What manner of phrasing is native to movement, in its disjunctively synthesizable difference from

9 Deleuze, "Un manifeste de moins," 96.

language and from other registers of experience? This is the core question around which Forsythe organizes his company's choreographic process, and to which he never ceases to return. He has returned to it now, in November of 2011, in pre-rehearsal for the upcoming season. A new piece is to be created from scratch, and old ones revived. Included in the revivals is *Woolf Phrase*, which will be performed after a long hiatus. New dancers have joined the company and are being inducted into its movement generation techniques. They likely already know the answer to the question, as would anyone familiar with Forsythe's dance vocabulary and practice: it "starts from any point" ... and stops at any other.[10]

This sounds like a non-answer of indifference: "whatever ..." (to phrase it like a teenager again). But when it is added that "any point within each piece contains the essence of the whole,"[11] it becomes apparent that far from being indifferent, any-point is a most singular point. In movement alone, in its own order, uncontrasted with language, any point is already overfull. The overfullness of movement with itself, on its own plane, Forsythe says, must be *felt* for movement to be made.

"Once you stretch one limb," Forsythe remarks, "all kinds of other places stretch and retract." (Forsythe 2011–11–10). Stretch it again in a different way, and the stretches and retractions shift. Every part of the body is a knot of different potential stretches and retractions radiating from that point as "so many vectors" (Forsythe 2011–11–10). So many lines of movement, potentially passing through each point. Each starting point of movement holds these potential passings-through in itself, together in their difference from each other. The move is less a point than a vectorial gestural nexus: a differential dynamic knot of potential variations on itself. A milieu of movement potential.

The dancers are instructed to "take the movement as far as it will go" (Forsythe 2011–11–10). But if you are attentive to where it is going, you feel it move in more than one direction. When the arm is stretching into a curve, the hip has already joined in throwing the curve. When the arm has finished moving, the hip is still going. Its movement radiates down into a flourish of the leg. A single movement comprises a differential "spray"[12] of movement lines akimbo, now beginning now ending, in the moment differently together. The "any-point" where a

10 Caspersen, Dana, "Decreation: Fragmentation and Continuity", in: Stephen Spier (ed.), *William Forsythe and the Practice of Choreography*", New York 2011, 93–100, 94; ead., "It Starts from Any Point: Bill and the Frankfurt Ballet," in: *Choreography and Dance* 5.3 (2000): 25–40, 33; Forsythe, William, *William Forsythe in Conversation with Zachary Whittenberg*, 2012, http://www.movementresearch.org/criticalcorrespondence/blog/?p=5213 (24.1.2013).

11 Caspersen, "Decreation: Fragmentation and Continuity," 94.

12 Forsythe, *Interview with John Tusa. Balletco Magazine.*

movement starts and stops is actually a spray, a splay, a spread: a point-*field*. It doesn't start at any *point*. Movement is not simply localizable in space.[13]

If the point-field of movement is non-localizable, when a dancer seeks to generate a movement it is not enough to get into position. Position is the gateway to movement. But it is not of its order. Position is of the fixed spatial order. Movement's order is of the always dynamically passing. "A move," Forsythe reminds his dancers, "is not so possessed of a place that you can't find another." "Keep in touch with the motion. Give up your position" (Forsythe 2011–11–8).

The question of dance is: "what is it about a position that made it motion?" (Forsythe 2011–11–10). Answer: the dynamic knot. The differential gestural nexus. Feel the spray, field the nexus, form the movement out of the feeling. Keep in touch with the forming motion, "chain the sensations rather than the positions" (Forsythe 2011–11–10). Giving up your position and getting in touch with the forming feeling of the motion triggers what Forsythe calls an "activation": the field throws a curve amidst a spray (Forsythe 2011–11–10). Movement's order is an ordering of formative feeling activation. Making movement concerns this fielding, from the middle, in the gestural milieu, more fundamentally than any notion of form in the completed sense.

This answer to the question "what is it about a position that makes it a motion?" raises another question. What do we see of what makes a motion? What is predominantly seen is indeed a line of movement: the arm describes a curve. But this visible form of the movement is only a certain salience of the field of motion. It is but the striking, to the eye, of a dominant vector. Form is a simplification of the field occurring between registers, flashing up from their difference to each other. The visible form of a gesture is a distancing of movement from itself. A movement seen is a visual effect of movement's envelopment in a larger differential field of different orders, different registers of experience, different keys of life, rhythmically rolled into the moment, in their difference from each other. Movement in itself, as itself, can only be felt.

The task of the dancer on performance night is to make visible what can only be felt. To make visible what can only be felt is not the same as performing a move. It is not to imitate or reproduce a movement form. "Things happen, but are not performed," Forsythe warns his dancers. "You don't want to perform your work." "Do it without portraying" (Forsythe 2011–11–10). You do it without displaying. You don't perform it in that sense. You just, like, enter into the sensation. Field the point, chain the sensations, and the audience also will feel the

13 The concept of simple location is famously critiqued in Whitehead, Alfred North, *Science and the Modern World*, New York 1967, 49–55, 58.

movement's making, in the flash-form of its visual effect. They will eye your differential gestural nexus at a distance. The activation of your point-field will activate the optical spray of their field of vision (what Gibson calls the optical array[14]). Dynamically, the visual activation will be analogous in its dynamic contours to the movement spray. They will share the same *activation contour*[15] – with a significant difference. Native to the visual system is the tendency to structure its field not at any-point, but at a focal point. What in the gestural nexus was a multivectorial knot now comes into retinal focus. The throwing of the curve becomes a stand-out, stand-alone, visually focused form. The stand-alone form of the movement belongs wholly to vision. But it can belong to vision only because it has come to stand out in this way from the shared activation contour that triggered its appearance. In this shared activation contour, movement and vision dynamically coincide. They are in a state of dynamic superposition across their difference from each other, each in its own manner, on its own register, according to its own order.

The sharing of the activation contour passes through the middle. Of itself, it is in neither register. It is in transit. It is trans-manner, trans-register, trans-order: transductive.[16] What is in trans-manner is amodal: neither in one modality of life nor another, but in the coming to pass of their disjunctive synthesis.[17] In the rhythm of the passing event that is life in this moment. The visual form is not displayed. The movement is not portrayed in it. Life splays into it.

Could not precisely the same be said of the relation between movement and language? Language and vision? The different planes of life come concertedly together in their disjunction, linked eventfully in an analogy that is grounded in their difference from each other, rather than any formal identity or resemblance. The dynamic analogy of amodal superposition is what Walter Benjamin, discussing the relation between language, gesture, and vision, termed a "nonsensuous similarity."[18] Just like that. Just like this. One, two, one, one-two one-two, sprayed multiply across the registers of an experientially splayed event.

---

14 Gibson, James J., *The Ecological Approach to Visual Perception*, Hillsdale 1986.

15 On the concept of the activation contour, see Stern, Daniel, *The Interpersonal World of the Infant*, New York 1985, 57–59, and Massumi, Brian, *Semblance and Event: Activist Philosophy and the Occurrent Arts*, Cambridge 2011, 107–114, 122–125.

16 On transduction see Simondon, Gilbert, *L'individuation à la lumière des notions de forme et d'information*, Grenoble 2005, 31–33, 107–110 and passim, and Combes, Muriel, *Gilbert Simondon and the Philosophy of the Transindividual*, Cambridge 2012, 6–9.

17 On the amodal see Stern, *The Interpersonal World of the Infant*, 47–53, and Massumi, *Semblance and Event*, 17–18, 109–113, 122–123).

18 Benjamin, Walter, "The Doctrine of the Similar," in: *Selected Writings* 2 [1927–1934], Cambridge 2004, 694–698; Massumi, *Semblance and Event*, 105–110, 123–124.

Escape Velocity

Chaining positions and postures is how we experience movement in its average, everyday functioning. We have our routines. The lunch bell sounds, a routine sequence of steps leads us, largely unthinkingly, to the cafeteria. Position after position, posture by posture, we proceed to the lunch-line. We see a vending machine on the way. Unthinkingly, our hand delves into the pocket and feeds the machine. It's habitual. The sensori-motor sequencing follows a logic: *if–then.* If the bell has rung, then ... it all follows. Recognition, logical consequence.

To the extent that it unfolds from a recognition according to a logic, a sensori-motor habit is a kind of thought in *action* automatically executed, below the threshold of conscious awareness. But action is of a different order than movement as such, as in dance. What defines action, as opposed to movement, is that its execution is swaddled less immediately in rhythm, than it is mediated by preestablished meaning. In habit, the body moves into what today's situation has recognizably in common with yesterday's, and what this sameness in the situation *means* for the stomach. A sensori-motor habit is a general idea of the body. The meaning is always *adaptive:* this repeated sequence fits the general situation, as it did before and will again.

The general idea of these habitual sensori-motor sequences is what embodied cognition and enactive perception theory (and earlier, Gestalt) have called body schema. Body schema are explained in terms of "implicit knowledge." These theories typically appeal to the implicit knowledge of body schema to account for the thought-like aspect of action. Body schema are presented as the thinking of the body, in the body. What is "embodied" *are* the schema. What is "enacted" is the general meaning of the adaptive situation. Maxine Sheets-Johnstone rightly observes that, far from grasping the manner of thinking incumbent in bodily movement as such, the concept of body schema implicitly construes what occurs, thought-like, at the heart of movement, moving on its own level, in its own register, as mental (why else call it embodied *cognition?*).[19] What other conceivable status could the element of general meaning have, but mental? Generality has nowhere else to be but in a "mind." The mental status of the body schema is confirmed by the fact that sequencing can be adequately expressed in logical form: *if–then.* It has the status of a syllogism. The syllogistic

19 Sheets-Johnstone critiques the concept of the sensori-motor in embodied cognition and enactive perception theory on similar grounds. See in particular Sheets-Johnstone, Maxine, *The Corporeal Turn: An Interdisciplinary Reader*, Exeter 2009, 219–252, 221, and ead., "Animation: the Fundamental, Essential, and Properly Descriptive Concept," in: *Continental Philosophy Review* 42 (2009): 375–400, 377, 394–395.

premise is the general meaning of the situation implicitly recognized in the triggering of the sequence.[20] That general meaning could just as well have been expressed in language as in movement. Language does syllogism to a "t," as the body does it on habitual cue. The concept of body schema brings language and movement together not in their dynamic differing, but in what they generally have in common, of a mental status.

Approaches relying on the concepts of body schema and implicit knowledge fail in their attempted anti-Cartesianism. The body's dynamism is implicitly returned to a dependency on a core mentality that can be adequately expressed in logical form and is in the element of general meaning shared by language. The singularity of movement's moving is unwittingly retrofitted into an implicit opposition between the mental and the physical, mind and body. The only way to avoid this dichotomizing refit, the only way to restore movement to the gestural nexus that is its native element, the only way to respect the dynamic spray of movement in and of itself, as well as in its form-flashing splay between other planes of experience, is to say that *movement embodies nothing but itself.* Movement never embodies anything. It just bodies-forth, at any-point. Embodiment is the wrong concept.[21] Just "bodying" is better. Movement goes a-bodying.

Where the bodying is a thinking in movement, most intensely and alive, is in movement *extracted* from action is this sensori-motor sense. "Subtract" the action, Forsythe tells his dancers (Forsythe 2011–11–17). "We start out with a process and try to collapse it," he continues (Forsythe 2011–11–23). Collapse movement back into itself. Subtract the general meaning from it. Undo it of its pre-established adaptations. Return it to its gestural nexus. Give it a more complex logic. Point-field it. Don't chain positions, chain sensations. Return the making of movement to the immediacy of its feeling. In that feeling, a different, more intense, utterly singular thinking will occur.

Where the body is most immediately thinking-movement is not in the unfolding of body schema. It is in its folding back into itself, back into the dynamic nexus of its native gestural element. Not unfolding, but infolding. A turning back around onto itself to begin again, never taking leave of itself, never ending. A turning, a twisting, in: inflection. Movement returns to itself in its each and every inflection. With every twist, it contracts back into its own element, to rebegin. Forsythe calls this the "centerpoint" of movement (Forsythe 2011–11–10). This centerpoint of movement is unlike the centerpoint of vision. It is not a

---

20 On syllogism and generality see Deleuze, Gilles, *Logic of Sense*, New York 1990, 294–297.
21 For Sheets-Johnstone's critique of embodiment, see Sheets-Johnstone, *The Corporeal Turn*, 219–252, 215, 220, and ead., "Animation," 390, 392, 393.

focal point but an any-point. Not a focused center but a many-vectored point-field. At this inflection point, movement contracts back into its own "dynamic difference" (Forsythe 2011–11–23), the differential gestural nexus where it is wholly and only in its element.

"*A body is that which folds.*"[22]

Movement's folding in on itself is not something the body *does.* It is what bodying *is.* Movement embodies only itself. Movement's making is corporeogenic: becoming-body.

Habit's adaptive action functionally unfolds. Dance movement inventively infolds. New moves come from movement's contracting back into its centerpoint, to re-splay as never before, throwing new curves for the seeing. Of course, a newly minted dance move is practiced, repeated, perfected. It becomes a habit, an acquired skill. But at the heart of every habit there is a *contracting* of the habit. Every adaptive unfolding comes from an inventive infolding. For this reason, there is a rhythm to dance creation. To avoid simply performing the movement sequences for the audience, to avoid simply displaying the visual form, and to avoid the dancers just portraying a character role in the performance – or just portraying themselves in the role of performers – the bodying has to be reactivated at every moment. All of the techniques that Forsythe has developed over the years are techniques for movement reactivation.

With respect to all of the techniques, it helps, Forsythe says, to go "extremely fast."[23] Because at extreme speed, the body "pushes the limit of its coordination."[24] Its habits collapse into the velocity of movement. The body is at the limit of its functional capacity to chain positions and postures meaningfully and adaptively. It has no choice but to surrender itself to its own order of sensation. Movement approaches escape velocity, where it returns to its own orbit. Collapse of the sensori-motor schema.

Take it to the extreme. Take movement to its limit – "and then go further" (Forsythe 2011–11–10). Movement doesn't actually stop. It subsides into itself. It relaxes back into its field, for the reoccurring. It collapses back into the nonlocality of its any-point, the unlocalized interval of no perceptible movement. In every movement's rise and fall, at any-point, there is an imperceptible interval where the rising and falling coincide. ***Waves collect. Overbalance. And fall.***

---

22 Forsythe, William, Conversation with the authors, Amsterdam, 3.6.2008.

23 Forsythe, William, "Observing Motion. An Interview with William Forsythe," in: ZKM Center for Media Arts (ed.), *Improvisation Technologies: A Tool for the Analytical Dance Eye*, Ostfildern 2003, 20.

24 Ibid.

The form of movement, in and of itself, is not a straight line. It is a pulse. It is a waveform. Forsythe: "Waves on the seashore, one coming in as the last is going out, overtaking, receding. Don't be where you are. Make yourself tidal" (Forsythe 2011–11–10). In the tidal interval, the waves cancel each other out for an uncountable beat. They collapse into each other, in dynamic superposition. At this any-point of superposition, there is no visible movement. The body, movement, is in suspense. But the interval is no less motional for being in suspense: a welling; already a next overtaking a receding. A one, a two, the interval is comotional. It doesn't stand for anything. It doesn't mean anything in particular – particularly not in general. It stands for what it is, which is what it *does:* a repotentializing interval of movement folding its sequential forms into its rolling on, the subsiding trace of the last a-bodying with the next to come. Movement never stops. It suspends itself in its own furtherance.

Movement always starts from a superposition, a mutual inclusion of sequential forms. The form of the movement cannot be reduced to the sequencing of positions and postures. Don't strike the pose: "put the activation in every part" (Forsythe 2011–11–10). Let your body subside into its movement pulsing, which will then throw itself a curve. The arc may well surprise you. Forsythe: "No movement is so possessed of its form that it cannot find another" (Forsythe 2011–11–10). Ride the wave of movement's occurrent variation.

Movement Moving

Movement only comes from movement (Spinoza). Movement is *sui generis:* all and only in and of its own order (its own nexus). Language is likewise *sui generis:* words only come from other words, in recurring waves, rising and falling from the linguistic any-point of the superposition of sound and speech, and of silence and noise. The same goes for vision, in its own difference, at its dedicated any-point (Klee's grey point where light and shadow, hue and illumination, enter into a zone of indistinction, and clarity of reflection clouds back into a spray of refraction, re-becoming pure optical array).[25]

25 On the "relation of non-relation" (65) between language and vision (their incommensurability and disjunctive synthesis) see Deleuze's analysis of "visibilities" and "statements" in the work of Foucault (Deleuze, Gilles, *Foucault*, Minneapolis 1988, 47–69): "what we see is not lodged in what we say" (64; translation modified). For Klee on the gray point as originary any-point of "primordial" motion which Klee places at the center of the color wheel, and thus of visual field, see Klee, Paul, *Notebooks*, vol. 1: *The Thinking Eye*, London 1978, 3–4.

Language cannot fully describe movement. Movement does not give itself over to the order of language, any more than it surrenders itself integrally to visible form. The orders of experience are incommensurable. There is always a residue, a holding itself in reserve, each in its own element. At the extreme, each suspends itself in its own reserve potential: in the trace of a last foretracing many a next, in a zone of indistinction. Each order is self-tracing. In the middle, they splay together in their difference. Absolutely amodally, in the dynamic concert of nonsensuous similarity. Modally incommensurable, they only relate outside anything in common – most especially outside the logical common of anything that can be attributed a core mental status.

There is no need to surreptitiously appeal to a common realm of implicitly mental status in order to account for the thought-like aspect of movement. Movement "feels like potential" (Forsythe 2011–11–17). Potential is abstract by nature, in the sense of not yet being this or that, here nor there. What is abstract feeling, if not thought? Movement-moving is thinking-feeling: sensation, integrally imbued with singular notional force. Potential is not of the if–then. Potential is allied to *what–if*. The thinking-feeling that is movement-moving is speculative: notional in the sense in which these two words can be used as synonyms.

Language and movement meet in the thinking-feeling. In thinking-feeling, both are notional, in analogous fashion: at the immanent limit of their ownmost centerpoint, wholly and only in their own mode. Each is a thinking-feeling after its own fashion. It is here, at the extreme point of their incommensurability, that they resonate. It's, like, they mimic each other. They notion to each other, at a distance, in their difference from each other. Each any-point, as immanent limit, twists around into the middle. In the middle, the immanent limits are in abstract superposition. *One, two, one, one-two one-two*, in many a multiple. Gestural, visual, aural, linguistic, in reciprocal transduction. Thinking-feeling is the transversality of all planes of experience, in the immanent twist. It is, like, life. Of the moment.

The Motional-Relational

In what manner is movement induced into furthering itself in the invention of new dynamic forms? The answer is quite simple. Movement only comes from movement. But movement does not come from movement only sequentially, in a rolling continuity of thinking-feeling motionally bodying. Movement is always triggered relationally. That is the answer: movement only comes from movement *relationally*.

Going extremely fast, as Forsythe suggests, is a key technique for making movement move. When a body folds in on its own gestures faster than the general speed limit of sensori-motor coordination, it self-generates an inflection of its line of movement. It activates to the extreme, and throws itself a curve, across the regenerative interval of its return to the any-point. You have to go so fast as to "be ahead of yourself" (Forsythe 2011–11–23). "Don't be adaptive, be predictive" (Forsythe 2011–11–10). That is, be *speculative*. Feel it coming at the speed of thought.

Even one body alone is dancing relationally. Its movement stands out in its difference against the background of all the other dance movements it potentially might have been. Or – why not? – of all the movements of dance history. One body alone is moving to the traces of movement's own passing, where it just was, absorbed in tracing where it is going. "Sense motion traces," Forsythe instructs.[26] This is what chaining sensations instead of positions is all about. When one, two multiple bodies are dancing, they are moving in relation to a sea of motion traces enveloped in the any-point. When there is one body alone, its potential makes two. When there are two, it makes three. When there are three, it makes ... a second one and a second three ... The count gets vague, because movement is always in rolling numerical excess over itself. The "thirdness" of all movement: the sea of motion-traces enveloped in the any-point in relation to which dancing dances.[27]

Movement is not sensori-motor. It is *motional-relational*.[28]

The motional-relational is in the speculative mode of the what–if. What–if the bodying, in its thirdness, folds now, extremely fast, too fast for any subject to think it? What–if one, two, five bodies do this at once? What–if? The what–if is moving thinking itself out, in an uncommon intensity of feeling.

26 Forsythe, "Observing Motion. An Interview with William Forsythe," 16.

27 On the thirdness of experience and the "sea" of traces as reserve of potential (specifically as regards sound and language), see Massumi, Brian, "Floating the Social: An Electronic Art of Noise," in: Michael Goddard/Benjamin Halligan/Paul Hegarty (ed.), *Reverberations: The Philosophy, Aesthetics and Politics of Noise*, London 2012, 40–57.

28 This term was suggested to us by Sheets-Johnstone's essay "Sensori-Kinetic Understandings of Language" (Sheets-Johnstone, *The Corporeal Turn*, 219–252). The term occurs on page 236. Sheets-Johnstone emphasizes the overlap (what we would term a "superposition" or "zone of indistinction") between language and movement in the motions of the body physically productive of speech. She uses the term "motional-relational" for that overlap and to argue that nonlinguistic dimensions of experience are not "prelinguisitic" but rather that language is "post-kinetic" (225). We take the concept of the post-kinetic in Sheets-Johnstone to imply a disjunctive synthesis between language and movement, rising from and returning to their zone of indistinction, as each come into themselves coming into and out of each other. For our own part, we reserve the term "motional-relational" for movement in its own order, as it co-composes disjunctive-synthetically with the "notional-relational" order of language, emphasizing how both orders, in their co-composing, overspill the sensori-motor (and even Sheets-Johnstone's far more adequate "sensori-kinetic"), into the suspension of nonsensuous perception.

What–If

Forsythe asks: "Can ballet actually produce ... something we don't know? We'd like to [...] investigate what it *feels* like to speak that way."[29] What–if? What–if this *épaulement*, this singular movement-formation, were turned in on itself, subverting the organization of the organs, of the limbs? What–if the gaze, normally turned askance toward the audience were turned in, turned back into the head in visual "disfocus"[30] so that "the body [...] becomes an ear, [...] an organ for listening" to the rhythmic milieu.[31] What–if this extension we call an arabesque opened less onto a posture or a pose, completed form, than onto the *feeling* of a passing balance?

Forsythe's work builds on the what–if? It is never a question of formally working something out in advance. Speaking of his collaboration with composer Thom Willems, he explains: "He always wants to know what are we thinking and I say nothing, we're just, we're just organising bodies."[32] *We* are not thinking, the thinking is in the moving.

Organizing bodies does not mean placing bodies on the stage and planning them into forms. It means reorganizing the body itself, in its commotion with other bodies. It means activating collective rhythm on the level of a relational movement, a level on which form does not single out. "When I go into the studio I want to be able to see what is in front of me, I want to not have another idea." Idea, here, is allied to form. What–if asks another kind of question, a question of technique: "the biggest challenge, is not seeing what I want to see but seeing what's actually in front of my eyes."[33] A spray.

This encounter with not-knowing – epitomized, while we were in Frankfurt, with the development of a new piece called *Whole in the Head* (2011) – keeps the process alive and uncertain. With *Whole in the Head*, weeks of rehearsal were completely unmoored two hours before the performance when Forsythe decided to change the piece, using only certain sections of what had been rehearsed and reordering them. In keeping with his idiosyncratic practice of the what–if, Forsythe then fundamentally altered the piece for each of the nine performances to come. This was not just staging. It was a rigorous engagement with "seeing what's actually in front of my eyes," the what–if of movement activated both

29 Forsythe, *Interview with John Tusa. Balletco Magazine.*

30 Forsythe, William, *Interview with Valerie Lawson. Balletco Magazine*, 2000, http://www.ballet.co.uk/magazines/yr_01/oct01/interview_william_forsythe.htm (24.1.2013).

31 Forsythe, *Interview with John Tusa. Balletco Magazine.*

32 Ibid.

33 Ibid.

on the stage-space of performance and in its co-composition with other performances just past, a movement precise with training but still open to regeneration.

The focus in Forsythe's work on the what–if is everywhere present in his ample arsenal of movement techniques, but can be felt particularly keenly in the techniques of cueing and aligning. In moving-to-cue, Forsythe asks the dancers to become sensitive, collectively, to how the dance as a whole moves. You cue to the trace of a movement-moving, to its repeatable tendency, yes, but even more so to the what–if of its emergent evolution. This means that you dance less into the movement-form as it happened in the past than into the speculative figuring of movement's incipient future. Aligning, then, means tuning to the what–if as it commotionally appears, in movement-transition. You move to the trace of a movement coming to a landing, you align to its capacity for transition, to its complex of potential, already-on-its-way. You align to the what–if of a movement's futurity – the not–quite of its having come into being. If–then necessarily follows. Gravity acts on the body in movement, moving it to land or flow in relatively habitual ways. But movement-moving has been opened up by the what–if to its own variation – to new collective sprays, new combinations, new "flow-matches," as the company calls them.

"Synching is not what's important, in the sense of matching already known timing," Forsythe explains (Forsythe 2011–11–8). What's important to the what–if of cueing and aligning is how position itself becomes movement. "Go slower, be in the other's past (right before they catch up to you), then move past them to their future – look for the moment – *aim* at it rather than going directly to it [...] then shift" (Forsythe 2011–11–8). Time-shift. You don't *always* move at extreme speed, you vary the rhythm. Remember the rhythmic milieu.

What–if is the "actually seeing what's in front of you" of movement's capacity to speculate.

What–if is the "body becomes an organism for listening" of a collective body's capacity to activate, and be activated by an "acoustic environment" (which we take for a synonym for the rhythmic milieu).

What–if is the collective movement-moving created, in mutual superposition, in the commotion of a complex that cannot be reduced to two: *one, two, one, two, one-two, one-two*, in thirdness.

What–if the one were always already more than one.

What does it *feel* like to speak that way?

The Notional in the Motional

"In a truly successful dance," Forsythe explains, the body "takes over at the point where you" have "no more idea."[34] Like the choreographer watching the dance unfold, you, the dancer, must dance with the unknowability of "what's actually in front of my eyes." You must dance the future in its visionary unfolding. In the motional intensity of "not knowing," operational, always, within the time-loop of the flow-match, it is the "body that dances you around":[35] the thought-body. Thinking in movement. "The body thinking itself into the flow of the world, and the world flowing into the thinking of the body."[36] The dance does not have to embody movement ideas, does not have to think the idea in advance of its commotion. Motionally, the body is already speculatively tinged with mentality, in its own mode.[37] In motion, the body is already notional, in its own order. "Corporeal-kinetic forms and relations are conceptual by their very nature."[38] There is no *having* ideas. *You* do not have ideas. The body itself, with its rhythmic milieu, *is* a motional-notion: a movement of thought. Dance that thought around.

Becoming-Visionary

One, two, one-two, one–two. He taps the mic, ***one–two, one–two.*** We sense a shift in register: he is preparing to tell us something, preparing to give us an image in a register of direct address we will never see/hear again throughout the piece. She continues to move slowly in the distance, movement-mewling. And then he speaks, walking back while holding the microphone, as though opening the acoustic environment to its furthest recesses, recesses that include her without in any way addressing her role, her place, in the story. He speaks directly to us: ***Did you hear the one about Virginia Woolf and the Dalai Lama?*** Pause. ***Well, Virginia Woolf and the Dalai Lama are walking down the beach, and there's something shining in the sand, and the Dalai Lama, he bends down***, he bends, ***and picks it up. It's a mirror. A little mirror. A pocket mirror. And he looks into the mirror, and he says, "hey, I know this person!" and Virginia***

---

34 Forsythe, *Observing Motion*, 24.

35 Ibid., 26.

36 Caspersen, "Decreation: Fragmentation and Continuity," 100.

37 Whitehead develops the concept of the mental and the physical as two modally distinct yet inseparably bound "poles" of every occasion of experience (Whitehead, Alfred North, *Process and Reality*, New York 1978, 239, 248–249 and passim).

38 Sheets-Johnstone, *The Corporeal Turn*, 221

Woolf says, "Give me that mirror." And so she looks into the mirror, and she says, "Well, of course you know this person, silly. That's me! That's you, one hundred years from now!" One hundred years from now.

When Forsythe speaks of not wanting to miss seeing "what's actually in front of my eyes," it seems that he is referring to vision in the standard sense. Look in the mirror, and see what's there, right in front of you. He wants to see what is happening, and to choreograph with the present in the mode of the just like that. No ideas in advance – let the process show you its own evolution. "Don't be afraid to let things reveal themselves," Forsythe tells his dancers. "Sometimes it [what the work's about] doesn't come until the end."[39] See it for what it is, in the middling of its coming to be. The problem is, the movement is aligning to futurity, its reflection askance. *One hundred years from now: that's me, silly!* How can he or the dancers work with "what's actually in front of my eyes" when it is landing onto the present from the unknowability of the future? How can they work with the mirror when the mirror at once reflects me and you, *one, two, one–two* in the uncertain field of vision that includes the now and the one hundred years from now? What does it mean to "really see" in the mode of the speculative?

Movement generation techniques in The Forsythe Company are enlivened by a paradox of double-vision. To be in the present in its unfolding is to be in the future-flow of a non-linear time-match. The sensori-motor is always in commotion with the motional-relational, in disjunctive synthesis. The body, articulating in the present the complexity of dance-form, is always also being asked to match-flow to a welling future movement. The sensori-motor is alive in the taking-form, led by the motional relational, poised at the limit where now and next are co-composing.

Forsythe Company dancer Dana Caspersen explains: "I experience two kinds of vision at once. The space around [is] alive with potential, invisible forms, and lines of movement. I am continually ambushed, caught up, in this welter of doubled vision."[40] This superposition of actively engaging with the just–now – just like that – of the complexity of the present presenting itself, is superposed with the futurity of how the what-will-come has already left its trace on the movement-welling.

---

39 Corbin, Patrick, *Patrick's Month in Frankfurt*, 2012, http://patrickcorbinsblog.blogspot.com/2012/01/january-25th-back-at-bockenheimer-depot.html (24.1.2013).
40 Caspersen, "Decreation: Fragmentation and Continuity," 97.

Forsythe describes this technique as having a vision.[41] Having a vision, in the moving, means moving beyond this body, this form, toward the unfolding of how movement will be affected by the middling which will flow into it from the future. A dizzying task, but one that cleaves movement from its habitual sensori-motor cues and alignments, inviting it to invent with a bodying. Two versions of cueing and aligning are co-active here, in the field of double vision. There is the habitual sensori-motor organization of the body in its causal flow – a forward step cueing to gravity, aligning to the slope of the floor – and the Forsythe Company's technique of cueing to movement-moving in a non-local field of co-action and collective aligning in counterpoint to the field's patterned interplay.[42]

Virginia Woolf's *Mrs Dalloway* also plays with vision, tuning it toward the visionary. It was this aspect of the text that most excited the Forsythe Company in the creation of *Woolf Phrase.*[43] Working through the text, Forsythe speaks of having the insight that a technique could be invented to make dancing visionary. The dancers could *technically* become visionary. The technique that he originally developed, and that was enhanced through the making of the piece, "had to do with identically remembering another person's variation, or sprays, and building a kind of architecture of movement around it." The caveat: "you had to keep seeing this other person dancing in order to perform it, so it was a way of having a vision."[44] A collective vision.

This evolved into two techniques for becoming-visionary, both of which were set in place for the making of *Woolf Phrase*. The first, "telescoping," involves actively working with the gaze. "We tried to focus our gaze, if looking at an arm movement, on the line proposed by the movement. The body follows the gaze to where the movement begins, turns around this point, and finally turns the gaze in another direction."[45] First, you follow the line of movement with your gaze. And then, still looking, you visualize where the line can go and move into it. You turn around that point, looking both for what you can see and for what you can't see. Double the vision. Then turn the gaze in another direction,

41 Forsythe, *Interview with John Tusa. Balletco Magazine*; Manning, Erin, *Always More Than One: Individuation's Dance*, Durham 2013, 86.

42 For an extended analysis of cueing and aligning in the work of Forsythe, see "Choreography as Mobile Architecture," in: Manning, *Always More Than One*, 99–123). On cueing more generally, see Massumi, Brian, "Power to the Edge: Making Information Pointy," in: *Ontopower: War, Powers, and the State of Perception*, Durham (forthcoming).

43 Forsythe, *Interview with John Tusa. Balletco Magazine.*

44 Ibid.

45 Lang, Prue, "Denken, Bewegung und Sprache," in: Gerald Siegmund (ed.), *William Forsythe – Denken in Bewegung*, Berlin 2004, 126.

in a splay of focus. Make the unseen felt, bring it into appearance through movement. The aim is not to replicate a movement, or even to follow a movement, but to continue it by other means. Activate the field by pulling the line of movement into a new direction. *Twist.*

"Metascoping," the more amodal version of the two techniques, works by feeling-seeing the time of movement. "We had to react to the movement of another body by allowing the gaze to move toward, away or around lines of movement."[46] You follow the movement as it unfolds, at the same time exceeding its arc, away and around, entering into its future. Time bends space: this is what it means to be visionary. Differentials of speed and intensity are created. "When the movement reaches a certain speed, these short dialogues become increasingly reactive and remind us of swarming bees" (Lang 2004: 126). Commotion. Moving into movement's incipient future, following its lines of intensity, means inventing the path movement can take in the untimeliness of its future-forming. Spacetime of movement-moving, as generative movement field.

The sensori-motor organization of the body in its sequential form-following is subverted by the force of the visionary. To generate a movement, the dancer must move beyond the form of the what-was and learn how to "sense motion traces." This opens up movement to its otherness, similarly to the way Proust talks about the necessity to open language to "a kind of foreign language within language, which is neither another language nor a rediscovered patois, but a becoming-other of language, a minoritization of this major language, a delirium that carries it off, a witch's line that escapes the dominant system."[47]

Double vision, in the end, is less about vision than it is about proprioception. To "visualize" a movement is to "see" it proprioceptively: in body-vision.[48] To move into the futurity of movement-moving is to have a body-vision in the dancing. It is to move with the future in its unfolding, it is to be moving in the very creation of spacetimes of experience, to be caught up, *just like that,*

---

46 Ibid.

47 Deleuze, Gilles, *Essays Critical and Clinical*, Minneapolis 1997, 5.

48 Forsythe has developed a unique approach to accessing proprioception for his choreographic process. He teaches his dancers to use their skin surface as an organ of proprioception more than of touch. The stretching of the skin registers the spray of the movement forming the differential gestural nexus. The dancers are asked to activate their body through their skin, making the skin what Deleuze would call the "surface of recording" or "abstract surface" of whole-body events of thinking-feeling. "Put the activation into every part. Think about where the movement starts and stops. If you raise your arm, where does your skin stretch? (Down the side). Activate the skin. ... Get feedback from the skin. Go further if it tells you something. What you feel is what you know. Look for the chain of sensations rather than the chain of positions" (Forsythe 2011–11–10).

in the simultaneity of the wake of movement perishing as it meets the welling of its futurity, dancing at the nexus of experience infolding.

Dance that thought around.

As–If

The motional-relational in the mode of the what–if is futurity in movement. What–if: the body folds, extremely fast, too fast for any subject to think it. What–if: cueing to movement sprays, we activate a collective bodying that alters the affective tonality not only of the movement itself, but of the rhythmic milieu. What–if: a curve is thrown that cannot be predicted in its particulars, precisely because it has been relationally imbued with futurity, fielding itself into novel form.

What must language do to become-visionary? What are its techniques for opening language to the foreignness, as Proust says, of its becoming-other? In Virginia Woolf, the as–if plays this role, folding as it does the present into its double: "She began to go slowly upstairs, with her hand on the bannisters, as if she had left a party, where now this friend now that had flashed back her face, her voice [*that's me, silly!*]; had shut the door and gone out and stood alone, a single figure against the appalling night, or rather, to be accurate, against the stare of this matter-of-fact June morning."[49] As–if, the holding pattern of a difference in time, time looped onto itself toward the impossibility of getting the story straight, once and for all. Not just "it could have been this way," but "it was this way, and that way as well, at the same time," disjunctive synthesis, *just like that*. As–if – "straight" in the middle.

As–if, he says. **As if things were just this bright, or ... as if things were just that loud**, or twisting at the hips **this fast** gesturing toward her **that far. As if there was just this one distance, this one**, marking a short distance between his hands **separation. One separation, one, two. One. Two. One, one–two**, tapping the mic, **one–two, one–two.**

As–if, Virginia Woolf writes, "as if this beauty, this scent, this colour [...] were a wave which she let flow over her and surmount that hatred, that monster, surmount it all"; "as if one will worked legs and arms uniformly, and life, with its varieties, its irreticences, had been laid under a pavement of monuments and wreaths and drugged into a stiff yet staring corpse by discipline."[50] Seventy-five times *as if* in *Mrs Dalloway*, each time touching on the impossibility of the

49 Woolf, *Mrs Dalloway*, 33.
50 Ibid., 14.

simple coupling of this and now. As–if, a rhythmic suspension that holds the presenting of the present to itself in itself, that denies *this* image its simplicity of unfolding, creating a knot in the discursivity of language. As–if: a mode that moves language to its narrative limit, pushing language to say not how if and then follow, but how "if" becomes its own limit, a transversal limit that cuts across the if–then.

When the Forsythe Company decided to work with *Mrs Dalloway*, the focus, once again, was not on retelling her story, but on "moving around the rhythm of Woolf's language." How to move into the text's visionary force? How to fold movement into the as–if of *Mrs Dalloway*'s speculative language? How to create techniques that were capable not of responding to language, or illustrating language, but of moving in counterpoint with language? Dog-like, seagull-like, not in the realm of representation, but in the elemental realm of formative feeling-tone. "Richard and I worked to dehumanize our bodies by giving them elemental qualities (like the ocean) or animal qualities (like a bird, a dog, or bees). Movement thus becomes entangled with sound or nature: circling and mewling, barking and jumping, the sound of waves surging."[51] Circling, jumping, surging: not forms but activation contours. As–if: in nonsensuous similarity.

Forsythe defines counterpoint as "a field of action in which the intermittent and irregular coincidence of attributes between organizational elements produces an ordered interplay."[52] Counterpoint of movement and language in *Woolf Phrase* is best understood as a coinciding of both at their limit, in the relational any-point of the motional and the notional. Here, the what–if of movement and the as–if of language co-compose. Together they unsettle the linearity of time in the tense of the if–then. In the speaking and in the moving, working in disjunctive synthesis, two speculative time-signatures interrupt one-another. What–if asks the dancer to "anticipate the other dancer's future." In this movement proposition, the dancer is being asked to shift time, make time in the tense of the future-felt. As–if asks the reader to place in an uneasy holding pattern more than this one image, more than this one narrative line.

As if there was just this one. Language interrupts the dancers who, a moment ago, were moving together, he drawing circles around her seagull-like movements in a contorsion that seemed to meet her movement halfway, never touching, always sidling. *As if there was just this one*, she, standing to his left, now immobilized as though holding her breath. Immobilized in a movement

---

51 Lang, "Denken, Bewegung und Sprache," 128.

52 Forsythe, William, *Synchronous Objects*, http://synchronousobjects.osu.edu/blog/introductory-essays-for-synchronous-objects (17.1.2013).

alive in its stillness, a lingering suspension. As–if this were the only way of moving the text, of speaking the dance. As–if it had to be language *or* movement. As–if the relational didn't already hold them together in disjunction, *just this way.* Superposed, what–if and as–if, moving together in their difference. Think the superposition as a flashpoint, felt as rhythm. Rhythm as the expression of movement and language's coming together.

"The perception of similarity," Benjamin writes, "is in every case bound to a flashing up. It flits past, can possibly be won again, but cannot really be held fast as can other perceptions."[53] The coinciding of the as–if and the what–if produces a flash of similarity. Just like this. Similarity at the flashpoint is nonsensuous because it is thought-felt at the any-point limit of expression where both movement and language become foreign to their general functioning, in "a delirium that carries [them] off, a witch's line that escapes the dominant system." Nonsensuous similarity: directly thought-felt in the register of the visionary.

Nonsensuous perception comes together with sensuous perception at all stages of *Woolf Phrase.* We sit and watch the work in its unfolding, alive to the minute changes in tempo, the rise and fall of voice, the undulating of the wall of sound. We also perceive, nonsensuously, the edgings into co-composition of rhythm-becoming-language, of rhythm-becoming-movement. The paradox of disjunctive synthesis makes itself felt at each stage of the commotion: language and movement are not one, and yet, at their limit, they flash up in a oneness of similarity. As–if.

Woolf Phrase: an ordered interplay that invents relationally at the limit of the notional and the motional. *Woolf Phrase:* a recurrence in lingering suspension of a rhythm in the making. The piece, Forsythe, says is always changing, ever generating variations on itself: no two the nights quite identical, but ... somehow ... yet quite the same.

When the notional meets the motional, it's not that we no longer understand the words. It's not even that we don't ally the words to the movement. It's that the words, in the mode of the as–if, contort to the movement they activate. It's that language, when in counterpoint with movement at the limit, tweaks into a rhythm with movement. At this any-point, language, like movement, proposes itself as "an organ for listening." We hear it sounding, its counterpoint with movement creating a joint rhythmic milieu. While we can differentiate the words from Willems' acoustic environment, while we know what is language and what is movement and what is sound – ***PAAF!*** – we are no longer certain

53 Benjamin, "The Doctrine of the Similar," 695.

where expression begins and ends. Notionally, we have “no idea.” Knowing and unknowing co-compose, dancing thought around.

And so we are left not with a story about a dog – he wags his leg – or about a sparrow – her movement-mewls – or about a woman and the Dalai Lama seeing each other cross-eyed – but with a dance that moves language in the middle. Just like that.

Reset

Midway through the piece, a shift occurs. The movement is cut by touch, touch taking on the role of the *all of a sudden*, resetting movement in the making.

The touch cues the return of the same moment, with a difference. ***And now the same moment, supposedly without happiness.*** As–if. He takes her hand. ***Still no happiness.***

Stricken once

He pulls her back, away from the microphone, away from language, his movement folding into hers, the gesture of touch leading her into the vortex of a commotion increasingly turned in on itself, their movements swarming, future-forming.

She moves back to the microphone, his body still connected to hers, his balance passing through an arabesque. She: ***Stricken twice.*** An arabesque, spiralling already into a circling, his body folding through her backward movement. She, reaching toward him. Hands held, a sense of push and pull. A suspended connection, wary of the measure of the touch, it seems. For while the movement looks to proceed by cause and effect, in fact what is active is a mobile relational interval – a co-constitutive suspension of movement in the moving. Touch strikes. It cuts into the relational interval. It does not stop movement so much as set it in motion, again, differently. It reorganizes, reorients, pulling movement-moving into knife-edge precision. A moment of clarity. Stricken.

The moment of clarity appears from the wash of movement-moving. The clarity was of course always already there, active in the midst, if eclipsed by the velocity, the agility, of the movement-moving. Clarity is a signal for the viewer. Look again, it says, think-feel the difference. Don’t forget: the touch comes between the moving and the speaking, recalibrating not only the motional but also the notional. Think touch as that which acts on the flow to pull from it a clearing of relation.

Touch resets relational movement not by bringing it to rest in this or that human body, but by activating a momentary pause in the frenzy that reorients the field as a whole.[54] Yes, he lays his hand on her shoulder. Yes, she stills. But really what is happening is less about him, or about her, or about them touching. It is about giving us a window into the wider field, giving us pause to encounter, with them, the counterpoint of movement and language in a field of experience continuously folding back on itself. No linear sequencing here: touch as the reset button for a rebeginning, a spiralling back.

Ok, we're back

The touching reset of relational movement creates a hiatus in movement-moving. This invites us to see dance moving not into form, but into the sprays, the traces, through which form spreads. Sprays: the folding through of movement becoming rhythmic milieu. Sprays: the affective surplus of the sensorimotor, the curve that exceeds this habitual stance, the moving at the point of inflection that tunes toward a collective individuation not only of the dancers, but of the whole field of activation, the field which includes, at its limit, the folds of *Mrs Dalloway* as well as every movement proposition that went into each of the choreographed sequences. Everything is there in abeyance, suspended in each movement, lingering in each utterance. But what makes the work work is not the everything, but the overflowing singularity of how *this* point of inflection tunes to *this* rhythmic fold at any-point, in the counterpoint of the notional and the motional. *Stricken.* Reset: from lingering to *just like that.*

Stricken thrice. PAAF! Willems' suddenly amplified soundwall fills the moment. ***PAAF!*** Same moment again, without the sound.

Touch resets the any-point of movement. By calling attention to the any-point of movement, touch makes visible the incommensurability of the two, *one, two, one–two.* Watch them move: note how the touch activates a slowing, an intensive attending, not only to through which relational movement moves, but to the shape of the work as a moving whole, to the shape of its sidling to language and contorting to movement. Watch how touch acts on movement, stilling its delirium. But note also how its stalling of movement works to make apparent what following Proust we might call movement's witch's line, its capacity to upset the dominant code that seeks only to generalize experience. The touch is unnerving – it undoes expectation, refusing to connect the two in an easy intimacy. Touch: giving the viewer, the dancer, the bearings from whence to start again, differently, in the middle. For what moves is not one dancer connecting

---

54 There are other resets besides touch. The many times repeated *PAAF!* is a vocal reset that operates on the cusp of language, where speaking, sounding, and gesture coincide. The Dalai Lama's mirror is another rest. In *Woolf Phrase*, the mirror plays a deviant role. It actually blocks vision from self-reflection, infusing otherness into identity.

with another. What moves is a moving in the middling, a collective moving-with that connects to movement's visionary potential, in a nexus with language.

We start out with a process and try to collapse it.

In the collapse, the witch's line of the dance is reactivated. The any-point beckons. We watch, the movement dizzying in its speed. No, start again. Watch again. Be in it, in the middle. Touch. *Stricken.* Any point in space resets to the any-point of movement. Learn to feel it. "It feels like potential" (Forsythe 2011–11–17).

Same moment. All of a sudden, again, somehow. PAAF!

Somehow, Circling

All of a sudden, you remember where you are. Pause. ***Goodbye.***

Ok, we're back.

Woolf Phrase's visionary quality creates a suspension, a lingering in the future-forming moment. She, movement-mewling. He, again sitting toward the edge, watching her, waiting. This holding pattern, cut by touch, reset – *PAAF!* – circles the piece on itself, somehow. Two velocities are continuously at work, two modes in attention, two levels, language and movement, in counterpoint. This doubling keeps the piece active, alive at the limit of what movement and language, in their differential, are capable of. One register alone, if given dominance, would dampen the force of what is at work here. We might look for Meaning, instead of heeding the call to attention to movement-moving. We might catch ourselves seeing Mrs Dalloway told, in movement. Illustration, representation redux. But this is not what is happening. A thirdness is coming into play, in the elemental force of the two, *one-two*, coming together. This thirdness contracted, always, through the creative disjunction of the match-flow and its reset.

What do language and movement do, at this limit where they refract into thirdness? They remake themselves at the edge of time, where time itself is at the limit of its spiralling. Time folds. ***In. Ever. Widening. Circles.*** Life and death in co-composition, in the shape of a wave cresting. ***Striken. Leaden. Circles.***

Fear no more, says the heart. Willems' sound acute, the movement frenzied, she at the microphone. ***Fear. No. More.*** He, on the ground, moving around her, the intense sound crescendoing in a semblance of the *PAAF!*, resetting the field of movement and language. She, still standing at the mic, her breath sounding, in and out, the effort of her words, of her movement, palpable. He, crouched, watching her from a distance.

She: ***Fear no more, says the heart.*** Breath: in, out, in, out. ***Committing its burden to some sea.*** In, out, in, out. ***And with a final twist.*** Twist. ***She bowed***

her head. Very politely. In, out, in, out. ***She went*** barely audible ***paaf.*** A quiet darkening. ***She. Had gone.*** She, had gone. She, the face in the mirror – *that's me! that's you!* – one hundred years from now, death folding into life, life folding forward. She, had gone. She, ***voice of no age, no sex*** – elemental, here and gone, of the wave. Overtaking and receding. Just like that. Somehow.

Across the co-composition of just like that and somehow, across the threshold of death and its lingering at the edges of life, the refrain returns: ***As if there was just this one.*** As–if, as if two could make one. But what if two made three? What–if the two, the one–two of language and movement were reset, just this way, into an inexpressible thirdness? What if just like that and somehow were the cues through which the piece aligned itself, again and again, to the force of its doubling? What if just like that and somehow were the terms of the work's limit, the mark of how the work must continuously move through the absolute difference of a vertiginous lingering in expression?

Somehow, a likeness. Somehow, as if two times, two events, this and its likeness, could linger in co-habitation, as–if. *Just like that*, the flashpoint of nonsensuous similarity, the cut into likeness that brings time's overflow into the moment, this moment. But what a strange moment, this specious present, a moment qualified not by its being-present but by its spread, by the span of time it includes, the past and future rolling into it. A visionary moment, a futurity in the making.

This is what language and movement can do in co-composition. They can take the lingering of the *somehow* and, with touch, reset it towards the *just like that* such that suddenly, the contorting of time is acutely felt across expression in the making. *Woolf Phrase*'s proposition: to allow both movement and language, at their counterpoint, to create a movement of thought that moves time differentially, that motions across speeds and slownesses, that sidles towards the disjunction of differential tendencies even while composing across their nonsensuous similarity.

Co-composition, suspended in the force of a recurrence.

The piece comes to an end. He, at the microphone. She, standing back, movement-mewling.

And as they looked, the whole world became perfectly silent. Pause. ***And a flock of gulls crossed the sky, first one gull leading, then another.*** Pause. ***And in this silence, in this purity, in this pallor, bells struck eleven times, their sound floating up there among the gulls.***

Thirteen seconds of silence. He, standing at the microphone, looking forward, but gesturing back in her direction. She, behind him, completely still. No more mewling, the acoustic environment full, once more, with Willems' sound. The scene is set.

And then, all of a sudden. The stage bathes in a darkness that lingers across the *all of a sudden.*

All of a sudden, a lingering. And so we are reminded that the suspension is everywhere speciously present, speculatively active at different velocities, notionally in different registers, relationally across every modality of experience. A lingering, a suspension of isolatable movement form, a suspension of language, as designation or narration, and also a suspension, within movement-moving, of its human characters, tuning to their elemental qualities, gesturing to dance's capacity to hold dissonant times together in movement.

As if there was just this one.

References

Benjamin, Walter, "The Doctrine of the Similar," in: *Selected Writings* 2 [1927–1934], Cambridge 2004, 694–698.

Caspersen, Dana, "Decreation: Fragmentation and Continuity," in: Stephen Spier (ed.), *William Forsythe and the Practice of Choreography*, New York 2011, 93–100.

Caspersen, Dana, "It Starts from Any Point: Bill and the Frankfurt Ballet," in: *Choreography and Dance* 5.3 (2000): 25–40.

Combes, Muriel, *Gilbert Simondon and the Philosophy of the Transindividual*, Cambridge 2012.

Corbin, Patrick, *Patrick's Month in Frankfurt*, 2012, http://patrickcorbinsblog.blogspot.com/2012/01/january-25th-back-at-bockenheimer-depot.html, (24.1.2013).

Deleuze, Gilles, *Essays Critical and Clinical*, Minneapolis 1997.

Deleuze, Gilles, *Logic of Sense*, New York 1990.

Deleuze, Gilles, *Foucault*, Minneapolis 1988.

Deleuze, Gilles, "Un manifeste de moins," in: Gilles Deleuze/Carmelo Bene, *Superpositions*, Paris 1979, 87–131.

Deleuze, Gilles/Guattari, Félix, *A Thousand Plateaus*, Minneapolis 1987.

Deleuze, Gilles/Guattari, Félix, *Anti-Oedipus*, Minneapolis 1983.

Forsythe, William, *Synchronous Objects*, 2013, http://synchronousobjects.osu.edu (17.1.2013).

Forsythe, William, *William Forsythe in Conversation with Zachary Whittenberg*, 2012, http://www.movementresearch.org/criticalcorrespondence/blog/?p=5213 (24.1.2013).

Forsythe, William, Forsythe Company Rehearsals, Bockenheimer Depot, Frankfurt, November 2011 (authors' studio observation notes).

Forsythe, William, Conversation with the authors, Amsterdam, 3.6.2008.

Forsythe, William, *Interview with John Tusa. Balletco Magazine*, 2003, http://www.ballet.co.uk/magazines/yr_03/feb03/interview_bbc_forsythe.htm (24.1.2013).

Forsythe, William, "Observing Motion. An Interview with William Forsythe," in: ZKM Center for Media Arts (ed.), *Improvisation Technologies: A Tool for the Analytical Dance Eye*, Ostfildern 2003.

Forsythe, William, *Interview with Valerie Lawson. Balletco Magazine*, 2000, http://www.ballet.co.uk/magazines/yr_01/oct01/interview_william_forsythe.htm (24.1.2013).

Gibson, James J., *The Ecological Approach to Visual Perception*, Hillsdale 1986.
Klee, Paul, *Notebooks*, vol. 1: *The Thinking Eye*, London 1978.
Lang, Prue, "Denken, Bewegung und Sprache," in: Gerald Siegmund (ed.), *William Forsythe – Denken in Bewegung*, Berlin 2004, 125–132.
Manning, Erin, *Always More Than One: Individuation's Dance*, Durham 2013.
Manning, Erin, *Relationscapes: Movement, Art, Philosophy*, Cambridge 2009.
Massumi, Brian, "Power to the Edge: Making Information Pointy," in: *Ontopower: War, Powers, and the State of Perception*, Durham (forthcoming).
Massumi, Brian, "Floating the Social: An Electronic Art of Noise," in: Michael Goddard/Benjamin Halligan/Paul Hegarty (ed.), *Reverberations: The Philosophy, Aesthetics and Politics of Noise*, London: 2012, 40–57.
Massumi, Brian, *Semblance and Event: Activist Philosophy and the Occurrent Arts*, Cambridge 2011.
Sheets-Johnstone, Maxine, *The Corporeal Turn: An Interdisciplinary Reader*, Exeter 2009, 219–252.
Sheets-Johnstone, Maxine, "Animation: the Fundamental, Essential, and Properly Descriptive Concept," in: *Continental Philosophy Review* 42 (2009): 375–400.
Simondon, Gilbert, *L'individuation à la lumière des notions de forme et d'information*, Grenoble 2005.
Stern, Daniel, *The Interpersonal World of the Infant*, New York 1985.
Sulcas, Roslyn, "Woolf Phrase and Bees of the Invisible", in: *New York Times*, http://www.nytimes.com/2001/12/09/arts/dance-using-forms-ingrained-in-ballet-to-help-the-body-move-beyond-it.html (24.1.2013).
Whitehead, Alfred North, *Process and Reality*, New York 1978.
Whitehead, Alfred North, *Science and the Modern World*, New York 1967.
Woolf, Virginia, *Mrs Dalloway*, New York 1992.

Gerko Egert

Movements of Touch in *MAYBE FOREVER*

> Touching: that is, giving free rein to the interplay of attraction and repulsion, integrity and intrusion, differentiation and transference. Leaving the whole to play as such, the entity and its dissolution, its diremption brush each other by.[1]

The stage is dark, slowly the outlines of two bodies become apparent: half lying, half sitting, Meg Stuart and Philipp Gehmacher begin stretching out their arms and feeling their way towards the other. The moment they touch for (probably) the first time is barely discernible in the dimness of the stage. Only when Gehmacher reaches for Stuart's arm, clasps it, and uses it to drag himself towards her, do the two dancers touch in a clearly visible manner. Now they lie, body against body, half next to, half on top of each other, almost motionless. Yet this bodily contact is only short-lived: already, they are both pushing their bodies away from each other, until they lie on the floor, back to back, separate, even isolated, curled up and self-contained, barely discernible in the dark. These movements of *drawing near* and *withdrawing* are repeated several times before they slowly sit up. Kneeling, arms outstretched in front of him, Gehmacher remains almost motionless until Stuart nestles in his arms, her body against his, clasping him, making them both lose their balance and fall. Again and again, the bodies touch, but just as often do they separate, calling and reaching for each other at a distance. Darkness. Light: the stage is large, slightly too large and too empty, lacking the intimacy generated by the dimness at the start of the performance. This emptiness establishes physical and emotional distance – between the dancers and between stage and audience.

When bodies touch – as described in this dance scene – they move, they draw close to and away from each other, maybe leaving traces – they transform each other. Instances of touch are possibilities of relating, implying the prospect of closeness, tenderness, and intimacy, but also of violence, restraint, and power. They are complex relationship configurations, displaying both an affective and an ethical-political element. Tenderness and violence, closeness and separation are no mutually preclusive determinants of touch, they are in fact im-

1 Nancy, Jean-Luc, "Der Lustkörper," in: *Ausdehnung der Seele: Texte zu Körper, Kunst und Tanz*, Berlin, Zurich 2010, 25–30, 29 (translation M.C.).

manent to touch as mutually conditioning motives. What violence does touch exert even in its most tender manifestation? What relation exists between these (seemingly) contradictory aspects of touch?

Drawing on Stuart's and Gehmacher's 2007 piece *MAYBE FOREVER*, this essay will consider different gestures of touch. It will advance the thesis that instances of touch between the two dancers are not mere moments of actual bodily contact, but also contain elements of separation and distance. Instances of touch become movements of the bodies, convergence and repulse. They shall not be examined as actions unfolding in time and space in a present and linear manner, nor is it the aim to differentiate them clearly from other, non-touching movements. Rather, movements of touch shall be understood as gestures with their own, inherent virtuality, their *not yet* and *no more.*

Subsequently, the dancers' bodies shall also be considered from the perspective of movement: in touching, the bodies change, they are moved by touch, and as moving bodies they are constantly de- and refigured. What forms of bodily relations arise between the dancers? How do instances of touch open their bodies to each other, without dissolving their form, but rather again and again marking their boundaries and their distinctness – de-fining them?

The emotionality of these tactile relations shall be investigated on the basis of *MAYBE FOREVER*'s representation of a fading love relationship. In what way are instances of touch between the dancers also moments of *being touched*, of emotional stirring? Is this restricted to the relation between the two dancers or does it also – and in what way – affect the audience? In what follows, the movements of touch shall be examined using the examples of three gestures in *MAYBE FOREVER: drawing near*, *tender touch*, and *withdrawing.*

Drawing Near

Slowly and haltingly, as though they were struggling against great resistance, Stuart and Gehmacher move towards each other. Yet their attempts at drawing closer to one another do not lead to a direct contact of bodies: they miss each other, walk by each other, and begin a renewed attempt to touch. Their bodies rarely meet, sometimes forming an embrace which is broken off unfinished because the other retreats from this touch and withdraws. Suddenly, the dancers hold each other tightly by the hands, as tightly as if they felt obliged not to let go of this fleeting moment. Yet these instances of touch also vanish after only a few moments, Gehmacher retreats, Stuart persists for a while – her hands stretched towards him. She follows him slowly, once again they hold hands, and once again they disengage after a few moments. For neither of

Video still from *MAYBE FOREVER*

Video still from *MAYBE FOREVER*

them is touch an original, primeval situation of protective intimacy, an expression of emotional closeness; rather, it is an act of overcoming.

These instances of touch cannot be understood as the mere moment of contact between one's hand and the other one's body, they are movements that begin with the body *drawing near* the other. These movements of *drawing near* do not precede touch, they are rather an essential part of touch itself. Erin Manning describes an element of potentiality in this movement towards the other, in the commenced but not yet completed touch: "Potentiality is at the heart of this reaching-toward."[2] According to Manning, it is potentiality that enables touch without it being enacted: touch is not yet carried into execution, it is not (yet) present, and yet it is already prefigured.

Stuart's and Gehmacher's movements are attempts at convergence, which question the notion of a potentiality of touch through their temporal elongation, their interruptions, their diversions, and their ever renewed postponement of bodily contact. In *MAYBE FOREVER*, touch is not always carried out, it is truncated, the dancers change their direction of movement, and thus draw apart once more. They approach each other with arms opened forward, but when their bodies touch, they do not put their arms around each other, instead turning away and withdrawing (with their arms still outstretched), thus interrupting the contact of their bodies.

Their *drawing near* is not the beginning of a linear movement whose aim it is to touch the other. Their movements could rather be characterized by elements of postponement, hesitation, and suspension. Gehmacher's movements in particular seem defined on the one hand by strong restlessness and frequent changes in direction, and on the other by constant interruptions and standstills. There is thus a recurrence of moments in which no purposeful sequences of movements or actions are discernible, or in which initiated actions are suspended. These movements of the dancers do not point directly to the moment of touch and can thus be seen as a potentiality which is always also a potentiality *not to touch.*[3] Through this dual possibility of potentiality, the moment of decision between touching and not touching appears in the process of the *drawing near* of the bodies, or rather the *drawing near* itself precedes the decision, delays it, and suspends it for the time being. The movement of *drawing near* could therefore be understood as a moment of non-decision, which maintains touch in the realm of potentiality.

The meeting of the bodies is the act in which the decision finally takes place, an act that is always also one of violence. Not only does it erase the potentiality *not to*, it also entails a decision whose consequences reach beyond the individual enacting the touch. Unlike the offer of touch, the other cannot elude bodily

---

2 Manning, Erin, *Politics of Touch: Sense, Movement, Sovereignty*, Minneapolis 2007, 7.

3 Agamben also describes *potentiality* as a *potentiality not to*. Understood in this sense, potency is thus less a "power to do something" than also the "possibility not to do something," to interrupt and refrain from it (cf. Agamben, Giorgio, "On Potentiality," in: *Potentialities*, Stanford 1999, 177–184).

contact. She/he is touched without being able to oppose this touch. The direct implication of the other, perhaps against her/his will, turns the decision to touch into an ethical decision. While *drawing near* contains a potential suspension of touch that does not infringe upon and violate the other in her/his otherness, this suspension always already points to the moment in which the decision to touch is made, it is indeed the very condition of the possibility of touch. Touch is not simply its own suspension, the pure potentiality of two eternally converging bodies, it is precisely the movement consisting of approach and contact, potentiality and actuality, thus forming a dual mode of touch.

In order to preserve this dual mode, the act of touch must not be conceived as the causal consequence of *drawing near*, the former supplanting and erasing the latter. Potentiality must persist in the moment of bodily contact. If *drawing near* is to be considered both as a *potentiality to* and as a *potency not to*, the question arises as to what happens with this potency *not to touch* in the moment of contact? Following Giorgio Agamben, one could conceive of a form of action that would not erase the potentiality, but would allow it to subsist within itself. Rather than a causal consequence of the potentiality, the act thus becomes the simultaneity of its execution and its own suspension. Touch becomes a gesture, of which Agamben writes that it is "a potency that does not pass into the act in order to deplete itself within it, but that persists as potency and dances within the act."[4]

In *MAYBE FOREVER*, the dual mode of touch – the potentiality and its act – is suggested in Gehmacher's and Stuart's wavering movements of *drawing near*. Again and again, the process of *drawing near* is interrupted and begun anew. In these unfinished movements, characterized by their own interruption, or in which touch misses its end and bodily contact remains partial, the potentiality of touch comes to light. Even in the bodily contact made possible by *drawing near*, this potentiality is not erased. When the bodies of the two dancers touch, begin to embrace, a potentiality remains in the incompleteness of the embrace: in the very moment in which the arms are not (yet) closed, a non-touch appears, which allows for an openness of the possibilities of movement.

---

4 Agamben, Giorgio, "Noten zur Geste," in: Hemma Schmutz/Tanja Widmann (ed.), *Dass die Körper sprechen, auch das wissen wir seit langem*, Cologne 2004, 39–48, 46 (translation M.C.).

Tender Touch

It is "the tender" in touch[5] that is on the one hand paradigmatic for *MAYBE FOREVER* and on the other hand repeatedly called into question. In *MAYBE FOREVER*, Stuart and Gehmacher make regular use of "ready-mades" of all too well known and everyday forms of touch in romantic relationships.[6] However, these movements do not accord with their normally ascribed association with romantic love, their caresses become "distracted gestures that go past an alleged end";[7] they break apart, break off, and thus expose the cultural formation of the human body and its movements. These are movements characterized by ambiguity: is the dancers' touch tender or violent? Is Stuart's forceful stroking over Gehmacher's body with extended arms an act of closeness, of contact, or a movement meant to establish distance? It is the fragility of their embrace that makes Stuart and Gehmacher lose their balance and fall. This embrace goes beyond itself. Was it too much? Was it too little? Must tender touch always be carried out in a measured way? Must it be tactful?[8]

Emmanuel Levinas describes the touch that carries within it an element of "the tender" as a caress.[9] According to him, the caress is the physical expression of a desire felt for the other, it is the calling and reaching for the other in her/his otherness. This otherness, which Levinas views as a fundamental condition of possibility both for eros and for ethics, is desired in love (and in the caress as the bodily movement of love), but must not be subsumed by it. On the one hand, the other is the end of this erotic desire, on the other hand, the desire must go beyond the other; it must desire something that is not (yet). The erotic movement of

5 The term "the tender" refers to Emmanuel Levinas. In contrast to *tenderness*, *the tender* is not a categorical description of specific forms of touch, but rather an element that may appear in various forms of touch. On Levinas' notion of "the tender," cf. also Derrida, Jacques, *On Touching – Jean-Luc Nancy*, Stanford 2005, 92 ff.

6 On the notion of the "ready-made" in Stuart's choreographies, cf. Brandstetter, Gabriele, "Figur und Placement," in: Hermann Danuser (ed.), *Musiktheater heute: Internationales Symposion der Paul-Sacher-Stiftung Basel 2001*, Mainz 2003, 310–326.

7 Brandstetter, Gabriele, "Körper-Transformationen in zeitgenössischen Tanz-Performances: Benoît Lachambre, Meg Stuart, Xavier Le Roy und Jeremy Wade," in: Alexandra Kleihues/Barbara Naumann/Edgar Pankow (ed.), *Intermedien: Zur kulturellen und artistischen Übertragung*, Zurich 2010, 343–357, 348 f. (translation M.C.).

8 The French word "le tact" (just as the English word "tact") means both "a form of the sense of touch" and "tactfulness." Both go back to the Latin word "tangere," a root that can also be found in the German word "Kontakt."

9 Cf. Levinas, Emmanuel, *Totality and Infinity: An Essay on Exteriority*, Pittsburgh 1969, 256–266.

the caress brings together the two bodies. Yet, in touch, the other must not be grasped and restrained, the caress also touches that which is not (touchable).

> The caress consists in seizing upon nothing, in soliciting what ceaselessly escapes its form toward a future never future enough, in soliciting what slips away as though it *were not yet*. It *searches*, it forages. It is not an intentionality of disclosure but of search: a movement unto the invisible. In a certain sense it *expresses* love, but suffers from an inability to tell it.[10]

The caress is a movement of search for something that lies in the future. This future is not simply something that is not yet but will certainly soon occur, but rather something that is "closed and dormant beyond the *future*, consequently dormant quite otherwise than the *possible*, which would be open to anticipation."[11] In this conception of the future – akin to the notion of the potential – no linearity of time is imagined; rather, the future is understood as something that does not belong to the order of the presence of the present, but stands in (non-linear) relation to the present, and in this case, to the presence of the other. Through the caress, the face of the other, which attracts desire, stands in close relation to its own absence, the "inversion of the face" (Derrida). In the caress, this realm of the unsayable, which does not simply lie beyond the sayable, but as its negation remains in relation to it, is desired just as much as the face of the other itself.

For Levinas, the caress as a movement of touch goes beyond touching the face or the body, and thus beyond "the tender": it touches upon something untouchable, something that is not (yet) there, but makes possible touching the other as an other. Touch that would only be aimed at what is present, at a graspable body, would not be tender, it would be violent or, to use one of Derrida's counter terms to "caress," a "blow."[12] This juxtaposition of caressing and striking must not be equated with the distinction between touching and not-touching, which in this context appears as an element of "the tender" in touch. Caressing, as a counter term to striking, is not only the omitted touching of the present

---

10 Ibid., 257 f. [emphasis in the original].

11 Ibid., 258 [emphasis in the original].

12 On the relation between caress and violence, Derrida writes: "The tact of a caress, [...] also plays with death, and not only with death as the absence of life or the negative absence of whatever it may be, but with murder. Death at one's fingertips – that is what the movement of this hand is. Far from saying 'You shall not commit murder,' the first of the commandments and the opening of the ethical according to Levinas, far from calling for respect of the face that is as one with this commandment, the femininity of the tender invites profanation, 'at the limit of the obscene,' 'disrespect,' in this 'inversion of the face' and in accordance with this 'disfigurement': it is not the absence of good, it is evil." (Derrida, *On Touching*, 88)

body. It is precisely the simultaneity of touching and not-touching, the possibility of "the tender" in touch. It cancels the opposition between these two terms and, in its dual mode, constitutes the opposite of a clearly definable bodily contact such as dealing a blow.

Many instances of touch in *MAYBE FOREVER* contain these elements of "the tender" that cannot be unequivocally determined as a caress or a violent blow. Take, for instance, an embrace performed by Gehmacher, which ends with him lifting Stuart, carrying her slightly to the side, and then dropping her, or conversely, the moment she pulls him to the ground in her embrace before withdrawing from him. These movements are characterized by "the tender," yet they make equally clear that the caress also contains elements of impending violence. The instances of caressing touch in *MAYBE FOREVER* are never unequivocal: moments of "the tender" are pushed aside by violence-filled movements, and there is a constant need for new movements that attempt to touch the untouchable. Yet this violence does not step in from an outside position, it cannot obliterate "the tender" as its opposite. As part of the caress itself, the violence always also stands in relation to "the tender." The dancers' violent dropping or pushing away would be something else if no tender moment had taken place before. And conversely, the caress would not be possible if it did not always also desire and elicit an element of violence.

Withdrawing

Stuart's and Gehmacher's movements rarely come to a standstill in the moment of touch, they are continued, go beyond the aspects of bodily contact. Let us consider as an example the last instance of touch between Gehmacher and Stuart in *MAYBE FOREVER:* on the edge of the stage, behind the open curtain, the two dancers embrace one last time. The interplay of rigid, almost violent clutching at the other one's body and mutual holding gives way to a slow, difficult distancing. In this movement of *withdrawing*, a dissociation occurs between the dancers' body parts: while they use their legs to move away from each other, their upper bodies (especially their arms) linger in the position of touch. Facing, arms stretched out towards the other body, they move away from each other. Slowly, lagging behind the rest of the body, the arms also loosen their position and join in with the other, new movements. Stuart's subsequent solo is thus initially still clearly defined by her outstretched arm, which as an 'echo' of the recently loosened touch informs and disrupts her new movements. In the movement of *withdrawing*, just as in *drawing near*, the bodies appear in taut relation to each other.

It is the touch that was just now still there, that is now vanishing – for, just as *drawing near*, *withdrawing* is not an add-on to touch, it belongs to touch itself.

In his interpretation of the “Noli me tangere” scene, Jean-Luc Nancy identifies two dimensions of *withdrawing:* withdrawing not only changes the spatial distance between two individuals, the body of the person withdrawing also changes. It vanishes and is present as a vanishing body.[13] This twofold movement is essential for touch: in touch, both the movements of the body in space and the changing body itself are at stake. Only if these two movements are not discontinued is touch possible. The vanishing of the body and the *withdrawing* are intimately connected: only the withdrawing body moving in space can also be a vanishing body.

In what way do the two dancers stand in relation to each other when withdrawing in *MAYBE FOREVER?* Their bodies are facing each other, and their arms are extended towards the other. This lingering of the movement of touch leads to overlaps with subsequent movements. Take Stuart’s final solo: starting from a (last) embrace with Gehmacher, she moves back to the center of the stage, her gaze and her body still directed towards him. As she withdraws, the fingertips of their outstretched arms still touch. In the subsequent solo, the position of the arm does not immediately vanish, it merely fades slowly and blends in with the new motion sequences. Tentatively, Stuart brings her arm back to her body, yet without giving up its extended position and her gaze along its length: arm and gaze still move synchronously. Then she starts bending the arm; she initiates a self-embrace, in which her gaze also lets go of her fingertips. In spite of what is now an almost entirely new position, the arm appears to resist her other, new movements, like a foreign body, an element of disruption. A moment of simultaneity occurs – several movements overlap, disrupt, and condition one another. Even though the moment of bodily contact has passed, its trace remains present in the movements of the solo. It is the imprint of the elapsed instance of touch, which leaves its trace and does not immediately give way to another movement. The trace of touch shows that touch does not come to an end with the discontinuation of bodily contact. In *withdrawing*, it holds present its own absence. The slow vanishing makes it impossible to define a clear endpoint of the movement of touch. Unlike *drawing near*, *withdrawing* is not based on a con-

---

13 In his study of the “Noli me tangere” scene in the New Testament, Nancy describes the body of Jesus, which as a vanishing and transforming body must not be touched: “The true movement of giving oneself is not to deliver up a thing to be taken hold of but to permit the touching of a presence and consequently the eclipse, the absence, and the departure according to which a presence must always give itself in order to present itself.” (Nancy, Jean-Luc, *Noli me tangere: On the Raising of the Body*, New York 2008, 50)

cept of the potential of touch. Here, touch has (perhaps) taken place, the non-touch of *withdrawing* is a *no more* of touch; Stuart's outstretched arm is – at first glance – pointing to the past.

These possibly irreconcilably different conceptions of non-touch – *drawing near* and *withdrawing* – cannot be reduced to a common term, nor are they independent of each other. They always appear in conjunction: only the person who has drawn near can withdraw. In *MAYBE FOREVER*, the dancers' *withdrawing* creates the very conditions for the next *drawing near*, which in turn gives way to retreating, in an ever renewed configuration of a process of movements that is brought neither to a standstill nor to an end during the performance.

Movements of Touch

These highly diverse moments of touch between the two dancers share a *being in movement*. The instances of touch are movements of bodies that draw near and withdraw, that find themselves in an interplay of closeness and distance, connection and separation. Lasting touch would immobilize the dancers and thus freeze the other body as well as one's own. In order to touch, one must always begin anew,[14] for the movements do not lead directly toward the moment of touch or away from it; neither do they establish touch as their end, nor do they establish themselves as touch (and thus as an end), they are "means without end" (Agamben).[15] These are movements that are possible only in the (accruing or vanishing) distance of the bodies; a distance that is not touch (in the strict sense of bodily contact) but non-touch. Touch is interrupted, postponed, or suspended in its movement by non-touch. At the same time, non-touch as *drawing near* is what enables touch in the first place. The movement of touch is not – as

14 Cf. also Erin Manning's concept of touch as movement: "My gesture toward you is a momentary one. There is no touch that can last beyond the first moment of contact. To touch longer, I must touch again: as my focus shifts elsewhere, my skin soon forgets to acknowledge yours. To touch me you must return the touch to and from yourself in an ongoing process of exchange. Because it is temporary and immediate, the gesture is never more than momentary. This is a political moment in the most ethical sense, for it demands a continual re-articulation rather than subsuming into the same." (Manning, *Politics of Touch*, 60)

15 As regards the concept of "means without end" cf. also Agamben's notion of the gesture: the gesture is neither a means directed towards an end, nor does it declare itself to be an end, rather "the gesture ... breaks with the false alternative between ends and means that paralyzes morality and presents instead means that, *as such*, evade the orbit of mediality without becoming, for this reason, ends." [emphasis in the original] (Agamben, Giorgio, "Notes on Gesture," in: *Means without End: Notes on Politics*, Minneapolis, London 2000, 49–60, 57).

the terms *drawing near* and *withdrawing* might suggest at first glance – linear, the relation between touching and not-touching is one that disrupts the linear configurations of movement and thereby makes possible non-touch in touch. Non-touch is thus not simply the opposite of touch, it is not *not touching*. Non-touch is itself part of touch. It is what makes touch possible, and yet it is not its equivalent – it is at the same time that which interrupts, disrupts, and suspends touch.

These descriptions of (non-)touch not only question the immediate contact postulated between two people touching, but also the possibility of its perception, in this case its perception by the audience. When does touch take place? At what point does a movement become *drawing near*, and for how long can it be described as *withdrawing?* Unequivocal identification of instances of touch by the observer seems to become impossible: perceptual framings dissolve in ever renewed configurations of the *not yet* or *no more* of touching. Linear space–time parameters with their clear subdivisions in presence and absence can no longer serve as unequivocal description of movements of touch, since in their simultaneity of touching and not-touching, these movements of touch elude unequivocal presence or, as Krassimira Kruschkova – quoting Derrida – phrases it (not only) regarding touch: "It takes place when it doesn't."[16]

Touching Bodies

Non-touch not only keeps suspending the act of touch, it also questions the two touching bodies as autonomous entities. In a dual mode, it interrupts the perceivable and definable forms of the bodies: touch is opening and closing, defiguration and refiguration of their boundaries.

In this context, the following scenes in *MAYBE FOREVER* could be described as moments of the de-finition of bodily boundaries: Stuart and Gehmacher are facing each other at arm's length after a series of movements of which it is not possible to say whether they are meant to keep the other one away or pull him/her close. Slowly, Stuart starts caressing his head with her hand, then lets her hand glide along the neck, the shoulder, the chest, the stomach, further down, until – having arrived at the legs – she slowly breaks off contact with Gehmacher and, in a similar manner, feels along her own body's boundaries. What is happening in this movement? While our first association is one of caressing, tenderly and lovingly, soon what arises is the impression of a much

---

16 Cf. Kruschkova, Krassimira, "The *Fault Lines* of Touching," 105–111, 106 (in this volume); Derrida, Jacques, *Dissemination*, London 1981, 213.

harder touch, of a stroking across the body. Distinctly and with precise movements, Stuart traces Gehmacher's body outline, delineates it one more time with her extended arm, which, while it establishes contact through touch, equally serves as a measure of the distance between herself and Gehmacher. The twofold touch of caressing and stroking establishes both contact and distance between the two dancers. In this instance of touch, the form of Gehmacher's body is also de-fined in a twofold manner: his body is linked to the other body and at the same time determined in its boundaries. The body, writes Nancy, is opened in touch, this opening being conceivable only in relation to something closed.[17]

Here, the body is thought from its exterior, its form, i.e. the skin. It is this boundary that – just as touch itself – is interrupted. The body's boundaries cannot be determined unambiguously, they are open, blurred – they are no longer a self-contained entity. Just as in touch, the opening of the boundary is not the dissolving, not the melting of the bodily form, but precisely its condition of possibility, yet at the same time the impossibility of absolute self-containment. The body is thus no fixed and rigid form, but rather the simultaneity of opening and closing.

For Nancy, form, and thus body, means not the delimitation of a self-contained entity, but rather the opening outward of the body: bodies engage in articulation with other bodies through touch. The articulation of bodies, which necessarily takes place between bodies, always comes from the outside; touch, including the touch of one's own body, is something that happens on the skin, on the body's boundary. "It's through my skin that I touch myself. And I touch myself from outside, I don't touch myself from inside."[18] From outside, touch creates the form of the body and thus the body itself. This form, this entity is unique, it makes the body – and thus the "I" – appear as a singularity. Yet, still according to Nancy, neither the body nor the subject are self-contained and exclusively self-referential; rather, through touch, they stand in relation to the outside.[19]

---

17 On the concept of the body used here, cf. Nancy, Jean-Luc, *Corpus*, New York 2008, especially his text "On the Soul" (122–135). On the relation of the open and the closed, Nancy writes: "[T]he body is the open. And in order for there to be an opening, something has to be closed, we have to touch upon closure. To touch on what's closed is already to open it. Perhaps there's only ever an opening by way of a touching or a touch. And to open – to touch – is not to tear, dismember, destroy." (122)

18 Ibid., 128.

19 Nancy describes the touched body as (impossible) unity of the subject: "The body is the unity of a being outside itself. [...] After that, it becomes instead a question of thinking the unity of being outside the self, the unity of the coming to self as a 'self-sensing' a 'self-touching' that necessarily passes through the outside – which is why I can't sense myself without sensing

In dance also, bodies and their forms are a constant topic. Both the ideal of a perfectly trained and shaped body and its critical calling into question determine the concepts (not only) of contemporary choreographers. "Prefigured" bodily forms and movements are questioned, interrupted, and shifted using the means of dance.[20] The entity of the body and its movements (but also the elements of change and transformation) form the "figure" of the dancer. As the outline of the representing body, writes Brandstetter, the figure describes precisely that boundary, that interface between inside and outside, one's own and someone else's, which de-fines the body from its outermost borders. In dance, this body cannot be separated from its movements: as a body, it moves through space, shifts its position, and with it its relations to others. Yet even the body itself, its form, does not remain static in movement – it is transformed: "However, the principle of these transformations in movement is not consistent with the model of metamorphosis. Rather, it is to do with monstrous, de-monstrative transgressions of body boundaries."[21] The movement of the figure as a process of "transformation/transfiguration" (Brandstetter) is not a movement that passes from one unitary figure into another, but rather a process of constant change, in which establishing individual familiar and self-contained figures becomes impossible. Figures of bodies and movements in dance are no static states, they are "processes of defiguration and refiguration."[22] Brandstetter inscribes within the notion of the figure its own dissolving, figure becomes a notion of movement, which ends neither in the self-contained form nor in absolute formlessness, which does not end at all, but pushes ahead to ever new figures and forms. The form of the dancing body is movement, it is performative (beyond linear conceptions of time), and as a figure it is involved in a process of change and transformation.

Here, it is the movements of touch that con-figure the body, open it to an outside, break open and interrupt its closure – de-figure it. Bodies, writes Nancy, are forms that are articulated in touch. In *MAYBE FOREVER*, the instances of touch between the dancers, the articulation and the simultaneous breaking open of bodily forms and figures, could be described in the touch of caressing/stroking. In this case, caressing/stroking the body of the other would be a gesture that articulates the body, and in this articulation makes apparent the opening and re-

otherness and without being sensed by the other. It involves thinking the unity of what a little earlier I called articulation, unity as a form, which is inevitably an articulation." (ibid., 133)

20 Cf. Brandstetter, *Körper-Transformationen in zeitgenössischen Tanz-Performances*, 343.

21 Ibid. (translation M.C.).

22 Brandstetter, Gabriele, "Defigurative Choreography: From Marcel Duchamp to William Forsythe," in: *The Drama Review* 42.4 (1998): 37–55, 38.

lationality of the bodies. The bodies of the two dancers appear as bodies of touch, which change and reconfigure themselves with each new touch.

Yet which touches, which movements enable the transformations of the bodies, and which freeze them? Instances of touch that exclude elements of non-touch – of *drawing near*, of *withdrawing*, of *the tender* – make the movements of the bodies impossible. In their ambivalence, elements of non-touch not only enable and disrupt instances of touch, but also the body itself: whilst the act of touching determines the body's boundaries, these are opened (again) by elements of non-touch, making possible a relation to the other, to the one touching. Touching is opening, is articulation, is the relation to the outside, and in order to be all these things, it requires an element of (non-)touch, of set bodies and boundaries, of its own interruption.

As tender caressing, Stuart's touch appears to be searching for what Gehmacher's body (form) is not (yet), and opening it for a forthcoming, potential re-figuration, yet one that in its very potentiality becomes formless or de-figurative. Caressing does not persist as a lasting state, in its very movements of search it brings about this re-figuration, actualizes it. Caressing tips over into stroking, Gehmacher's body boundaries are determined (anew) and the initial tenderness now appears as an act of violent freezing of his body.

Here, instances of touch are more than mere movements. They (and not the holding) are what make the body boundaries of the two dancers appear open, and their bodies thus not as self-contained entities, but as figures in motion, as "processes of de- and refiguration." And it is also this openness that makes the movement of touch appear not as an act or as a completed entity. No instance of touch in *MAYBE FOREVER* is exclusively tender or violent, none is only a caress, none only a blow – and yet they differ from one another. The configurations of non-touch and touch are what make every instance of touch different from the other. Even an extreme touch such as a blow can merely ignore the element of non-touch, but not erase it. And on the other hand, no touch can make do without its execution, which in this case is to be thought essentially in both its ethical and its erotic dimensions. Beyond classifications, touches are movements that disrupt and break open themselves, but also the bodies of those touching, without entering these breaches and without closing them again.

Relations of Touch

MAYBE FOREVER repeatedly represents fragments of a (fading) love relationship. The advances, the shared experiences, the closeness, the (small) fights, the struggles, and lastly the separation, all of it repeatedly interspersed with text fragments that seem to have come from another era, just as the music, the costumes, the carpet, and the sepia-toned image of two dandelion flowers as a stage backdrop. All these elements remove the action to an inaccessible past, from which it seems equally impossible to escape. It is the representation of a downright stereotypical heterosexual relationship,[23] yet one that goes beyond a simple image or representation, keeps exceeding and displacing it.

The emotionality of love, just as desire, is in no way the natural basis of relationships, it can only be analyzed in its socio-cultural context. Yet at the same time, it constitutes the element that regularly breaks open and questions these constructs.

How do these breaks occur within the love relationship depicted in *MAYBE FOREVER* and the emotional configurations it implies? In what way is it the instances of touch between the dancers – especially the abovementioned moments of hesitation, drawing near, tenderness, or withdrawing – that make possible elements of affect, of potentiality, and thus of *being touched?*

The little gestures between the two dancers in *MAYBE FOREVER*, those instances of touch that within a love relationship establish intimacy in daily life, look a bit too large on the empty stage – they seem displaced, amiss, discontinuous, or aborted. The dancers' movements are tense, too tense to suggest tenderness. The loving caress of the other one's arm does not open up possibilities of further contact, but leads to a distance between the two bodies. Just as the tender clasping of the wrists, which at first looks like an intimate play between lovers, soon appears violent; a violence that destroys any preceding associations with tenderness. Stuart's extending of her arms in order to touch Gehmacher's body remains mainly one thing: indeterminate. Do her arms establish contact with his body or keep him at a distance? A few scenes earlier: the attempt of an embrace, in which Stuart pushes her head and upper body into the arms Gehmacher is extending towards her, accepting his offer of an embrace, and in turn laying her own arm around his neck. Yet she does not interrupt the movement of *entering the embrace*, she does not remain in his arms, but continues her movement, until their bodies have once more moved away from each other.

---

23 The concept of the heterosexual relationship is neither natural, nor to be described as the social organizational form of natural lust and desire. Since its emergence in the 18th century, the (lifelong) partnership (mostly in its institutionalized form of marriage) has been one of various powerful apparatuses, eliciting our structures of desire and inscribing them within the heterosexual and patriarchal discourses of our society as biopolitical modes of operation and subjectification of a (patriarchal) power, which Michel Foucault and Judith Butler have analyzed at length. The notion of romantic love, which since the 18th century has been intimately linked to physical lust, can be understood as the motive providing meaning and coherence to the justification of heterosexual partnerships, and thus as an essential part of this patriarchal apparatus.

However, these moments of separation, which keep appearing in the instances of touch between the two dancers, are not breaks in their relation, they question the conventional relationship model that is being represented, and are at the same time the condition of possibility for a love relationship beyond oneness and immediacy. The relation established by the two dancers in *MAYBE FOREVER* is determined by those moments described above as characteristics of touch: moments of hesitation, of doubt, of separation, of distance, and – to put it in the terminology of emotions – of shame: "Shame is the feeling of the distance that imposes itself in closeness, and this distance derives from the remove that is endless – and endlessly free – it derives from the remove that love opens up in the existing."[24] What is the shame that interrupts so many instances of touch in *MAYBE FOREVER* and keeps pushing the dancers to retreat? The shame of getting *too close* to the other and hurting her/him? Or the fear of revealing *too much* of one's own body and its intimacy, of opening the body *too much?* According to Nancy, the distance in the shame of touching, or rather of not-touching, of hesitating and retreating, is constitutive for the love relationship, whose love is always already shattered,[25] but which precisely in this fact finds the possibility to open up to the other. This opening creates the possibility for singularities to come together without fusing into a whole that is always predicated on oneness, appropriation, and the exclusion of difference. In her book *I Love to You*, Luce Irigaray shows that this imagined *becoming one* is not only a phantasm of male universalism and claim to absoluteness, but that it equally secures the patriarchal domination in our relationship concepts. Irigaray demands a mode of love that, just as for Nancy, is not predicated on becoming one (subsuming of the lovers by the male), but is understood as the communication between two individuals. "*I Love to you* means I maintain a relation of indirection to you. I do not subjugate you or consume you."[26] Irigaray thus no longer understands love as a state of oneness and unambiguousness; in fact, love no longer even denotes a state. From a distance, the two lovers turn *to* each other, they articulate, and thus draw closer to each other (a movement that, as shown above, always requires a distance to the other). It is this movement of *turning to* described by Irigaray that makes possible a relation of love to the other, without subsuming her/him and destroying her/his singularity. Such descriptions of

---

24 Nancy, Jean-Luc, "Ereignis der Liebe," in: Nikolaus Müller-Schöll (ed.), *Ereignis: Eine fundamentale Kategorie der Zeiterfahrung. Anspruch und Aporien*, Bielefeld 2003, 21–36, 30 (translation M.C.).

25 Nancy, Jean-Luc, "Shattered Love," in: *A Finite Thinking*, ed. Simon Sparks, Stanford 2003, 245–274.

26 Irigaray, Luce, *I Love to You: Sketch for a Felicity Within History*, London, New York 1996, 109.

love as a movement can also be found in Nancy, though not as a movement of two individuals (a man, a woman), but rather as a movement that cuts across such concepts of masculinity and femininity, or in more general terms, concepts of individual entities.[27] According to Nancy, the love relation is

> nothing other than the coming-and-going. The other comes and cuts across me, because it immediately leaves for the other: it does not return to itself, because it leaves only in order to come again. This crossing breaks the heart: this is not necessarily bloody or tragic, it is beyond an opposition between the tragic and serenity or gaiety. The break is nothing more than a touch, but the touch is not less deep than a wound.[28]

This relation of love is a movement, a movement of touch, which always also breaks open the other one's body (and heart), yet without fully destroying her/him in her/his form and thus in her/his singularity. The movement of the love relationship is a rhythm of *drawing near* and *withdrawing*, a movement that neither has reached its end, nor is moving in a linear fashion towards an existing end; love is always a movement of arriving, "maybe even when it leaves, and also when it comes back."[29]

In this sense, the instances of touch between the dancers in *MAYBE FOREVER* could also be described as movements of a love relationship, although or perhaps precisely because they elude unequivocal ascriptions of closeness, intimacy, or joy. Their movements of touch cannot be assigned unequivocally to specific emotions, for it is the distance between both dancers necessary for both the love relationship and the movements that breaks open this unambiguousness and makes it possible, e.g., for their touch to contain not only affection and tenderness, but also aspects of shame. While terms such as joy or grief can always serve to describe instances of touch, a categorical attribution becomes impossible if instances of touch are not simply understood as expressions of affection for the other or fear of losing her/him. Movements of touch such as they have been described here are not an expression of distinct emotions,[30] but rather

---

27 Nancy describes love as something that cuts across and thus questions the subjects' boundaries, especially their gender determination. "It [, the love, G.E.] is sexual, and it is not: it cuts across the sexes with another difference (Derrida, in *Geschlecht*, initiated the analysis of this) that does not abolish them, but displaces their identities. Whatever my love is, it cuts across my identity, my sexual property, that objectification by which I am a masculine or feminine subject." (Nancy, "Shattered Love," 266)

28 Ibid., 262.

29 Id., "Ereignis der Liebe," 21 (translation M.C.).

30 Various approaches in the theory of emotion take as their starting point a series of human emotions that can be read off the human body on the basis of distinguishable facial expressions. For example, Paul Ekman describes emotions such as fear, anger, joy, disgust, etc. as distinct

complex emotional constellations between the dancers. This emotional indeterminacy shall not be described here as a conceptual deficiency or a pathological disturbance; rather, it is the expression of the affect that exceeds and breaks through patriarchal concepts of conventional love relationships.

Just as in movements of touch, in love relationships also it is precisely the affect that does not yet possess a form that on the one hand breaks open the "formedness" of the emotion as a potentiality, and on the other hand is what makes it possible in the first place.[31] In the distance to the other, an opening is created, which makes it possible to be touched by her/him without the breach being closed and the relation being formed. Yet, as moments of potentiality, these breaches never last, they push for form, for actualization, they are therefore never beyond dominant discourses. Neither the love relationship nor the dance are to be understood here as possibilities of escaping these heteronormative practices, and yet these are modified, fractured, and their social configurations of emotionality are displaced.

When Stuart and Gehmacher move towards each other, initiate an embrace, change their movements, push each other away only to reunite elsewhere, these instances of touch create relations not only between the two of them, the audience is also touched. Yet how can touch between the dancers touch beyond itself, how – in the absence of bodily contact – can it establish a relation to the audience? The instances of touch on stage do not represent emotions of the dancers, they even disrupt the structures of representation in *MAYBE FOREVER*, and yet they do not necessarily lead to a termination or a suspension of the audience's aesthetic experience. The audience's *being touched* becomes possible in the very moments in which no familiar emotional states are represented and recognized, but rather in which the audience's perception is momentarily irritated, delayed, and diverted from what is being represented. The audience is able to open up to being touched in the sense of a potentiality of touch, a *being touched* that perceives and establishes a relation that possesses no emotional form (yet). It is precisely in interruption, in distance, and in formlessness that the instances

---

basic emotions (cf. Ekman, Paul, "Basic Emotions", in: Tim Dalgleish/Mick Power (ed.), *Handbook of Cognition and Emotion*, Sussex 1999, 45–60).

31 I follow Brian Massumi's notion of emotion, which following Spinoza describes it in simultaneous demarcation from and relation to the affect. The affect belongs to a different logic than the emotion, it cuts across the emotion, can disrupt, diminish, or reinforce it, yet at the same time it is its condition of possibility: "Formed, qualified, situated perceptions and cognitions fulfilling functions of actual connection or blockage are the capture and closure of affect. Emotion is the most intense (most contracted) expression of that *capture* – and of the fact that something has always and again escaped." (Massumi, Brian, *Parables for the Virtual*, Durham, London 2002, 35)

of touch between the dancers, but also between the dancers and the audience, allow the emergence of emotional configurations that do not (yet) possess a form, but first create the possibility of being touched. These instances of touch move, draw near, and put at a distance both the bodies of the dancers and the audience; they are movements of which Stuart writes:

> It is the ongoing practice of connection, accepting the unpredictable present without escaping or forcing an outcome. It is not easy and demands constant practice – moving, letting the invisible bones be a guide, buffering the inevitable projections of discomfort and emotion that are present in any challenging new connection and meeting.[32]

Translated by Maud Capelle

References

Agamben, Giorgio, "Noten zur Geste," in: Hemma Schmutz/Tanja Widmann (ed.), *Dass die Körper sprechen, auch das wissen wir seit langem*, Cologne 2004, 39–48.
Agamben, Giorgio, "Notes on Gesture," in: *Means without End: Notes on Politics*, Minneapolis, London 2000, 49–60.
Agamben, Giorgio, "On Potentiality," in: *Potentialities*, Stanford 1999, 177–184.
Brandstetter, Gabriele, "Körper-Transformationen in zeitgenössischen Tanz-Performances: Benoît Lachambre, Meg Stuart, Xavier Le Roy und Jeremy Wade," in: Alexandra Kleihues/Barbara Naumann/Edgar Pankow (ed.), *Intermedien: Zur kulturellen und artistischen Übertragung*, Zurich 2010, 343–357.
Brandstetter, Gabriele, "Figur und Placement," in: Hermann Danuser (ed.), *Musiktheater heute: Internationales Symposion der Paul-Sacher-Stiftung Basel 2001*, Mainz 2003, 310–326.
Brandstetter, Gabriele, "Defigurative Choreography: From Marcel Duchamp to William Forsythe," in: *The Drama Review* 42.4 (1998): 37–55.
Derrida, Jacques, *On Touching – Jean-Luc Nancy*, Stanford 2005.
Derrida, Jacques, *Dissemination*, London 1981.
Ekman, Paul, "Basic Emotions", in: Tim Dalgleish/Mick Power (ed.), *Handbook of Cognition and Emotion*, Sussex 1999, 45–60.
Irigaray, Luce, *I Love to You: Sketch for a Felicity Within History*, London, New York 1996.
Kruschkova, Krassimira, "The *Fault Lines* of Touching", XX–XX (in this volume).
Levinas, Emmanuel, *Totality and Infinity: An Essay on Exteriority*, Pittsburgh 1969.
Manning, Erin, *Politics of Touch: Sense, Movement, Sovereignty*, Minneapolis 2007.
Massumi, Brian, *Parables for the Virtual*, Durham, London 2002.
Nancy, Jean-Luc, "Der Lustkörper," in: *Ausdehnung der Seele: Texte zu Körper, Kunst und Tanz*, Berlin, Zurich 2010, 25–30.

32 Stuart, Meg, "What does it mean to make contact?," in: Jeroen Peeters (ed.), *Are we here yet? Damaged Goods: Meg Stuart*, Dijon 2010, 52f., 53.

Nancy, Jean-Luc, *Noli me tangere: On the Raising of the Body*, New York 2008.
Nancy, Jean-Luc, *Corpus*, New York 2008.
Nancy, Jean-Luc, “Ereignis der Liebe,” in: Nikolaus Müller-Schöll (ed.), *Ereignis: Eine fundamentale Kategorie der Zeiterfahrung. Anspruch und Aporien*, Bielefeld 2003, 21–36.
Nancy, Jean-Luc, “Shattered Love,” in: *A Finite Thinking*, ed. Simon Sparks, Stanford 2003, 245–274.
Stuart, Meg, “What does it mean to make contact?,” in: Jeroen Peeters (ed.), *Are we here yet? Damaged Goods: Meg Stuart*, Dijon 2010, 52f.

Katherine Mezur

Cold Burn (Teion Yakedo)

On Touch in Contemporary Japanese Performance

> Loving a disappearing image means finding a way to allow the figure to pass while embracing the tracks of its presence, in the physical fragility of the medium.[1]

The project of this 'cold burn' investigation is to bring new media theory and practice into the messy and fleshy interactive arena of the cultural politics of 'different' dancing bodies in our already transnational visceral and virtual worlds. I want to place this analysis and these performances in the churning, evolving, shape-shifting action at the fulcrum of touch theory and practice. Nothing is stable; everything is in flux, particularly any corporeal condition, nation, or state. Instability is uncomfortable.

Touching and being touched are deeply cultured acts. By this I mean that touch is gendered, classed, raced, and controlled by religious, ethical, social, and bio-medical systems within micro-cultures such as families and macro-cultures such as nation states. Touch is political. The media have further contributed to these complex patterns and controls of touch perception, adding screened, digital, 3D, and animated forms to our visual/kinesthetic systems. Imagining through our bodies, viscerally and virtually, could facilitate an engagement with the complexity of touch perception and its still unexplored knowledge.

This analysis includes the action of performer bodies and the spectators, who may be kinesthetically 'touched' by the performing or projected bodies. I consider both no-tech and hi-tech dance works, but focus on particular moments of choreographed 'almost-touch,'which radically shift and transform our visual/kinesthetic modes of perception through doing less and less with greater intensity. This 'less' is not necessarily reductive or abstract. Instead, in the case of the Japanese artists we will consider, the 'less' of cold burn strikingly expands and intensifies with smallness, stillness, distance, and vagueness in 'touch.' That is, the power of cold burn lies in its particular manipulation of touch sensation that is partially controlled by Japanese rules of behavior and the contemporary mo-

1 Marks, Laura, *Touch: Sensuous Theory and Multisensory Media*, Minneapolis 2002, 96.

ment of subverting those rules of 'touching' discourses. The new media,[2] even with their own culture and politics of gesture and touch, reveal new technologies of perception that further illuminate the complexities of the cultural codes governing sensations, especially touch. I use 'cultural' codes because 'cultural' can include the political, corporeal, spiritual, ethical, and aesthetic codes that shape and produce certain modes of behavior which, in turn, become effective in maintaining a way of life: a culture.

I analyze moments of 'touching' in these works by the Japanese artists in the companies dumb type and chelfitsch, and the solo artist Naoko Tanaka, which deploy a particular kind of 'cold burn' in relation to their individual techniques and outcomes. I also suggest placing these works within the larger context of live and mediated works of the Wooster Group and the video installations by artists such as Shirin Neshat, Bill Viola, and Kim Sooja. I do not want to isolate these artists in some special 'Japanese' cell, but instead reveal the complexities of 'other' touching.

In all three cases presented here, the touch of the performer does not touch another live human. The circuitry of touching and being touched is almost incomplete. This almost-touching is repeated in different ways with different intensities. These different intensities of proximity and energy and duration of touch are at the heart of my theory of 'cold burn.' Their indirectness, distance, and briefness make them quite different from what Jennifer Fisher suggests in her analysis of deliberate skin-to-skin confrontations in certain artworks of Chris Burden or Marina Abramovic that

> produce enlivened zones of tactile intensity, often impacting skin to skin [...] performances featuring gestures of touching other human beings enact particular affective socialities and close-range transference. Yet, while often intimately implicating the audience, these performative touches exceed the codes of personal relationship.[3]

2 There is often an assumption that new media are universal. Yet screens, platforms, and devices in their specific cultural milieu are all in the hands of a user. Similarly, the creators of new media themselves have their own cultural background and are implicated in their own cultured space. The new media artist is situated. New media also have their own culture and politics of gesture and touch: new media are situated. Then the new media artist is situated and re-situates new media with regard to his work. The complexities of the codes governing sensations, especially touch, from personal belief systems to elite corporate or government censorship, are dense and activated in works that cross senses and technological reflections of those senses.

3 Fisher, Jennifer, "Tangible Acts: Touch Performances," in: Sally Banes/Andre Lepecki (ed.), *The Senses in Performance*, New York 2007, 166–178, 168.

Quite differently here, dumb type, chelfitsch, and Tanaka do not "exceed," but instead their intimacy/touch recedes, becomes less.

Cold Burn 2000

In 2000, when media works inundated museums and stages, I went to the millennial opening exhibit at the Museum of Contemporary Art in Tokyo. The exhibit was entitled *Teion Yakedo* (低温 火傷), *Land/mind/body-scapes in the Age of Cold Burn.*[4] Room after room was filled with artworks using mediated images and objects: moving, framed, or sculpted. I experienced this sense of slight, yet constant pressure from the discomforting repeated stimuli: muffled radio sounds, flashing lights, out-of-focus images, glowing neon, all kinesthetic teases. Most works used repetition or excess, like a section of McDonald golden neon arches and neon tubes arranged in the over-familiar patterns of chain fast food and convenience stores. Seven Eleven's murky white, green, and red neon bars glowed along the walls. These works called on the daily, repeated exposure to visual, aural, and kinesthetic signals, photographs, video installations, and interactive works; all used repetition, mediated imagery, and an exaggeration or diminution of the everyday (urban sensorium). The exhibition as a whole left me with this vague sensation of discomfort, dissatisfaction. I sensed through my visual/kinesthetic exposure what the exhibition brochure suggested, "Cold Burn."

Okamura Keiko, the head curator for the MOT (Museum of Contemporary Art Tokyo) Annual 2000 show, curated and formally named the exhibition *Teion Yakedo, Land/mind/body-scapes in the Age of Cold Burn.* She explains how it works physically and then how she experienced this sense theory in the art works:

> A cold or low-temperature burn is a burn that results from over-exposure of the skin to a relatively mild heat source [...] that is not hot enough to cause immediate burning if it is held next to the skin for long enough period of time. The pain emerges only gradually as the victim is enjoying a pleasant sensation of warmth, the heat gradually works into his or her skin with eventually painful results. [...] The problem with a cold burn is that treatment may be delayed because the effects are felt only gradually, so it may actually take lon-

4 Okamura Keiko, *Teian Yakedo, 'Cold Burn' or Land/mind/body-scapes in the Age of Cold Burn*, Museum of Contemporary Art Tokyo Annual, Tokyo 2000. Okamura Keiko is also the curator for the Tokyo Metropolitan Museum of Photography. I should be clear from the start that I am very aware of the fact that I am an outsider, but I hope that my transiting between my US perspective and Japanese 'experience' will cause a kind of productive friction that unsettles the possible essentializing.

> ger to heal. Such burns are seldom fatal, but they cause real pain. [...] This situation can be compared to life in today's society, where the uneventful repetitions of everyday life can be transformed into sources of pain.[5]

When I saw the *Cold Burn* exhibit, I was stuck by these sensations. I felt something uncomfortable on the edge of my consciousness, a vague aching. I found that "cold burn" resonated with various performance works that I had seen. I had so often thought: what is going on here? Why is no one really touching each other? Why are they doing this sequence over and over again? Several artists included in the exhibit had created interactive video installation/performance spaces that were reminiscent of performances where media and live performers intersect through gestures and projections. This cold burn 'touch,' was directly connected to my investigations of body and media interface theories and practices in Japan. This concept was very different from my experience of the direct or enveloping touch in Euro-American contemporary dance, which loads on layers of visual and tactile stimuli, saturating our proprioceptive grid (sometimes even to the point of shutting down from the onslaught). Only the slightest attention is paid to the possible sensitivity of the individual spectator, cultural coding, or prescriptions for viewing. One of the purposes of this study is to suggest other ways of touching and perceiving, which may expand the entire field of kinesthetic empathy.

The Japanese touch performance works presented here are at the other end of the scale, dominated by 'less,' by distance and restraint. Even if the load of sensation is increased, as in some cute girl performance and media, touch at the heart of sense worlds remains slight, repetitive, and disturbing. As Okamura suggests in her comments on works in the 2000 exhibition: "All of these works directly express a sense of strangeness, discomfort, or mystery [...] discovered in the inconsequential events of daily life."[6]

What is cold burn and how does it work? How does cold burn work with touching and being touched? The methods of the artists examined here vary, but in each case their work produces in the viewer/experiencer and performer a kind of 'hyper' sensation that results in a low-intensity dull pain, a longing, a hurt. I suggest that these works produce what Laura Marks in her book *Touch: Sensuous Theory and Multisensory Media* refers to as "haptic images": "Haptic images do not invite identification with a figure so much as they encourage a bodily relationship between the viewer and the image [...] a dynamic sub-

---

5 Ibid., 88f.
6 Ibid, 90.

jectivity between looker and image."[7] Here in the Japanese works, our interaction of touching and being touched is between ourselves and the moving images/gestures. Perhaps this 'not-identifying' but interacting with the digital or flesh figures is part of the cold burn, constantly touching, at a distance.

Cold Burning: Culture and Touch

Cultures control and shape acts of touching and being touched. Gender, highly regulated by state and culture, codes the rules of touch. Belief systems manage touch and sensation. The modern state overrides all local regulations and delineates public, private, and criminal touch. Touch is politically complex and sensitive, and the media may amplify or repress those distinctions. When visual media intersects with touch, for example when we see a projection of an embrace over a live body, matched perfectly in size and shape, or screened flames flickering over a slow motion dancing body: what happens? Do different sensations overlap and filter the 'cultured' touch? Do the projected light, shadow, and image change mediate touch?

Touch is at once visceral and virtual and plays with those complications of media and skin. This analysis includes a variety of actions of touching: live performers touching other bodies and themselves, projected performers touching and being touched, and spectators kinesthetically feeling as if they are 'being touched' by the performing or projected bodies. I consider the range from no-tech to only-tech performance works in order to better assess these situations of touching. We experience our proprioceptive senses differently depending on our cultural upbringing and touch, seen, and felt phenomena. I believe that touching and being touched is fundamental to contemporary performance and visual cultures where narratives have given way to an abstracted, charged seeing-touch. The increased violence in visual news and the entertainment media has also added new codes to our touch censors.

Issues of culture and race mark these cold burn touches. These works are created by late 20th- and early 21st-century artists from Japan, riddled with its own racism, fragmented by waves of Americanization, its own global pop culture of *manga* (comics), *anime* (animated films), and games, and a cinema culture of saturated sensualities. Japan has its own set of patterns and codes of touching and being touched. Yet this is not an analysis of these codes, I just want to foreground this frame of culture over and around the phenomena that

7 Marks, *Touch*, 3.

I analyze. I want to bring forward cold burn, because it foreshadows one of the ways in which the media runs interference with our bodies' senses, and these Japanese artists have picked up on this: a touch that is slight, repetitive, and vanishing, which drives our desire for engagement, but refuses that engagement, always just out of reach. Yet this mediated touch with its light constant touching may also relentlessly press and eventually wound us.

I am always struck by how the codes and practices of public touch in different cultures change depending on the medium. In Japan, in contemporary performance and art practices, touch is often intensified by both distance and proximity in a kind of rebellion against the status quo. The power of cold burn lies in its particular manipulation of touch sensations and the particular intensity and pain associated with it. With touch only partially controlled by Japanese rules of behavior on the one hand and the contemporary moment of subverting those rules of 'touching' discourse on the other, new media technologies amplify this sensational performance landscape even further.

In this study, I focus on particular moments (not entire works) of choreographed touch, which radically shift and transform our visual/kinesthetic modes of perception by doing less and less. This 'less' may be reduction or abstraction. Yet 'less' can intensify through small repeated acts. The 'less' in cold burn requires the insignificant: the small, the slight, the light burn though the repetitive 'touching and being touched.' In each instance, the performer's own touch is emphasized. Each touching moment begins with its particular skin, culture, visuality, politics, and technology. These works do not build up to a powerful crescendo conclusion. Rather, each carefully delineated cold burn moment within the larger performance gets its flashlight and microscope moment: to be seen, touching and touched, one touch at a time.

In the works that follow, the light repetitive touch tangles with different virtualities. There was no 'flesh-to-flesh' touching. What occurs is 'almost-touching,' barely touching, or the touching and being touched is caught in a freeze frame before sensation occurs. Because of the slight space in between, we imagine the reach across time and space to complete the touch. We can be frustrated from that act of incomplete contact. We long for the touch. This 'longing for' or, in Japanese, *nastukashii* will become one important outcome and action in the examples discussed. *Natsukashii* relates to 'nostalgia' and a whole field of longing for a Japanese 'past' or 'tradition' or future that might restore that past. Marilyn Ivy in *Discourses of The Vanishing: Modernity, Phantasm, Japan* refers to these "longings" and "a nostalgia for a Japan that is kept on the verge of vanishing, stable, yet endangered."[8] This kind of affect, longing for the vanishing, has a long history, which continues to be embedded in popular culture as well as in the contemporary performances of Kabuki and Noh. Remembering and imagining the past are like virtual acts of touching over time and space. Repetition is also key to this cold burning of virtual or imaginary touch. It must occur over and over again. This practice of burning for or with the longed-for touch can be found in countless dance dramas in Kabuki, in Noh, and in the forms of poetry where the visual and tactile intersect with intensities of proximity. *Natsukashii kimochi* or "a longing for feeling" is an affect that arises in the present moment, to bring that past or future imaginary into this moment here.

I keep these complexities present in each of the case studies: the properties of cold burn, the light touch, the repetitive and continuing touch, the vanishing or phantasmic touch bring to mind the subtlety that drives this process of cold burning.

Burning Examples

Each one of the artists deploys the media and the time/space of gesture, touching, and being touched. They also work with the micro-intensities of 'almost-touching.' Each artist manipulates his or her cultural prescriptions (also unfixed) differently. Each deploys touch and media in different ways. In the dumb type installation, all the performers are projections, but the live bodies of the spectators/experiencers activate their movement and touch gestures. In chelfitsch, the performers are all live, but their small gesture/touch vocabulary is digitally performed. Tanaka's live object and doll 'play' become mediated through light, in shadows, new technologies of touch. All three share certain parameters of touch: 1) repetition, 2) small, light repeated gestures of touch, and 3) touch memory and disappearance.

To better sense the 'differences' in touching, my examples will draw on the variety of intersections of performing bodies with virtual and visceral technologies, the tensions between live and screened bodies, and how the performer's and the viewer's consciousness of virtual and visceral presence and touch changes and evolves over time. I explore a visceral and virtual concept of visuality and kinesthetics that I call 'cold burn,' which comes from my earlier research on live and media performances in Japan. Here, 'cold burn' is used to denote a visceral sensation, like a bruise of a burn that we experience from slight but constant pressure on our skin/membranes (a rub on your heel from an ill-fitting shoe that almost becomes a blister). In live and media performance, digital images press on bodies,

---

8 Ivy, Marilyn, *Discourses of the Vanishing: Modernity, Phantasm, Japan*, Chicago 1995, 65.

and the bodies press back onto the digital images. Through a performance or experience of an interactive installation, performer and viewer experience a kind of 'cold' pressure that 'burns' or creates a sensation that is visceral, virtual, and emotional: a sense of longing for something touching, but just out of touch, yet touchable. This is one way of considering how our bodies and new technologies intersect and challenge our assumptions on corporeal sensations and their broader implications in human cultures, encounters, and communications.

This study aims to bring forward more processes of affect, driven and subtle intensities, which flicker on and off, like neon signs or a bad connection, messages of simple and mysterious codes connected to touch. Different from 'in-your-face' physical encounters or the grasp, lean, and press of contact dance, these slight, distant almost-encounters and imagined illusory encounters repeat themselves and reflect on themselves. They create different worlds of perception and affect. I believe that these different touching encounters reflect a sense of contemporary Japan that lies below the radar. As I exclusively discuss the work of Japanese artists, this study has a specific cultural turn, although these works (dating from the mid-1990s to 2011) and their artists are internationally renowned. The 2010 work, *Die Scheinwerferin* (*A Search/flashlight*) by Naoko Tanaka, was created and performed in Germany. These are examples of how touching and being touched in the works of a variety of Japanese artists reveal and (in their very own way) require a revision of our kinesthetic grid in order to expand our range of perception required by particular cultural practices of performance. Each example differs in its use of touching and being touched. Each performance example challenges and shifts theories and sensations, with varying intensities experienced through repetition, over time.

I wish to connect affect with touch here in order to better understand how cold burn works in these examples. Drawing on Erin Manning for her connection and reminder of movement as a theoretical force, I will use this idea in my analyses (as a background that may cut across the cultural divides). I have said that touch is political, and I want to consider touch/affect as a kind of moving unit here and understand how the small gestures of Miffy in *5 days in March*, Tanaka's reaching doll/puppet's arm, and Furuhashi's open but missed embraces touch with cold burn and create, in that moment, "affective politics." According to Manning,

> Affective politics are not moral politics. An emphasis on movement does not promise an emancipatory politics. Affective politics are what we preconsciously make of them. They are of us and with us: we recompose with them. In this way they are much more dangerous and much more powerful than content-driven politics.[9]

9 Manning, Erin, *Relationscapes: Movement, Art, Philosophy*, Cambridge 2009, 137.

The "of and with us" of affective politics seems to resonate accurately with what we have been finding in different cultural codes for touching and being touched. If we can keep this "of and with" and affect in motion, then the cold burn examples might counter a simpler reading of 'distance' or what may seem barely touching if at all. Perhaps I am asking theory to grow a more sensitive layer of corporeal insight. I read Manning from another perspective, outside the fold of movement, in a paused or stilled moment of suspension (if that is possible, when cold burn has reached a certain level of pain, then it hurts, the area of skin burns). The sense of not quite touching could be a non-sensuous touch, a virtual touch but, according to Manning, "[a] sensing body in movement is activated both sensuously and non-sensuously. Perception occurs on a continuum of relation."[10] Then we could direct our attention to this ribbon of relation, a continuous path of sensuous material loading and unloading on the Möebius strip.

Haptic Politics and Kinesthetic Power

Laura Marks refers to this sense of reading bodies from the screen and camera techniques, as "haptic perception,"

> ... usually defined as the combination of tactile, kinesthetic, and proprioceptive functions, the way we perceive touch both on the surface of and inside our bodies. In haptic visuality, the eyes themselves function like organs of touch. [...] Because haptic visuality draws on the other sense, the viewer's body is more obviously involved in the process of seeing that in the case of optical visuality.[11]

Marks goes on to explain how visual media may also offer "haptic images" which "do not invite identification with a figure so much as they encourage a bodily relationship between the viewer and the image."[12] While Deleuze, Guattari, and Bal have all made use of the haptic gaze in different ways, Marks brings forward that haptic viewing and haptic techniques are useful strategies to broaden and complicate the theories of gaze identification with representations and narrative, alongside cultural differences.[13]

10 Ibid., 73.
11 Marks, *Touch*, 3.
12 Ibid.
13 Cf. Deleuze, Gilles, *The Movement Image, Cinema 1*, Minneapolis 2001.

In *The Skin of the Film*, Marks emphasizes how in haptic visuality the eyes can function like organs of touch.[14] She suggests that haptic looking moves over the surface and is inclined to move, and to blur rather than to focus. She demonstrates how different ways of viewing and sensing invite viewers to distance and/or come so close that what it is disappears. I would like to deploy this kind of viewing into the matrix of my cold burn touch/gesture performances. To go along with that, I will also borrow from what Deidre Sklar describes as "'embodied schemata' that are neither perceptions nor representations but cross-modal recognition patterns, whether of form or quality."[15] Sklar emphasizes how multiple sensorial operations are at work in our readings of bodies and that "[i]magination then is not merely a mental operation that works reproductively to duplicate or reflect experience; it is a perceptual/cognitive process that works productively and creatively to configure experience."[16] Sklar's explanation that a synthesis of imagination and sensorial perception is involved in culturally configuring experience, supports how touch activates cold burn in these performances.

I will move from a screened tech performance (dumb type) to a body-focused performance (chelfitsch) and then to a light body image performance (Tanaka), suggesting a variety of 'cold burns' from gestural differences through (in process) choreographic kinesthetics. That is, each artist carefully constructs a movement and technology score, which enables our sensorial grids to perform in a variety of ways within that score. In each case, there are wide variations in this process and what I am highlighting are only a few instances, although repeated, in the larger variegated performance work.

In each descriptive example I will suggest how the 'Japanese-ness' of touching might be part of the small, repetitive, insistent, and distancing gestural implications that fill these works. The energy is also indirect, not unlike the language structure that privileges indirect verbs without subjects: actions float, the source of the acts seems to be elsewhere. Further, in the world of new technologies perhaps we should consider just how much and how far we are within our own mediated bodies: how do we perceive touch, skin, sensation in any way that is not already always mediated in the worlds in which we work?

---

14 Marks, Laura, *The Skin of the Film: Intercultural Cinema, Embodiment, and the Senses*, Durham 2000, 162.
15 Sklar, Deidre, "Remembering Kinesthesia: An Inquiry into Embodied Cultural Knowledge", in: Sally Ann Ness/Carrie Nolan (ed.), *Migrations of Gesture*, Minneapolis 2008, 85–112, 94.
16 Ibid.

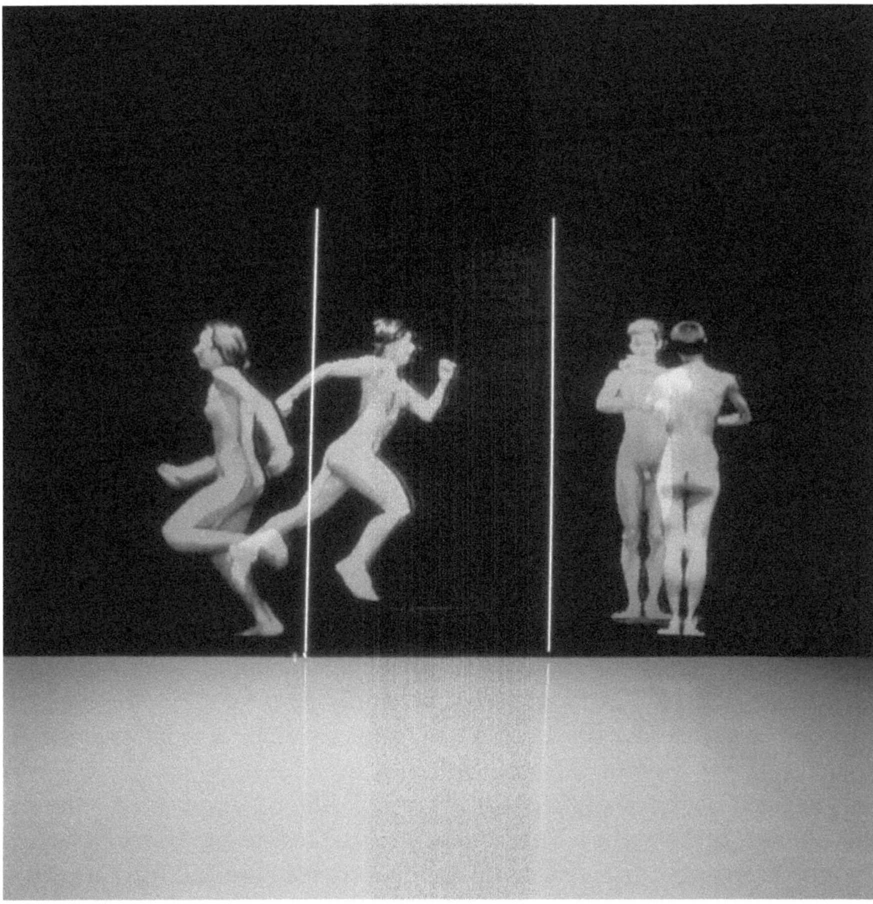

Lovers by dumb type, 1997, photo by Emmanuel Valette

Cold Burn *Lovers* by Furuhashi Teiji and dumb type

The media performance group dumb type (usually written with low case letters and sometimes as one word, "dumbtype"), an interdisciplinary multimedia performance and installation art collective, began creating what might be called visual dance theatre in 1984. Based in Kyoto, Japan, the artists use a number of combinations of projections, bodies, objects, and architectural spaces. They always position the audience in particular ways to set the conditions for seeing and perceiving. They frequently use strobe and pulsing pinpoint lighting to fracture the bodies moving in space and to disturb the viewers. Viewers have remarked that the light seems to 'attack' the bodies, making it painful to watch. To complement their use of 'violent lighting,' they work with repeated and episodic movement sequences, over and over again. They deliberately create a fragmented environment where bodies and projections meet and react against each other. In dumb type, the bright flashes of projections and strobes erase performing bodies. Skin is digitalized by flashing projections. Their stages are architectures of 'cold burn' touch.

Lovers, an interactive video installation, was supported by a research project grant from Canon in 1994. The installation is abstractly based on the work *S/N* (signal/noise) created and performed by the entire collective dumb type. *S/N* toured internationally and was dumb type's most overtly political work, specifically focusing on gay and sex worker rights, safe sex, and AIDS awareness. Furuhashi Teiji, one of the founding members and artistic directors of dumb type, who died from AIDS in 1995, was known for focusing on the political engagement of all dumb type works. While he considered *Lovers* to be a collective piece, it is his personal signature work, which employs only virtual performance figures and only operates through the movement of spectators.

In this circular performance/installation, the beams of a central projector unit (with stacks of data projectors) react to the spectators who wander in the circular screened area. There are set repeating taped loops of performers doing a movement sequence. The sequences or choreography are set for the group of six performers: they may run forward, they may turn and run in the opposite direction. They may walk quickly, they may stop, spread their arms as wide as possible and fall backwards (disappearing), and finally they may slowly hug the space in front of them. The have the options to put these in any order and timing. The performers were then videotaped and their dances were looped. There were also text loops that could appear across and around the moving and stilled bodies: "Do not cross the line. Don't fuck with me, fellah, use your imagination." When you moved around the space, the passing body triggered different projectors at different times. Each passing around or back through the circular space created a different sequence.

I believe that about ten people were given access to the space at one time. This meant that you did not run your own encounter with the images alone. Other bodies in the circular space could trigger the projectors simultaneously. This also caused the viewers to tune to each other. There was only a slight 'pinging' sound when the projected figures moved into presence or disappearance.

The figures were always soft, ghostly, like in an old fading analog video. They were naked but ghostly, a kind of flesh that floats with indistinct facial features. This softness, the distancing of the then analog image, already started the cold burn: you could not get close enough and they moved away, again and again. The running figures were lovely, passing you by, or you could catch up, but the stilled slow wide-open arms to the hug and fade really caught me by surprise: I did not want them to let go, or fall, or fade. Because of the loop, I kept trying to move through a sequence, to make the narrative of the run, hug, touch, fall work, but it always broke, interrupted. Left with this soft bruising affect, I wanted to try again, to get hurt. I kept trying to get inside the hug or even close my eyes and fall away into that grayed, black abyss: take me with your disappearing touch. This kind of cold burn, made before so much digital work, might have had a close-touch illusion, but I think the intention of Furuhashi was what the text said, "Do not cross the line, or jump over. Limit. Z." You have to stay distant, reaching, and slipping out of that hug, cold burn. People pass through me, the very substance of my flesh becomes the screen, passing, passed through, almost touching, and fading.

In one of dumb type's collective performance/media works, *OR* (1999), cold burn touching arises between the viewer and the screen, with a live performer as the interface. This coupling of the live performer and the live viewer doubles the burn/touch. In *OR* a woman stands utterly still dead-center stage in a long white dress, the strobe flashing fades into a panoramic video of a highway shot from the windshield of a speeding car. It is double-projected so that the image of a highway runs like a vortex up the center of the stage, making the woman's body the road and the pinpoint horizon. Her flesh body becomes the road, and the digital road is her body that at times disappears in the curves of the pavement and the scenery passing by. Layers of images spin by touching the eye, her body, (that becomes) my body. A moment of cold burning: I am the driver, the driven, and the body on the road. It is an exhilaration that is mixed with discomfort when I identify with the body, take on the body kinesthetically, and transfer the feeling of the road running over me like an electronic exhilarating zap. Finally the road burn erases the body and the woman, and myself: on the screen, 'cold burn' whites out, into ashes: white on white.

I think there is a productive tension in these different kinds of 'burns' that actually hurts us, a dull aching sensation/emotion, that is both distant and intimate 'touch.' This is not melodramatic emoting. Rather spectatorial experience borders on a physical/visual witnessing that repeats and loops with the soft gray figures and the woman disappearing, just out of reach.

Chelfitsch's Cool Touch

Okada Toshiki is one of the up-and-coming director/authors of new performance in Japan, and his group *chelfitsch* (from the English 'selfish') is internationally

Miffy in *5 days in March* by Chelfitsch, photo by Thomas Bremondo

renowned for their experimental performance. In the first decade of the 21st century in Japan, Okada emerged as both a lauded playwright and choreographer, winning several national choreography awards, including the "Toyota Choreography Award 2005: Discover the Choreographer for Next Generation." Okada himself was a little baffled that he was considered a 'choreographer' even though he sets all movement and gesture patterns, which he places into a particular space, time, and object configuration. His texts are fragmented conversations or short stories, like memory flashes, whose characters speak without narrative content or emotional direction.

In Okada's *5 days in March* and one of his newest works, *The Sonic Life of a Giant Tortoise*, a lone young woman moves about, in what I call 'frames' of light and architecture, performing spoken and corporeal gestures in a state of soft focus, or as if to herself. There is a broken 'narrative' or dialogue that might be between two or three people who are not on the stage but somewhere else. The physical gestures and the verbal vocal gestures have no accent: instead they prod, like a point finger into the silence, almost words-in-gestures.

Miffy in *5 days in March* and A Woman in *The Sonic Life of a Giant Tortoise*, create solo touch dances in which they only touch themselves, the space around them, and real and/or imaginary objects. While there is no new media intersecting their touch, their small, repeated gestural movements seem broken and fractured, or stilled in time. Further, their texts, on international tours, are projected overhead or sometimes on the walls behind the characters. The repeated gesturing and positioning of performers drives the minimal dialogue into the background, but also sharpens and abstracts these gestures.

In the following example, I will examine at a short sequence of gestures by the female characters in each of these plays. I suggest we consider how the small gestures are displacements in time, space, and energy. These displacements create tiny patterned puzzles that *touch* on meaning, but do not produce meaning. Not unlike text messaging or tweeting, these tiny gestural patterns communicate a visual and sensual scattering and patterning of partial meaning, which we feel: their repetition (in one case of jerky irregular gestures, in the other a sustained 'repetition' of one sustained touch gesture) deeply impresses our sensate surfaces. Perhaps these small gestures indent corporeal memory by both irritating the surface and touching deeply, burning in, and inscribing gestural resonances on the surface of the spectators' bodies.

At one point in *5 days in March*, Miffy is isolated by the 'frame' of a triangle of blue light, and she rocks from foot to foot and gestures with her hands as if she was typing on a computer. Then she just drops her arms, then she shakes her arms and hands, then folds them around herself. She gazes off, looking for something. Her small gestures that are so carefully lit and placed in the blue triangle become tiny proddings and irritations (on the skin, in the muscles, in the imagination). Miffy, who has just left an encounter with other characters, now argues (alone, with herself) for staying in her room in her own world, or possibly on Mars. She flaps her arms, points sharply into outer space, without an object, they flicker and dab. Her fingers curl for no purpose, her hand touches her arm, rubbing and chafing it.

When the light goes blue, the Miffy/performer enters her room, behind her the text projects her barely audible narrative of how, since childhood, she has studied in this room but even without studying she loves being in this room, doing things like reading comics or writing her homepage. For a while she gestures shaking arms and hands in the air. She tucks her hands in her boots. Then her ragdoll-like arms seem to flap and fly away from her torso. She keeps rocking from foot to foot. At times she stands still and works her fingers on an imaginary keyboard. After her missed encounter with this young man, she says she was "so embarrassed I decided to take a rocket ship to Mars."[17]

---

17 All quotes from *5 days in March* and *The Secret Life of a Giant Tortoise* are used by permission from Okada Toshiki from his unpublished English and Japanese play scripts and my notes from performances.

"In her own room Miffy orbits." Miffy, according to the young woman performing her 'voice,' has started to study the planets and space, and in this way she can forget her own social mistakes. Out there in space (in her room), she is safe. She is bent over, her eyes cast down. Her hands make small loose fists that uncurl, touch, open and close. She touches her elbows, her own hands, and then pulls at her hair. Her fingers dig into the front of the boots. Her hands smooth down her thighs, then she jerks sporadically with sharp quick stops and starts. She repeats these phrases: typing on the imaginary keyboard, shoving her hands into her boots, rocking from foot to foot, looking back and curling over.

What characterizes her gestures is this minutely careful repetition of fragments: the repetitive sequences agitate and shake off any sense of meaning attached to the gestures. Watching the DVD, the performances, and playing the gesture patterns in my head, I see over and over again, traces of incomplete gestures, persistent, irritating, and slight. The gestures keep alive the discomfort of the performer and the viewer.

Uchino Tadashi, a professor of performance studies at Tokyo University argues that,

> his/her language and his/her bodily movements do not agree as in a realist fashion, but always in discordance. It is not that Okada choreographed the movements as such, but when he let the actors speak the lines, he wrote during the rehearsal, they came up with those seemingly strange, 'displaced' movements. [...] Okada likes to say that their bodily reality resides only in those dance-like repetitive movements, and fading, gradually crouching gestures.[18]

While I agree with Uchino on the displaced sense of movement to text, I think there is a need to analyze those 'dance-like' gestures further if Okada regards their repetition and fading postures as the reality of the characters. I would take Okada's meaning of reality as the corporeality of these performer/characters, which is all that remains with the "postmodern-sequel fracturing of narrative, subject, and representation and presentation."[19] I suggest that we consider the time, space, and energy of these small gestures to be tiny patterned puzzles that touch on 'meaning,' in their very repetition. On top of that, I believe that young people today, who use social media sites, blogging, YouTube, and cell phone messaging, have a sense of a visual and sensual scattering and patterning

---

18 Uchino, Tadashi, DVD Program Note, *5 Days in March*, chelfitsch, excerpted from Theatre in Japan Contemporary Theatre.
19 Okada DVD Program Note.

of partial meaning. The fragments do not need to come together; they only need to be repeated. Thus these gestures indent corporeal memory with irritating surface touching that becomes burning.

Video still from *Die Scheinwerferin* by Naoko Tanaka, copyright: Naoko Tanaka

Die Scheinwerferin by Naoko Tanaka

Die Scheinwerferin, workshopped in 2010 and premiered in Berlin in 2011, was created and performed by Naoko Tanaka. Tanaka performs a 'solo with objects' about touching and being touched within a world made of the repetitive 'pressures' of light, shadow, skin, and things which contact, collide, and burn in her light, shadow, darkness repeated sequence. The performance is in its own constructed space within the performance hall: two walls of white fabric hang like curtains forming an angle in which a small table (with objects underneath it) stands with a chair, and one hanging ceiling lamp hanging overhead. On the table there is a doll, almost a replicant of Tanaka, lying on its back, face up, in repose. Tanaka is seated at the table with her back to the audience, her arms and head resting on the table. This small cornered space seems like the private space of *hikikomori*, the condition of young people in Japan today who stay inside all day and night, only occasionally leaving their rooms. In their rooms they are occupied with visual media such as DVDs and games and online worlds of anime, YouTube, and blogs.

Tanaka was born and raised in Japan and received her visual art training (BA and MFA) at Tokyo National University of the Arts in the late 1990s. Having received a scholarship that enabled here to work at the Düsseldorf Academy of Arts in 1999, Tanaka began to develop her interdisciplinary visual/installation/performances in collaboration with visual, dance, and sound artists. Tanaka has never performed this work or any of her 'performance works' in Japan. While I am including her in my thesis on 'other' ways of touching in Japanese performance, I realize that her works, in a sense, problematize the idea of both 'other' and 'Japanese.' I was immediately struck by her use of small repeated actions, gestures, and the intersection of media (here light and shadow) and the objects that come to life through touching and being touched by light, shadow, and touch. Furthermore, Tanaka performs herself as the *Scheinwerferin* (the person who manipulates the spotlight or flashlight) and the doll, which only has one gesture: a reaching arm. This format with the white fabric curtains for projections is reminiscent of the Japanese forms of doll theatre and other, older forms such as *kamishibai*, performed by solo storytellers who told narratives using scrolls and pictures lit by lamplight. Her touch and manipulation of objects with the flashlight's light and shadow sets up the conditions for cold burn touch.

The cold burn touch arises in numerous instances in Naoko Tanaka's performance work *Scheinwerferin.* I saw the production in a small space at Sophiensäle in early 2011. Tanaka had created the work the year before, the first in which she stars as a solo performer with low-tech technologies of flashlights and music, old analog film rolls, and her object world of an old dollhouse, kitchen utensils (forks, knives, spoons) and wires and strings next to tree branches and a toy railroad track. The moment I have chosen to demonstrate cold burn here is from the beginning of the work at the moment of transformation from the overworld (table top) to the underworld (under the table), and the beginning of the spectator's slippery switching between seeing the materiality of the performer performing and being inside the shadowland's touching-dream travelogue.

Tanaka begins this second section of the work with an inspection of the Tanaka-like doll, which sets up a strange identification between the doll, ourselves, and Tanaka who might become a surrogate or avatar. Tanaka touching her doll-like self displaces our sense of object and person. Tanaka inspects the doll's limbs, the periphery of her entire body, up to her arms. Tanaka carefully lifts the tiny arm so it stretches reaching, stiffly, to the ceiling. It remains in this beseeching position until the end of the play: the tiny arm raised as if asking for help, suspended, out of touch. Hutched over the doll, Tanaka moves methodically, almost medically, examining details slowly and smoothly, and then quickly, as she combs the light through the doll's hair that mingles with her own long cascading hair, and the strands of hair seem like veils of light and shadow covering the whole room. Tanaka, going for a 'shock' moment, plummets with the tiny flashlight into the dark macabre underworld under the table. We leave the doll reaching up into darkness, touchless. This is the first of our cold burn shocks. We are taken there suddenly but gracefully with sounds amplified. We hear trains on tracks, outdoor sounds, and traffic, close by and ominous. This first time, the shadows fly up around us: where are we, train tracks, a train whistle, giant knives, forks and spoons, stuck standing up, swelling up around us on the sheets, like legions of looming ghosts, or trees, or armies?

The cold burn touch is not just the dark-on-white images, but the loss of gravity, the float through Tanaka's hair that works like a mesh or screen or net to save our fall. We flail inside the hair. We navigate between identifying with the holder of the light, our imaginations and the room's 'reality.' And this happens again and again until, I suggest, the spectator takes pleasure in the strange or estranged touch, in the cold burn, to a certain point. The distance touch is Tanaka's (who we can barely see behind the objects and table on the floor) in her light and shadow Alice-in-Wonderland world of small objects, which outgrow their own shapes. We see from our seats but we are 'touched' by those shadows over and over again, cold burn.

Naoko Tanaka's *Scheinwerferin* beautifully demonstrates the tiny touch of a doll and the strange abyss of shadows that touch, and press, and brush on our skin, from far away. When she combs the doll's hair, we are plunged into her hair, and then we fall into the neverland under the table, into the shadows of everyday objects. The shadow shapes also touch our consciousness: railway tracks or whirling tornado? A table fork or a giant pitchfork? Trees bent over in a storm, or broken branches? We are moved and urged on in through imaginary and 'real' object imagery. You cannot be touched by a shadow, but, scarily, we feel the shadow's touch. The doll's hand is still reaching blankly into the sky. Touch me?

When Tanaka appears to be out of the shadow labyrinth, she stands, facing us, with her *Scheinwerfer*, the only light in the room. She looks at the tiny beam of the *Scheinwerfer*, suddenly she stuffs it into her mouth and her face glows like bloated red fish at the bottom of the dark sea. Or like a mask of a child's face, lit by flames. Then, just as suddenly, she snaps the light off, plunging everyone and everything into darkness. No touch, cold burn.

Towards a Conclusion: Cold Burning, Distance Touching and Catalyzing Gesture

Perhaps this is strategic essentializing: to say there is something 'other' about touch in these very different works and that touching and being touched are acts of complex learned practices, which are also dynamic and transformative. I agree with Deidre Sklar, who suggests that anthropologists have carefully debunked the myth of racial gesture/behavior links, but, as I suggest here, the degree of deep control by cultural practices[20] can differ in each and every specific cultural milieu. In this case, I argue that both imposed 'cultural' practices, especially from stylized patterns and daily life gender acts/regulations and mediated daily practices carve deeply into bodily gesture patterns. Small gesture patterns inscribe their range, duration, and frequency of movement onto our bodies, no matter what the culture.

Sarah Rubidge suggests that her own interactive installation works "through [their] interface between the body and technology, are intended to draw attention to the more subtle aspects of our being in the world, and in doing so implicitly, to sensitize participants to the 'hidden' sensibilities inherent in their responses to the environments."[21] Gretchen Schiller writes that her own work draws on the witness's "dynamic subjectivity," referring to Laura Mark's theory/practice ideas of "haptic cinema," where there is a "mimetic relationship between the perceiver and a sensuous object." Schiller breaks down her transdisciplinary moments, where the production of haptic/felt experience transforms from 'identification' to something 'else.' She argues that there is a play between "what is body what is space and what is image and what is screen" that transcends the figurative body into a sensual and moving dynamic.[22] Certainly the passing virtual/visceral bodies in *Lovers* and the close-up of the woman's eyes, or Miffy's blue light in chelfitsch's quiet gesturing, and Tanaka's tiny doll who seems to dream in the shadowland of strange forests on the curtain screens, may be part of this transcendence: moving us to different de-limits of touch.

In this media performance arena, Susan Melrose suggests a "triggering" between spectator, performer, and artist. This connects strongly to Furuhashi's run-

---

20 Here, Sklar refers to Judith Butler in her *Undoing Gender.*

21 Rubidge, Sarah, "Sensuous Geographies and Other Installations," in: Susan Broadhurst/Josephine Machon (ed.), *Performance and Technology: Practices of Virtual Embodiment and Interactivity*, London 2006, 112–126, 124.

22 Schiller, Gretchen, "Kinaesthetic Traces Across Material Forms," in: Broadhurst/Machon (ed.), *Performance and Technology*, 100–111, 107.

ning, hugging, falling *Lovers*, Okada's Miffy in her small gestures into outerspace orbit, and Tanaka's changes in perspective when the shadows leap out from her shadowland becoming ours: "whereby what is present but incomplete, triggers [...] something operating below the level of representation."[23] While we can accept or reject these triggered moments, these works suggest a different kind of consciousness to nuance, detail, and fine-tuning to our mediated corporeality. Cold burn asks more of us than interactiveness, instead a dynamic participation in its touch and burning.

Gretchen Schiller comments that

> [e]very technically mediated action we undertake or with which we are engaged is embodied. [...] Technical systems are not separate from movement perception but entangled in the manner in which we come to experience, understand, and perceive movement: they constitute our perceptual kinaesthetic systems.[24]

I would propose more entanglements in our art/science making processes, and, as Schiller suggests, with accepting this "entanglement between human-created techniques and movement as a dynamic structural and relational event, then we replace discussions of the 'body and space' or body and machine with the Fluid surprises of relational dynamics."[25]

In conclusion we could say that these live/mediated performances are a kind of "new sensory syntheses" in which, according to Mark Hansen and Derrick de Kerkhove, "new speeds and new perceptions" arise from "catalyzing those bodily senses – proprioception, interoception, affectivity," that is "through the internal, intensive space of our affective bodies." These performers/creators use precise exacting and repeated gestures, touching and being touched. Or, as Furuhashi's ghostly digital figure beckons, Miffy taps into the air, and Tanaka shadows: when I go into my room (or under the table), I can go to Mars or touch with shadows or fall away from your digital, ghostly embrace, over and over and over again, less and less: cold burn.

23 Melrose, Susan, "Bodies without Bodies," in: Broadhurst/Machon (ed.), *Performance and Technology*, 1–17, 15.

24 Schiller, "Kinaesthetic Traces Across Material Forms," 109.

25 Ibid.

References

Deleuze, Gilles, *The Movement Image, Cinema 1*, Minneapolis 2001.

Fisher, Jennifer, "Tangible Acts: Touch Performances," in: Sally Banes/Andre Lepecki (ed.), *The Senses in Performance*, New York 2007, 166–178.

Hansen, Mark, *New Philosophy for New Media*, Boston 2004.

Ihde, Don, *Bodies in Technology*, Minneapolis 2002.

Ivy, Marilyn, *Discourses of the Vanishing: Modernity, Phantasm, Japan*, Chicago 1995.

Lenoir, Timothy, "Foreword," in: Mark Hansen, *New Philosophy for New Media*, London 2003, xii–xxviii.

Manning, Erin, *Relationscapes: Movement, Art, Philosophy*, Cambridge 2009.

Marks, Laura, *Touch: Sensuous Theory and Multisensory Media*, Minneapolis 2002.

Marks, Laura, *The Skin of the Film: Intercultural Cinema, Embodiment, and the Senses*, Durham 2000.

Melrose, Susan, "Bodies without Bodies," in: Susan Broadhurst/Josephine Machon (ed.), *Performance and Technology: Practices of Virtual Embodiment and Interactivity*, London 2006, 1–17.

Morse, Margaret, *Virtualities: Television, Media Art, and Cyberculture*, Bloomington 1998.

Okamura Keiko, *Teion Yakedo, 'Cold Burn' or Land/mind/body-scapes in the Age of Cold Burn*, MOT Annual, Museum of Contemporary Art, Tokyo 2000.

Reiner, Miriam, "The Role of Haptics in Immersive Telecommunication Environments," in: *IEEE Transactions on Circuits and Systems for Video Technology* 14.3 (March 2004): 392–401.

Rubidge, Sarah, "Sensuous Geographies and Other Installations," in: Susan Broadhurst/Josephine Machon (ed.), *Performance and Technology: Practices of Virtual Embodiment and Interactivity*, London 2006, 112–126.

Schiller, Gretchen, "Kinaesthetic Traces Across Material Forms," in: Susan Broadhurst/Josephine Machon (ed.), *Performance and Technology: Practices of Virtual Embodiment and Interactivity*, London 2006, 100–111.

Sklar, Deidre, "Unearthing Kinaesthesia, Groping among cross-cultural models of the senses in Performance," in: Sally Banes/Andre Lepecki (ed.), *The Senses in Performance*, New York 2007, 38–46.

Sklar, Deidre, "Remembering Kinesthesia: An Inquiry into Embodied Cultural Knowledge," in: Sally Ann Ness/Carrie Nolan (ed.), *Migrations of Gesture*, Minneapolis 2008, 85–112.

Uchino, Tadashi, DVD Program Note, *5 Days in March*, chelfitsch, Theatre in Japan, Contemporary Theatre.

Krassimira Kruschkova

The Fault Lines of Touching

In the tense breathless silence, the motionless restlessness at the beginning of the installative performance *The fault lines* by Meg Stuart, Philipp Gehmacher and Vladimir Miller, something is foreclosed and revoked at the same time – something that will have been. *Fault lines:* something will have happened, and the tectonic fissures between the bodies, between the media – reconstruct, remember it just because of their paradoxical emptiness. From the outset, a dispositive of the past shimmers through what could have been, will have been totally different. A scene under the banner of a farewell which will pass into the strangely melancholic restlessness of technical challenges – choreographical and medial ones. Neon lamps flash up, glittering, which paradoxically mark a kind of ramp for the white cube and simultaneously engulf the scene in laboratory light. No black box, an exhibition space as the location of this installative performance.[1] A dance performance which turns into a kind of video installation that, however, happens frontally towards the audience. There is a curtain, too, yet it does not serve as a partition between stage and auditorium, but stretches along the one white wall of the stage area – a curtain behind which there is nothing. And nothing will stand behind the performative gestures either, they will stand for nothing – and disclose more and more nothingness, wistfully uninvolved.

We hear the room's breathing, the whispering of Vincent Malstaf's sound installation which will later turn into the stuttering acoustics of a film soundtrack. We see Gehmacher, Stuart and the video artist Miller (also on stage), far apart from each other at first. Abandoned. And exposed. Exposed, too, the projection equipment, the golden beamer, the golden cables. The tectonic fissures. Then, Stuart and Gehmacher will not fall into each other's arms but rather attack each other, in a fighting embrace, violently, repeatedly, with absentminded resolve – and fleetingly, an undecidable tenderness, an inactive solicitousness will yet arise. Already under the banner of parting, on the verge of a farewell the performers, the man and the woman, will turn towards each other for evanescently brief moments only in order to immediately turn away again. The movements will pause in a *still*, in a picture, but not as a picture. Idiosyncratic. A thwarted expenditure which cannot be actualised in any act. A movement which would rather be none, which prefers not to. Rapt touches bordering on violence, fierce, unrestrained and at the same time casual, oblivious. Amnesia of gestures, contingency of touch.

---

1 The Tanzquartier Wien performance took place at MUMOK on January 4th and 5th, 2011.

Later, the two will remain on the floor for some time as if they were about to fall asleep, and Miller's drawing directly on the projection area which shows a virtual double of the two bodies remaining in still will not wake them up, it just caresses their images. These drawings will not trace or continue the movements that took place, but rather continue to imagine dreamed touches. As if they were small whirlwinds and tongues of flame, proliferations in all the contingency of caresses or landscape structures, the drawings note down the *fault lines* of contact between chaotic structures, their contingent, maybe provisional and not retroactive traces – as a "memory of that which was not".[2] Stuart will briefly turn around on the floor, smile lopsidedly, tickled by the video artist's pencil – not she herself but her image. Later Miller will leaf through the room slightly changing the projection's position so that he virtually takes along the performers' live body, turning it into two dimensions. The paradoxical, trembling flatness of the live bodies, that in the beginning seem to miss each other so intensively, is taken apart when they converge in reality, dissected more and more by the medial manipulation. The projections of the figures that are actually positioned near each other are separated virtually – as if the medial event were articulating something the live event is not able to formulate even if it could only happen 'live'.

It takes place when it doesn't. Like touch. Between the bodies, between the bodies and their images, between the singular body surfaces and their plural projection areas. In *Être singulier pluriel*, Jean-Luc Nancy writes, "The law of touching is separation, and even more, it is the heterogeneity of the surfaces touching each other [...] insofar as the actual power of a body consists of its capacity of touching another body (or touching itself), which is nothing else but its de-finition as a body."[3] Thus Miller's medial de-finition, definalisation of the bodies, his articulation, his medial touch with the live event – especially in the heterogeneity of this contact of different presentation and projection areas – will not be illustrative but strikingly illusive, in all its openness and its apparentness of illusion, which exposes itself to its own *techné* and thus, in a literally potential manner, turns reality into possibility. Literally exposed illusion made visible and in spite of this – or rather, just because of it – magical. The video artist who performs his apparatus on stage becomes part of it. The exposed path of the images and electronic impulses through the golden cables will paradoxically oscillate between illusion and disillusion, simultaneously present and absent, visibly illusive: real virtuality instead of virtual reality. The whir of projections will make the projected live bodies (and not only their projections) tremble.

---

2 Cf. Agamben, Giorgio, "Bartleby o della contingenza", in: Giorgio Agamben/Gilles Deleuze, *Bartleby, La formula della creazione*, Macerata 1993.

3 Cf. Nancy, Jean-Luc, *Être singulier pluriel*, Paris 1996.

This is comparable – if in a different way – to Stuart's *Alibi* (2001) and *Visitors Only* (2003).

A brief reminder, something like *fault lines* between works: Stuart's *Visitors Only* begins with the vibration of the bodies with which her previous work *Alibi* ended. This time there are bodies clothed in transparent raincoats whose long trembling unsettles the scene's visibility. Like trickling, vibrating raindrops the bodies in transparent coats fold the transparency of sight. The virtual veil of rain translates the scenic air into another state of aggregation. Intension instead of intention. The trembling choreography makes the room vibrate and fold: real movement that, however, virtualises real space. And when at the end of the scene the vibrating bodies jump on the spot now and then, as spring-back ball-point cartridges, it is as if the gestural tension of the sequence were critically whipping the writing utensils out of the hands of the choreography.

In *The fault lines*, too, there is critical optioning instead of clinical representation; here, too – even if with an entirely different medial implementation, we witness a kind of virtual rainbow after the virtual veil of rain – a choreography in italics instead of boldface – as if it were only quoting the dancing body whose outline resists any presence like bristling skin, like goose bumps, as if the bodies were merely trembling quotes of themselves, put between quotation marks, as if they were not there at all. This is the strong mutual affinity between the choreographies by Stuart and Gehmacher, the author of *in the absence* (2003), *Mountains are Mountains* (2003), *incubator* (2004), *like there's no tomorrow* (2007), to name but a few of his works. What do the choreographies of two of the most interesting protagonists of contemporary dance try to present so passionately, long after having conceptually committed themselves to the un-presentability of passions? What shakes the bodies on stage, what makes them tremble like this – searching for an *Alibi* for their own movement, their own being moved, for grasping their own emotion? What may still touch them when every kind of solid ground withdraws from under their feet as if they were floating – like at the end of *Visitors only* – over an abyss? As a place of medial ascriptions, the motivation of touching becomes increasingly harder in contemporary dance and performance practice, and it is all the more interested in the emotive fall of the body which keeps evading the idea of its dancing weightlessness, even lightheartedness. As if this practice were asking again and again where the customary oppositions of conceptual/emotional, minimalistical/affective come from, by letting these oppositions fall. Instead of rehabilitating affects or opposing emotions to concepts, it tries to dis- and reassemble the ever emotive texture of choreography especially in the course of minimalism. It tries to defigurate the illegible figurations of feeling, to deconstruct its all too blind constructions – and to persevere, knowing about its referential imponderability.

The question of the potential of touch also deals with the main rule of the scenic – visibility. This investigation of the preconditions of a medium also has to be seen politically – i. e. against the ideology of sentiment, against the slogans of a positivistic view that postulates the evidence of visibility. The scenic emotion, however, stays in the trembling, the oscillation of potentialities – it is thus never actual, never present, but potential, in marked absence. "There is no falling in love, no falling out of love", it says in Meg Stuart's and Benoît Lachambre's *Forgeries, love and other matters* (2004). The fault lines of lacking, failing, falling: falling in love, falling out of love. "There is no dance in this place, there is no reason to stay in this place", it says in *Forgeries, love and other matters*. And yet the piece closes with the words: "I'm staying here forever."

Maybe forever is the name of Stuart's and Gehmacher's first joint performance created in 2007 which they continued in 2010 with *The fault lines* – to draw further confused fault lines and lines of distortion, of touching the other, prone to fault and missing, measuring, impudent, missed. Joint artistic research, too, between the video artist Miller and the choreographer Gehmacher: in the choreographic video installations *dead reckoning* (2009), *at arm's length* (2010) and the group piece *in their name* (2010). Here, too, choreography and video installation, body and images go along with each other – toppling and diving into each other, immersing and submerging. What seems to separate not only the live figures but also the various projection areas actually connects them – if they are to be connected at all. Bodies and their stories, put down by themselves but not anywhere else either, that linger at the fringe of their mirror-image without breaking the glass. "One reaches a border not by crossing it but by touching it", Nancy writes in *Corpus*.[4]

Fault lines: bodies that touch the border between each other without crossing it – that *are*, in fact, this border. Exposed bodies, exposed to touch, in all their immeasurability, incalculability – and vulnerability. As if they were phantom pain, a painful nothingness, completely exposed to the other. Touching each other as attention and distance. For it is necessary "to interrupt the immediacy and continuity of touch", says Jacques Derrida in *Le toucher: Jean Luc Nancy*.[5] It is this chance of possible interruption, of interrupted immediacy in all the anchorless melancholy of every gesture in *fault lines* that endows the scene with the optics of the optional, of openness. The camera makes the eye alert for the live event, the invisible distances within the live touches. Doubles, reflections, surfaces, layers. Bodies disrobed by themselves, bodies on withdrawal, which

4 Cf. Nancy, Jean-Luc, *Corpus*, Paris 2000.

5 Cf. Derrida, Jacques, *Le toucher: Jean Luc Nancy*, Paris 2000.

are, at the same time, quoting themselves, setting themselves in italics, every gesture resisting itself – and merging into the pixel-like goose bumps of Miller's projections. Medial replays that only play the live bodies back into their real virtuality. Stills that always assert the choreographical and medial movement. What remains is the never-shown, the performative residue of absence, the performative and medial gesture of the undeliverable. Gestures that are too big and too small at the same time, marking the rest of the inexpressible and only articulating with restraint – if at all.

The reserved manner of pathos and melancholy, so typical of Gehmacher's choreography, here deals with the incommensurability of the other with the utmost aesthetic strictness. The too much/too little of scenic gestures as a residue. The rest is silence. And the melancholic absentmindedness of these gestures that evoke the exceptional circumstances of dance, ecstatically immobile or stutteringly bespoken, existential and exhaustive. Gestures so small that they touch their absence, as if they were not even there yet. Gestures so big that they tear apart. *Fault lines*. The bodies of Stuart and Gehmacher will leave each other and themselves – while touching. What will remain will be their outlines. Even after the two fighting/embracing bodies separate, one of them will stay in the interrupted gesture of touch. An embrace with empty hands. And Stuart will not so much caress her partner but rather retrace the contours of his body – a line along his body, almost as if tenderly outlining a dead body on the crime scene. Nor is this gesture accidental in *Maybe forever.* Choreography as an epitaph, as touching the ephemeral. At some point in *The fault lines* HE will push her 'corpse', her unmoving body along in front of himself. And again SHE will caress his outlines, touch her border to him, cut out not so much the body but the touch. The peephole projection, too, with which Miller will softly spy on and sample the two bodies, inert again – entirely different, cuts out the live bodies or rather the distance of their touches in order to focus on them: however, as *punctum*, as a crossfade of something invisible, in the sense of Roland Barthes' *punctum* of photography, the incalculably interrupting and simultaneously painful punctuation of the ephemeral that records the literal withdrawal of the figurative, as a kind of blind spot in the eye of the hurricane.[6] In "Les morts de Roland Barthes", on the other hand, Derrida, on the occasion of Barthes' death, specifies the *punctum* of transience as "incompleteness made visible", as "punctuated yet open interruption".[7]

6 Cf. Barthes, Roland, *La chambre claire: note sur la photographie*, Paris 1980.

7 Cf. Derrida, Jacques, "Les morts de Roland Barthes", in: *Poétique* 47 (1981): 269–292 (republished in: Derrida, Jacques: *Psyché*, Paris 1987, 273–304).

Interruption once again, narrative spots instead of narrative plots: the peephole of projection, its spotlights virtually punctuating the live event. The virtual touch of reality punctuates, isolates, focuses, interrupts, hurts it. It, too, is a touch in the mode of *I prefer not to.* Like the performative violence in the beginning, when the two performers touch each other to become separated – along their opposing fault lines. When they attack each other in order to then let go of each other, to desist; when they go towards each other in order to part. Bodies parting. They turn towards each other only to turn away from each other. The live touch raves, goes up the wall, is played against the wall, in the beginning a brutal and painful touch of the two bodies literally throwing themselves at the wall, letting their embrace fail intensively – and later also cast their projection on the wall, the projection that vibratingly repaints the live figures. The foil which Vladimir puts in front of the projector like a curtain makes the projected bodies shimmer and lets them immerse in virtuality. The rainbow of movements passing into each other is virtually doubled by the lyrical rainbow of glittering colours Miller will cast on the wall. Like a "drop of sky" (Friederike Mayröcker), the touch will camber down to the other – without actually touching. And always, shortly before the bodies raving with and at each other throw themselves against the wall, his body will cushion her body's blow. He will protect her. Too much? Lyrical film soundtrack. The performance's *making of* by Miller. And time and again one will come to lie in the other's arms. Untouched.

Near the end Gehmacher and Stuart will sit on the floor together in front of the curtain and the white wall behind which there is nothing, and draw big circles around themselves with their arms. Two embraces without object, drawn embraces, two circles intersecting. The empty intersection of an embrace. The arms are folded, but in the frontal drawing of an embrace that never happened. An entanglement of two semaphores, two clockworks ticking peculiarly instead of signalling. Miller will pull the glittering foil over the images' projection, let the figures glimmer pixel-like, thereby transporting them somewhere else entirely, uncannily enlarging the pointillist distance between them. And once again he will virtually isolate only Stuart's projection which now – depixelated – will seem to inhabit a parallel world. The interrupting, painful, invisible *punctum* of touch in the image of the finale, the parallel worlds of touch will pause – in a downright transcendental longing for each other. Meanwhile, Gehmacher will have quoted his *long arms*, his self-referential gesture of a singular absentee's outward tension.

In the choreography – as a punctual temporisation and spatialisation of touches – rather the untouchable is inscribed. The untouchable in figures of touching, figures without shape. Choreography as a technique of borders. And the borders as the figures of touching. Where the choreography splits the scenic bodies and glances with its sense of rhythm and touch, no body and no gaze will have stayed intact. In the fissure of this impaired and longing seeing and feeling, optic and haptic contact with each other, contaminate without ever becoming one. The 'touchingness' of a scenic touch will have been its potential, its strong weakness of touching without touch; without transgressing any limits, without mingling surfaces, but rather touching the borders, affecting, tangential, contingent: in all the contingency of a contact that occurs, happens, is imparted – only in separation, only in the non-intactness of tactile experience which does not concern unimpaired subjects, which takes no immediacy as given, which aesthetically, ethically, politically opens and closes the quotation marks for "touching" – as if they were the eyelashes of an ever distant, interrupted gaze. No immediacy, uninterruptedness, continuity, symmetry. The technique of touching rather concerns the caesuras, the syncopes, the *fault lines*. As if our world were built on *fault lines*, on those subterranean fissures and crevices in deep rock strata that are supposed to be responsible for our aggressions and depressions, for our violent *stills* and tender distances. *Fault lines* – perhaps those fissures, disturbance areas, lines of distortion at which we always abide, anchorless and restrained, in our mutual inverse desires, the lines at which – only in our inconsistency, our brokenness – we can touch each other.

References

Agamben, Giorgio, "Bartleby o della contingenza", in: Giorgio Agamben/Gilles Deleuze, *Bartleby: La formula della creazione*, Macerata 1993.
Barthes, Roland, *La chambre claire: note sur la photographie*, Paris 1980.
Derrida, Jacques, *Le toucher: Jean Luc Nancy*, Paris 2000.
Derrida, Jacques, "Les morts de Roland Barthes", in: *Poétique* 47 (1981): 269–292 (republished in: Derrida, Jacques, *Psyché*, Paris 1987, 273–304).
Nancy, Jean-Luc, *Corpus*, Paris 2000.
Nancy, Jean-Luc, *Être singulier pluriel*, Paris 1996.

This text was first published in: *Scores 1: Touché* (2011). We would like to thank Tanzquartier Wien and the editors of *Score*, Walter Heun, Krassimira Kruschkova, Sandra Noeth and Martin Obermayr, for their kind permission to reproduce the text.

II Kinesthesia

Mark Paterson

On 'Inner Touch' and the Moving Body

Aisthêsis, Kinaesthesis, and Aesthetics

Introduction: The Aesthetics of Movement

The *idea* of movement has been of interest within phenomenologically-influenced studies in cognitive science and embodied cognition.[1] But what about the *feeling* of movement, the subjectively-felt qualitative dynamic, utilizing the so-called 'muscle sense,' the somatic senses that include kinesthesia and proprioception? Explorations of these sensations of movement occur unsystematically across the humanities and social sciences, in areas like cultural studies (e.g., Brian Massumi, Erin Manning), dance and performance studies (e.g., Susan Leigh Foster), philosophical aesthetics (Barbara Montero), even cultural geography. The Canadian phenomenologist David Morris has noted how interest in the philosophy of embodied cognition has met with empirical validation, not only through conventional TMS and fMRI scanning techniques but also through arts–science collaborations in dance and the performing arts, borne out by funded projects like Dee Reynolds' *Watching Dance: Kinaesthetic Empathy* (2008–2011).[2] In other words, the arts and sciences of bodily sensations of movement are coming together in unusual and experimental ways.

Yet that divide between historicized idea and qualitative feeling of movement is not straightforwardly bridged by empirical means of measurement, no matter how technologically sophisticated the equipment employed. In other words, the difficulties of articulating felt qualities of movement remains, despite any purported ability to 'read' brain states *in media res.* In this short chapter I offer selected highlights of an ongoing research project on developments in the history of neurophysiology that have informed our contemporary understanding of somatic sensa-

---

1 Cf. Sheets-Johnstone, Maxine, *The Primacy of Movement*, Amsterdam 1999; Sheets-Johnstone, Maxine, "Body and Movement: Basic Dynamic Principles," in: Shaun Gallagher/Daniel Schmicking (ed.), *Handbook of Phenomenology and Cognitive Science*, New York 2010, 217–234; Gallagher, Shaun, *How the Body Shapes the Mind*, Oxford 2005; Berthoz, Alain/Petit, Jean-Luc, *The Physiology and Phenomenology of Action*, New York 2008.

2 Cf. Morris, David, "Empirical and Phenomenological Studies of Embodied Cognition," in: Shaun Gallagher/Daniel Schmicking (ed.), *Handbook of Phenomenology and Cognitive Science*, New York 2010, 235–252. See the website of Dee Reynolds' *Watching Dance: Kinaesthetic Empathy:* http://www.watchingdance.org/ (19.12.2012).

tions of muscular movement or, in other words, how our bodies became sensori-motor. These are initial steps for what will become a 'felt' aesthetics of movement. The roots of such an idea are found in Aristotle's treatment of bodily sensations as *aisthêsis*, a generalized sense faculty in *De Anima* ('On the Soul,' 350 BC). This generalized *aisthêsis* works as both the philosophical and etymological starting point, since 'aesthetics' derives etymologically from the stem *aesthe*, 'to perceive, hear,'[3] and later medicalized variants include 'coenaesthesia' in the 18th century and 'kinaesthesia' in the early 20th.[4] Throughout the larger unfolding story, from *aisthêsis*, via kinaesthesis, to aesthetics, we consider how *aisthêsis* potentially binds sensation into aesthetic evaluations.

This chapter is divided into four parts, firstly an historical section that tracks the neurophysiological discoveries of movement and the sensory-motor into the early 20th century, effectively from *aisthêsis* to kinaesthesis. Here I investigate how movement and, more particularly, kinaesthesia as a modality and as a metaphor have become of interest at the intersection of phenomenology and dance. Situating the relationship between *aisthêsis* that collectively constitutes the bodily or somatic senses, and kinaesthesia as the sense of movement, the second section briefly highlights how *aisthêsis* develops into the more medicalized language of a distinct 'muscle sense' from Charles Bell (1833) and the emergence of the concepts of 'kinaesthesia' from Henry Charlton Bastian (1869) and 'proprio-ception' from Charles Sherrington (1906). The third part attends to some qualitative features of sensations that arise within sensori-motor activity, especially the pleasures of movement. The fourth section 'kinaesthesia to aesthetics' indicates how earlier historical treatments of *aisthêsis* connect into more recent empirical work on movement and the ethnographic treatment of so-called kinaesthetic cultures, potentially widening the focus from individual bodily sensation to the charting of collective movements and reading the movement of others as kinaesthetic empathic practice.

3 Cf. "aesthesis," in: *Oxford English Dictionary*, third edition, December 2011; online version June 2012. http://www.oed.com/view/Entry/3234 (11.8.2012).

4 There is no standardized spelling of this term, with *Gemeingefühl* being variously translated as 'coenesthesia', 'coenesthesis', and 'cenesthesia', along with US and UK versions. For consistency I refer to the Anglicized spellings for this and related terms from the Oxford English Dictionary ("cœnaes'thesia, n.". OED Online. December 2012. Oxford University Press. http://www.oed.com/view/Entry/35680. 6.1.2012).

From Aisthêsis to Kinaesthesis: the *Idea* of Movement

For a kind of proto-history of haptic knowledges we turn briefly to Plato and Aristotle to establish the necessity for touch, and the pervasiveness of a generalized 'inner touch' faculty into the Middle Ages.[5] In classical Greek scholarship the word *aisthêsis* (αἴσθησις) is translated by most, including the standard Greek-English Lexicon by Liddell and Scott (1843) and Chappell (2004), simply as 'sensation' or 'sense-perception.' At another point, Liddell and Scott elaborate upon it as "perception by the senses, especially by feeling, but also by seeing, hearing, etc. [...] also of the mind, perception, knowledge of a thing."[6] However, Michael Frede rightfully takes issue with the vagueness of this translation, arguing there are three senses of *aisthêsis* that need to be distinguished.[7] Firstly, an ordinary or more general sense like 'awareness' not necessarily connected with sensory perception. In both Aristotle's *Politics* and Plato's *Symposium*, for example, *aisthanomai* is translated as 'I notice.'[8] Secondly, a narrower sense as used in *Phaedo* and *Republic*, in which *aisthêsis* necessarily involves the body, does not equate to knowledge as such and is therefore allied with *doxa* (belief), and consequently not strictly sense perception. Thirdly, the narrower sense as used in *Theaetetus* in which *aisthêsis* comes to mean an entirely "passive affection of the mind" and so, more ordinarily, "sense-perception."[9] If there is an affection and therefore an alteration, what would be the cause? In *Theaetetus* it is not simply the perception of objects in the outside world, but also dreams and illusions.[10] This suggests that "the primary objects of *aisthêsis* are *internal to the mind*,"[11] to be categorized as immediate experiences, impressions – in modern parlance, 'sense-data.'

If *aisthêsis* involves both the awareness of sensation as well as bodily sensation itself, this implies a model of bodily perception as contact-like impressions based on a form of inner touch. However, touch is problematic for Aristotle. Capable of fine discrimination in the aesthetic evaluation of objects like a piece of sculpture, nonetheless touch remains at the bottom of the sensory hierarchy as

5 For extended discussion see Paterson, Mark, "Movement for Movement's Sake? On the Relationship Between Kinaesthesia and Aesthetics," in: *Essays in Philosophy* 13.2 (2012): 471–497.
6 Chappell, Timothy D.J., *Reading Plato's Theaetetus*, Indianapolis 2004, 53.
7 Cf. Frede, Michael, *Essays in Ancient Philosophy*, Oxford 1987.
8 Aristotle, *Politics*, 1267a29; Plato, *Symposium*, 220c7.
9 Plato, *Theaetetus*, 184–187.
10 Ibid., 157–160.
11 Chappell, *Reading Plato's Theaetetus*, 54.

we share it with beasts.[12] Aristotle correctly moves away from the locus of skin in order to argue "the sense-faculty of touch is within," like internal organs, rather than "without,"[13] like skin, eyeballs, or ears. Wearing a glove, we may still stroke an animal or imprecisely sense an object's texture; similarly, when walking with a stick we apprehend the roughness of the ground. The fleshy medium is corporeal then, and extendable through prosthetic means: "so it is necessary that the body be the ongrown medium of the touch-faculty and that the sensations (which are indeed many) take place *through* it," says Aristotle.[14] At times straightforwardly cutaneous, at other times Aristotelian touch is more diffuse or a generalized contact.

We return to *aisthêsis* here because of the persistence of a generalized form of bodily perception based on contact, a form of inner touch. For example, in his 1778 treatise *Plastik* ('Sculpture'), Herder postulates that if the student of art were to encounter something doubtful or contradictory in their interpretation of a work then they should make use of "the fingers of [their] inner sense (*den Finger seines innern Sinnes*) in order to discover that which [they] could not otherwise identify: the shape of the spirit within the form (*Gestalt des Geistes in dieser Form*)."[15] Along with his project of a sustained reexamination of tactility within aesthetics, contra Hegel, his invitation to discern the immaterial spirit (*Geist*) of a material work by using the metaphorical 'fingers' of a putative 'inner sense' invokes a striking tactile metaphor, a virtual probing, an imaginative grasping or palpating of an aesthetic work the better to discern its 'true' or intended form. This conceptualization of the inner senses as a kind of metaphorical or imaginative touching involves a somatic reflexivity, a knowing 'grip' on the body and its movement that forms a long-running strand in philosophy and medicine. It is consistent with the pervasive idea of "a kind of inner touch, by which we are able to grasp ourselves," as Heller-Roazen summarizes.[16] To reconsider the historical and etymological *aisthêsis* is to elaborate also upon 'common sense' (*aesthesis koine*) and later the medicalized term 'coenaesthesia,' so within the concept of *aisthêsis* and its derivatives we find a productive wellspring for reconsidering the type and nature of experiences that arise from perception through the body. This long trajectory originates in classical scholarship, but is something the Abbé de Lignac will term "the sense of the coexistence of the body"

12 E.g., Aristotle, *Ethics*, 1176a1–2.
13 Aristotle, *De Anima*, 423b.
14 Ibid., 423a.
15 Herder, Johann Gottfried, *Sculpture: Some Observations on Shape and Form from Pygmalion's Creative Dream* [1778], Berkeley 2002, 90.
16 Heller-Roazen, Daniel, *The Inner Touch: Archaeology of a Sensation*, New York 2007, 241.

in his *Elements of Metaphysics Drawn from Inner Experience* of 1752, and a series of sensations which Turgot in his *Encylopédie* entry of 1755/56, acknowledging Lignac, wished to place under a special class, "inner touch."[17] Variously understood by their contemporaries in terms of an expanded notion of touch as bodily self-perception, this includes Lamarck's "feeling sense" which comprises touch both inside and outside the individual, "self feeling," "the tonality of the sensory nerves" (Henle), or "the consciousness of our sensory condition" (Weber), amongst other formulations, all revealing the preponderance of conceiving the experience of manifold inner senses specifically in terms of some kind of touch. Recently, the philosopher Richard Shusterman has written extensively on what he terms 'somaesthetics' or "body consciousness" in the performing arts.[18] Continuing the trajectory of inner touch he explains this as "the embodied consciousness that a living sentient body directs at the world."[19]

The continuation of a medicalized form of inner touch, and physiological discoveries in the 19th century, start to differentiate the roles of the nervous system and the musculo-skeletal body, so now we focus on muscle and movement.

The Body as an 'Organ of Sense'

Between Aristotle and 19th-century physiologists like Bastian and Bell, Condillac and his followers describe a so-called 'active touch'[20] which predates Gibson's 20th-century use of the phrase.[21] But attention to the particularity of what is sensed through the muscular body becomes of scientific interest in Germany as the *Muskelsinn* (muscle sense) by George, Bernhardt, and others.[22] Similar terminology was first introduced into Britain as early as 1820 by Thomas Brown in his *Lectures on the Philosophy of the Human Mind*, where sensations were char-

17 Cf. ibid., 242.
18 Cf. Shusterman, Richard, *Body Consciousness: A Philosophy of Mindfulness and Somaesthetics*, Cambridge 2008; Shusterman, Richard, "Body Consciousness and Performance: Somaesthetics East and West," in: *The Journal of Aesthetics and Art Criticism* 67.2 (2009): 133–145.
19 Shusterman, "Body Consciousness and Performance," 133.
20 Jones, E.G., "The Development of the 'Muscular Sense' Concept during the Nineteenth Century and the Work of H. Charlton Bastian," in: *Journal of the History of Medicine and Allied Sciences* 27.3 (1972): 298–311, 299.
21 E.g., Gibson, James J., "Observations on Active Touch," in: *Psychological Review* 69.6 (1962): 477–491.
22 Cf. George, L., "Der Muskelsinn," in: *Archiv für Anatomie, Physiologie und wissenschaftliche Medicin* (1870): 251–263; Bernhardt, M., "Zur Lehre vom Muskelsinn," in: *Archiv für Psychiatrie und Nervenkrankheiten* 3.3 (1872): 618–635.

acterized as "an awareness of muscular contraction."[23] The new body of research was then reported to the British scientific community by Sir William Hamilton, writing in 1846, and subsequently William Hammond in 1871 provides an abbreviated report of George's 1870 paper in a section entitled "The Muscular Sense (*Muskelsinn*)" in *The Journal of Psychological Medicine*.[24] The portmanteau *Muskelsinn* persisted in both Britain and Germany, involving later uses such as Schaeffer in 1889, Hocheisen in 1893 on this sense for the blind, and Goldscheider in 1898.[25]

Muskelsinn is conceptualized as something distinct from cutaneous touch (*Tastsinn*), generalized bodily feeling (*Gefühlssinn*) as Kirchner had termed it, or common sense (*Gemeinempfindungen*) for Wundt as Titchener reminds us.[26] In Brown's lectures XX–XXII on touch for example the sensations particular to a muscle sense are recognizably distinct and worthy of attention:

> The feeling of resistance is [...] to be ascribed not to our *organ of touch*, but to our *muscular frame*, [...] as forming a distinct *organ of sense* [...]. The sensations of this class, are [...] commonly, so *obscure*, as to be scarcely heeded [...] but there is no contraction, even of a single muscle, which is not attended with some faint degree of sensation, that distinguishes it from the contraction of *other muscles* or from *other degrees of contraction* of the same muscle.[27]

His focus on a form of sensation which is distinctly muscular in origin, yet actually imprecise and vague is retained, so that shortly afterwards he states: "each motion of the [...] limb, whether produced by *one* or *more* of the [...] muscles, is accompanied with a *certain feeling* [...] which we distinguish from every *other feeling* accompanying every other *quantity of contraction*."[28] In other words, a more general awareness of bodily position is maintained through more localized and distinct muscular tensions and contractions. In line with his forebears, this is interpreted as a form of bodily touch that extends beyond the cutaneous and proposes the whole body as an organ of sense, where "our muscular frame is not merely a part of the living *machinery* of *motion*, but is also *truly an organ of*

---

23 Brown, Thomas, *Lectures on the Philosophy of the Human Mind*, Edinburgh 1820, cited in: Jones, "The Development of the 'Muscular Sense' Concept," 299.

24 Hamilton, W., "The Muscular Sense (Muskelsinn)," in: *The Journal of Psychological Medicine: A Quarterly Review of Diseases of the Nervous System, Medical Jurisprudence and Anthropology* 5 (1871): 396–398.

25 Cf. Baldwin, James Mark, *Dictionary of Philosophy and Psychology*, New York 1905, 1173.

26 Titchener, Edward B., *Lectures on the Elementary Psychology of Feeling and Attention*, New York 1908, 158.

27 Brown, *Lectures on the Philosophy of the Human Mind*, vol. I, 496.

28 Ibid., 497.

sense."[29] In this way, Brown, George, Wundt, and their contemporaries could be considered as re-articulating the ongoing inner touch of *aisthêsis*, while focusing on its manifestations through muscle fibers and the muscular frame.

The particular tensions and contractions that comprise the *Muskelsinn* become foregrounded in the absence of other sensory stimuli, especially sight. Attention to the musculo-skeletal frame of the body, albeit perhaps with the potential for amplified acoustic acuity and resonances, is heightened in the case of blindness, as Hochsein's "Über den Muskelsinn bei Blinden" (1893) evidences. Hochsein's paper was summarized in the British journal *Mind* of that year:

> The perception of passive movements – the perception of the 'muscle sense' in general – are more than normally acute in the case of practiced blind subjects. The causes of this are mental: attention and practice in the interpretation of sensations of slight intensity. The movement-sensibility of children is better than that of adults. The differences of 'right' and 'left' have but little to do with acuteness of movement-discrimination. The cutaneous 'sense of place' is not always provably more than normally acute in the blind: where it is so, the fact is referable to practice.[30]

Not long afterwards, Murray's 1909 essay "Organic Sensation" provides an historical overview of what she collectively describes as the "sensory contributions from the internal tissues."[31] Reviewing neurophysiological work from the 19th century onwards, she remarks upon "the least developed and systematized sphere of our consciousness" in order to "throw new light on processes of localization and attention."[32] Digestive, muscular, and respiratory systems each produce their own sensations, what Ebbinghaus in 1902 terms *eigenartige Empfindungen* ('strange sensations'). Again, Meumann in a 1907 article "On the sensibility of the internal organs" celebrates a multiplicity of "organic sensations" (*innere Tastempfindungen*) derived from distributed organs and tissues, but actually a direct translation of Meumann's phrase would offer the now familiar 'inner touch sensations.'[33] The unusual combination of qualitative diversity yet indistinctness of these sensations together obscures "the indefiniteness of localization of the sensations, and their deficiency in correlated visual images by

29 Ibid., 501.
30 Anonymous, "Philosophical Periodicals," in: *Mind, New Series* 2.8 (1893): 553 f.
31 Murray, Elsie, "Organic Sensation," in: *The American Journal of Psychology* 20.3 (July 1909): 386–446, 400.
32 Ibid., 402.
33 Meumann, E., "Zur Frage der Sensibilität der inneren Organe," in: *Archiv für die Geschichte der Psychologie* 9 (1907): 26–62.

which qualitative isolation might be facilitated," Murray argues.[34] In other words, these sensations remain vague and unsystematic, and cannot be compared to the kind of clarity that vision enjoys. At the end of one section Murray summarizes this indistinctness in frustrating terms:

> The fusability, absence of memory images, unanalyzability, lack of cohesiveness with other sensations, unlocalizability, capacity for eluding the attention, and other features ascribed guardedly or confidently in various quarters to our organic experience, demand critical verification.[35]

The terminological difficulties noted above extend to the laboratory, of course, as that legacy of Wundt's 'common sensation' (*Gemeinempfindung*) amongst experimental psychologists interested in internal or, as Murray puts it, 'organic' sensations, is problematic. The 'common' element across these sensations seems to be the "pain-pressure-temperature equipment common in his belief to the exterior and interior of the body alike."[36] Meanwhile, Luciani's volume on the physiology of the sense organs of 1917 acknowledges allied frustrations since "[m]any of the bodily feelings thus classified escape physiological analysis owing to their vague and obscure character."[37] From Wundt's *Gemeinempfindung* to Henle's *Gemeingefühl* ('common sensation' or 'coenaesthesia'), this is likewise "the sum, the confused chaos of the sensations which are instantly transmitted to the brain from all parts of the body."[38] Weber in *Der Tastsinn und Gemeingefühl* (*The Sense of Touch and Common Sense*, 1905) identifies possibilities within the chaos of sensation for the body to provide sensory resonances that accompany complex sensory impressions, such as encountering variously localized pain sensations or a combination of colors. Again, this diffuse yet almost synaesthetic model assumes an almost contact-like model of the senses not dissimilar from Aristotle's *aisthêsis* and its alteration, *alloiosos*.

Against this diffuseness of inner touch, a more refined answer that distinguishes a 'muscular sense' occurs in Charles Bell's *The Hand* (1833), where he considers the interaction between touch and movement. His earlier lectures on anatomy proffered a distinct muscle sense, in his words a "consciousness of

---

34 Murray, "Organic Sensation," 400 f.
35 Ibid., 402.
36 Ibid.
37 Luciani, Luigi, *Human Physiology*, vol. 4: *The Sense Organs*, ed. G.M. Holmes, New York 1917, 59.
38 Cf. ibid., 69.

muscular exertion," akin to a sixth sense.[39] Bell's earlier anatomical discoveries reported to the Royal Society of distinct specialized sensory and motor nerves had led him to investigate the mechanisms of the nervous system that governed and regulated muscular activity. Given that we customarily have a sense of muscular coordination, an awareness that heightens with exertions and spasms or even the estimation of weights through the use of our hands, Bell states his goal thus: "*I shall first enquire, if it be necessary to the governance of the muscular frame, that there be a consciousness of the state or degree of action of the muscles?*" This can be asked since "[w]e are sensible of the most minute changes of muscular exertion, by which we know the position of the body and limbs, when there is no other means of knowledge open to us."[40] In passing he offers the example of a rope-dancer or a blind man balancing his body, finding an explanation in neuroanatomical terms. Referring to prior discussion of the interactions of sensory and motor nerves, he summarizes in deceptively straightforward language: "Between the brain and the muscles there is a circle of nerves; one nerve conveys the influence from the brain to the muscle, another gives the sense of the condition of the muscle to the brain."[41] Likewise, this bidirectional nervous mechanism is referred to in a later book: "there is a nerve of sensibility to convey a sensation of the condition of the muscles to the sensorium, as well as a nerve of motion for conveying the mandate of the will to the muscles."[42] Conceptualized therefore as a circuit of sense in the nervous system and musculoskeletal movement, the body is indeed becoming sensori-motor.

The new modality of neuromuscular feedback was later termed 'proprio-ception' by Sherrington (1906), based on his laboratory research conducted on muscular reflexes between 1892 and 1894. Sherrington integrated these experimental observations into a prestigious lecture series at Yale, published much later as *The Integrative Action of the Nervous System* (1906). In his centenary appreciation of Sherrington's book, Burke notes how Sherrington identified "afferent feedback," nerve impulses returning to the brain from diffuse muscle tissue, and in Lecture III Sherrington considered how these afferents were 'proprioceptive' because they are caused by the organism's own movements, in contrast with 'exteroceptive' afferents that convey more distal information from the environment. In Lecture IX, Sherrington steps back from the details in order to offer a more

---

39 Bell, Charles, *The Hand: Its Mechanism and Vital Endowments as Evincing Design*, London 1833, 195.

40 Bell, Charles, "On the Nervous Circle Which Connects the Voluntary Muscles with the Brain," in: *Philosophical Transactions of the Royal Society of London* 116 (1826): 163–173, 167.

41 Ibid., 170.

42 Bell, *The Hand*, 196.

panoramic view of the evolutionary development of nervous systems through the phylogenetic sequence. The implication is that, for a range of anatomically sophisticated 'higher' animals, the sensory is invariably coupled with the motor, so 'sensory-motor.' He notes that "by its branching the motor neurone [*sic*] obtains hold of many muscle-fibres,"[43] and this diffusion of nerve endings through muscle fibers feeds back to a ganglion within the cerebellum, what Lidell and Sherrington will later term the 'motor unit.'[44] This allows Dewey in his famous essay "The Reflex Arc Concept in Psychology" of 1896 to speak of "sensori-motor coordination" in an almost proto-phenomenological way, uniting an initial sensory act or stimulus, say visually noticing an object or person, with an associated movement to achieve an overall action, for example steering a car around a raccoon or waving to a friend:

> [W]e begin not with a sensory stimulus but with a sensori-motor coordination [...] [I]n a certain sense it is the movement which is primary, and the sensation which is secondary, the movement of body, head and eye muscles determining the quality of what is experienced.[45]

Dewey continues in terms that reinforce the unity of perception and action: "The sensory quale [the property or value of a thing] gives the value of the act, just as the movement furnishes its mechanism and control, but both sensation and movement lie inside, not outside the act."[46] Our attention having been focused on the muscle senses, Dewey here raises a potential difficulty for aesthetically evaluating such movements in the arc of perception and action, given the unity of the sensory-motor circuit. Given the admittedly indistinct and potentially confused nature of neuromuscular feedback (strange sensations, inner touch), and the varying levels to which the embodied subject is consciously aware of them, the felt qualities of one's own movement (*aisthêsis*) is consequently problematic to evaluate in aesthetic terms. For a dance performance such feedback might be required to better effectuate and coordinate particular bodily movements, evaluated by both dancer and audience. We might then ask: what is the nature of the sensory qualia that arise in addition to the autonomic circuits of the sensori-motor? What obtrudes, or stands out in the consciousness of the body-subject? The answer might involve pain, fatigue, and pleasure as proprio-

---

43 Sherrington, Charles, *The Integrative Action of the Nervous System*, New Haven 1906, 309.
44 Cf. Burke, Robert E., "Sir Charles Sherrington's 'The Integrative Action of the Nervous System': A Centenary Appreciation," in: *Brain* 130 (2007): 887–894, 892.
45 Dewey, John, "The Reflex Arc Concept in Psychology," in: *Psychological Review* 3 (1896): 357–370, 358; see also: Pfeifer, Rolf/Bongard, Josh/Grand, Simon, *How the Body Shapes the Way We Think: A New View of Intelligence*, London 2007.
46 Dewey, "The Reflex Arc Concept in Psychology," 359.

ceptive markers within a series of coordinated actions. For the remainder of this section we concentrate on pleasure within movement.

The Pleasures of the Sensori-Motor

Expressed in more contemporary neurophysiological terms from original research by Mense and Stahnke in 1983, the afferent pathway for the sensori-motor body involves "small afferents arising in the muscle and joints," consisting of "small unmyelinated nerve fibres which arise within muscle. Most may be involved in signalling pain, contraction and temperature, but some appear to relay information related to fatigue," explain Cole and Montero.[47] Just like we did, Mense and Stahnke pondered whether these afferents reach consciousness, and what perception they might produce since the system could signal not just fatigue or pain but also pleasure. Such low-threshold touch, as research by Olaussen et al. reveals, involves unmylenated CT fibers which, unlike conventional touch through large myelinated fibers, is not localized or clearly perceived, yet occurs in the same brain area (*insula cortex*) as the monitoring functions of internal bodily functions like hunger, pain and discomfort.[48] In other words, this low-threshold touch is continually present as part of the sensations of embodiment, Sherrington's 'proprioception,' and perhaps Merleau-Ponty's noncognitive "motor intentionality," part of the *schéma corporel* ('body schema').[49] But low-threshold touch is also capable of producing pleasure from gentle caressing or stroking in a way that registers differently from a higher-threshold cutaneous touch *per se*. It supports the mixed sensations Bell had described in another section of *The Hand* (1833) entitled "The pleasures arising from the muscular sense":

---

47 Cole, Jonathan/Montero, Barbara, "Affective Proprioception," in: *Janus Head* 9.2 (2007): 299–317, 302.

48 Olaussen et al., "Unmyelinated Tactile Afferents Signal Touch and Project to Insular Cortex," in: *Nature Neuroscience* 5 (2002): 900–904.

49 Merleau-Ponty, Maurice, *Phenomenology of Perception*, London 1962, 110. Space does not allow more detailed exposition of links with Merleau-Ponty's influential work on the "spatiality and motility of one's body", and connections with Husserl on the 'kinestheses.' This is treated in a forthcoming book manuscript. However, Carman usefully re-examines 'body schema' in compatible terms to this discussion of inner touch: "The body schema is the crux or reference point that establishes a stable perceptual background against which I perceive and respond to changes and movements in my environment, and thereby opens me onto a world of other selves [...], the bundle of skills and capacities that constitute the body's precognitive familiarity with itself and the world it inhabits" (Carman, Taylor, "The Body in Husserl and Merleau-Ponty," in: *Philosophical Topics* 27.2 [1999]: 205–226, 220).

> The exercise of the muscular frame is the source of some of our chief enjoyments. The beautiful condition of both body and mind shall result from muscular exertion and the alternations of activity and bodily repose [...] This activity is followed by weariness and a desire for rest, and although unattended by any describable pleasure or local sensation, there is diffused throughout every part of the frame after fatigue a feeling almost voluptuous.[50]

Cole aptly refers to this passage to remark on "the rise of kinaesthetic-related pleasure" such as jogging or dancing,[51] and a notable characteristic of such pleasure is that it deepens with practice. The "simple ineffable pleasure of, and of being in, action"[52] exemplified through dance, as Cole and Montero put it, is enriched by moving in a way that 'feels right,' that is judged or feels beautiful, so Cole and Montero consequently argue that dancers enjoy a more "cognitively enriched" pleasure as a result.[53]

In proprioceiving a difference between a static body state at one point in time and anticipating the possibility of future movement in that body, Bell touches upon a kinaesthetic element in describing this muscle sense, to the effect that touching of any kind involves a proprioceptive component, the sense of one's own body, particularly when movement is involved. "We could not command our muscles in standing, far less in walking, leaping or running, had we not a perception of the condition of the muscles previous to the exercise of the will," he says.[54] Touching, especially through the prehensile organ of the hand, also betrays a kinaesthetic element since it is the "combined perception"[55] of touch with movement of the hands, arms and fingers that are necessary to embrace objects in the active register necessary for an engaged sense of touch. Here Bell anticipates, or perhaps acknowledges, Weber's 1834 psychophysical experiments on so-called "active touch" recounted in *Der Tastsinn:*

> So it is affirmed by physiologists [...] that the sense of touch differs from the other senses by this circumstance – that an effort is propagated towards it, as well as a sensation received from it. This confusion obviously arises from considering the muscular agency, which is directed by the will during the exercise of touch, as belonging to the nerve of touch properly. We proceed to show how the sense of motion and that of touch are necessarily combined.[56]

---

50 Bell, *The Hand*, 205 f.
51 Cole, Jonathan, *Pride and a Daily Marathon*, Boston 1995, 144.
52 Cole/Montero, "Affective Proprioception," 303.
53 Ibid.
54 Bell, *The Hand*, 200.
55 Ibid., 205.
56 Ibid., 197.

This observation, tying in a somatic, active tactility to the principle of movement, again expands tactility from mere cutaneous contact, and by invoking the idea of 'muscular agency' it echoes Herder's formulation of the inner senses as more active, prehensile feelers, an enlivened form of tactility. One may consider for example a choreographer tentatively sketching movements for a dance piece, or how a painter approaches a blank canvas with a series of broad brush strokes in mind, successively accreting further movements and gestures with paint and so sympathetically 'feeling' the textures inscribed. There are accordingly positive affects that accompany the mastery and refinement of movement, not only for a dancer or an accomplished musician but also a cook or a runner.[57] The pleasure which arises from muscular activity derives in part from a gratification that accompanies any refinement or active shaping of activity, "as that which mere dexterity, successful pursuit in the field, or the accomplishment of some work of art may give," Bell explains.[58] Part of the pleasure of the generalized muscle sense, and more specifically kinaesthesia in this instance, Bell explicitly identifies in aesthetic terms. For the next and final section we therefore consider the relationship between *aisthêsis*, kinaesthesia and the aesthetic values of movement for movement's sake. The aesthetic and even joyful qualities of touch and movement are praised in a rather romanticized way by Bell, nevertheless exhibiting a certain familiarity to modern sensibilities:

> We owe other enjoyments to the muscular sense. It would appear that in modern times we know comparatively little of the pleasures arising from motion. The Greeks, and even the Romans, studied elegance of attitude and of movement. Their apparel admitted of it, and their exercises and games must have led to it. Their dances were not the result of mere exuberance of spirits and activity; they combined harmony in the motion of the body and limbs, with majesty of gait. They consisted more of the unfolding of the arms than of the play of the feet – "Their arms sublime that floated on the air."[59]

From Kinesthesis to Aesthetics: Collective *Feelings* of Movement

At this juncture we can only gesture towards more practical implications for expressing the sensory-motor aspects of the body within the reflexive research process, in ethnographies of dance or so-called kinaesthetic cultures.

57 Cf. Cole, *Pride and a Daily Marathon.*
58 Bell, *The Hand*, 206.
59 Ibid.

Wacquant's invocation to research where the body serves "as a tool of inquiry and a vector of knowledge"[60] reminds us that the body itself has increasingly become "an instrument of research."[61] Yet rarely has this involved anticipative sensations of muscular embodiment, an attendance to motor intentionality. As we have seen, musculo-skeletal sensations become foregrounded and heightened in the non-visual spatial apprehensions and navigation of the blind[62] and visually impaired, but evidently also in dancing subjects. Acknowledging the complexity of neurophysiological explanations for somatic sensations or background embodiment is a necessary first step, before opening this concept out to the appreciation of other dancing bodies. For example, contra to this chapter's emphasis on kinaesthesia, Barbara Montero's work on dance and aesthetics has explored the role of proprioception as an aesthetic sense[63] and how one might "proprioceive another's movement."[64] But attention should also be paid to how these sometimes novel and sometimes familiar sensations are represented, given their transitory nature, flickering across thresholds of conscious awareness and preconscious states, sometimes visceral and sometimes vague, yet (almost) always directly experienced through the body.[65] One clue as to how this might occur is in ethnographic treatments of kinaesthetic bodies and kinaesthetic cultures, especially within non-Western traditions.

Previously we had noted the kinaesthetic orientation of a choreographer tentatively sketching out a new dance piece, or a painter approaching a blank canvas with a selection of kinaesthetic intentions and gestures. A kinaesthetic predisposition, a motor intentionality, a background capability that might then be played out through skillful movement. Likewise, Carrie Noland begins her book *Agency and Embodiment* (2009) by observing a graffiti artist spraying paint on the wall, and notices how movement and gesture were an integral part of the appreciation of that art form. Between the original script or ideogram,

---

60 Wacquant, Loïc, *Body & Soul: Notebooks of an Apprentice Boxer*, Oxford 2004, viii.

61 Longhurst, Robyn/Ho, Elsie/Johnston, Lynda, "Using 'the Body' as an 'Instrument of Research': Kimch'i and Pavlova," in: *Area* 40.2 (2008): 208–217.

62 E.g., Hocheisen, Paul, "Über den Muskelsinn bei Blinden," in: *Zeitschrift für Psychologie und Physiologie der Sinnesorgane* 5 (1893): 239–282.

63 Montero, Barbara, "Proprioception as an Aesthetic Sense," in: *The Journal of Aesthetics and Art Criticism* 64.2 (2006): 231–242.

64 Montero, Barbara, "Proprioceiving Someone Else's Movement," in: *Philosophical Explorations* 9.2 (2006): 149–161. A more detailed analysis of her arguments and their shortcomings can be found in my article "Movement for Movement's Sake?".

65 For more detailed considerations of haptic sensibilities within ethnographic fieldwork see Paterson, Mark, "Haptic Geographies: Ethnography, Haptic Knowledges and Sensuous Dispositions," in: *Progress in Human Geography* 33.6 (2009): 766–788.

through a mixture of movements, repetition and improvisation, and the final product, a painted script on a wall, "a body was afforded a chance to feel itself moving through space,"[66] and so a performance emerges. Like any performance which requires deliberate movement, this is simultaneously a "repetitive routine and improvisational dance," observes Noland. What she terms the "sensate motor body," what we have earlier considered *aisthêsis* and later the 'sensori-motor' body after Dewey, forms the medium upon which the "gestural regime" builds, she argues.[67] Overall, while Noland is more interested in questions of the 'subject' and 'agency' that result, she acknowledges how gestures belong to the domain of movement. Such iterative and performative gestures "provide kinaesthetic sensations that remain in excess of what the gestures themselves might signify or accomplish in that culture."[68] Much as in the discussion of *aisthêsis* above, the body's awareness of its own position and movement, its own interoceptive ability or kinaesthetic awareness, helps structure the range and indeed styles of movement that result.

References

Anonymous, "Philosophical Periodicals," in: *Mind, New Series* 2.8 (October 1893): 552–556.
Baldwin, James Mark, *Dictionary of Philosophy and Psychology*, New York 1905.
Bell, Charles, *The Hand: Its Mechanism and Vital Endowments as Evincing Design*, London 1833.
Bell, Charles, "On the Nervous Circle Which Connects the Voluntary Muscles with the Brain," in: *Philosophical Transactions of the Royal Society of London* 116 (1826): 163–173.
Bernhardt, M., "Zur Lehre vom Muskelsinn," in: *Archiv für Psychiatrie und Nervenkrankheiten* 3.3 (1872): 618–635.
Berthoz, Alain/Petit, Jean-Luc, *The Physiology and Phenomenology of Action*, New York 2008.
Brown, Thomas, *Lectures on the Philosophy of the Human Mind*, Edinburgh 1820.
Burke, Robert E., "Sir Charles Sherrington's 'The Integrative Action of the Nervous System': A Centenary Appreciation," in: *Brain* 130 (2007): 887–894.
Carman, Taylor, "The Body in Husserl and Merleau-Ponty," in: *Philosophical Topics* 27.2 (1999): 205–226.
Chappell, Timothy D.J., *Reading Plato's Theaetetus*, Indianapolis 2004.
Cole, Jonathan, *Pride and a Daily Marathon*, Boston 1995.
Cole, Jonathan/Montero, Barbara, "Affective Proprioception," in: *Janus Head* 9.2 (2007): 299–317.

66 Noland, Carrie, *Agency and Embodiment: Performing Gestures/Producing Culture*, Cambridge 2009, 1.
67 Ibid., 2.
68 Ibid.

Dewey, John, “The Reflex Arc Concept in Psychology,” in: *Psychological Review* 3 (1896): 357–370.
Frede, Michael, *Essays in Ancient Philosophy*, Oxford 1987.
Gallagher, Shaun, *How the Body Shapes the Mind*, Oxford 2005.
George, L., “Der Muskelsinn,” in: *Archiv für Anatomie, Physiologie und wissenschaftliche Medicin* (1870): 251–263.
Gibson, James J., “Observations on Active Touch,” in: *Psychological Review* 69.6 (1962): 477–491.
Hamilton, W., “The Muscular Sense (Muskelsinn),” in: *The Journal of Psychological Medicine: A Quarterly Review of Diseases of the Nervous System, Medical Jurisprudence and Anthropology* 5 (1871): 396–398.
Heller-Roazen, Daniel, *The Inner Touch: Archaeology of a Sensation*, New York 2007.
Herder, Johann Gottfried, *Sculpture: Some Observations on Shape and Form from Pygmalion's Creative Dream* [1778], Berkeley 2002.
Hocheisen, Paul, “Über den Muskelsinn bei Blinden,” in: *Zeitschrift für Psychologie und Physiologie der Sinnesorgane* 5 (1893): 239–282.
Jones, E.G., “The Development of the ‘Muscular Sense’ Concept during the Nineteenth Century and the Work of H. Charlton Bastian,” in: *Journal of the History of Medicine and Allied Sciences* 27.3 (1972): 298–311.
Longhurst, Robyn/Ho, Elsie/Johnston, Lynda, “Using ‘the Body’ as an ‘Instrument of Research’: Kimch'i and Pavlova,” in: *Area* 40.2 (2008): 208–217.
Luciani, Luigi, *Human Physiology*, vol. 4: *The Sense Organs*, ed. G.M. Holmes, New York 1917.
Merleau-Ponty, Maurice, *Phenomenology of Perception*, London 1962.
Meumann, E., “Zur Frage der Sensibilität der inneren Organe,” in: *Archiv für die Geschichte der Psychologie* 9 (1907): 26–62.
Montero, Barbara, “Proprioception as an Aesthetic Sense,” in: *The Journal of Aesthetics and Art Criticism* 64.2 (2006): 231–242.
Montero, Barbara, “Proprioceiving Someone Else's Movement,” in: *Philosophical Explorations* 9.2 (2006): 149–161.
Morris, David, “Empirical and Phenomenological Studies of Embodied Cognition,” in: Shaun Gallagher/Daniel Schmicking (ed.), *Handbook of Phenomenology and Cognitive Science*, New York 2010, 235–252.
Murray, Elsie, “Organic Sensation,” in: *The American Journal of Psychology* 20.3 (July 1909): 386–446.
Noland, Carrie, *Agency and Embodiment: Performing Gestures/Producing Culture*, Cambridge 2009.
Olausson, H./Lamarre, Y./Backlund, H./Morin, C./Wallin, B.G./Starck, G./Ekholm, S./ Strigo, I./Worsley, K./Vallbo, Å.B./Bushnell, M.C., “Unmyelinated Tactile Afferents Signal Touch and Project to Insular Cortex,” in: *Nature Neuroscience* 5 (2002): 900–904.
Paterson, Mark, “Movement for Movement's Sake? On the Relationship between Kinaesthesia and Aesthetics,” in: *Essays in Philosophy* 13.2 (2012): 471–497.
Paterson, Mark, “Haptic Geographies: Ethnography, Haptic Knowledges and Sensuous Dispositions,” in: *Progress in Human Geography* 33.6 (2009): 766–788.
Pfeifer, Rolf/Bongard, Josh/Grand, Simon, *How the Body Shapes the Way We Think: A New View of Intelligence*, London 2007.

Sheets-Johnstone, Maxine, "Body and Movement: Basic Dynamic Principles," in: Shaun Gallagher/Daniel Schmicking (ed.), *Handbook of Phenomenology and Cognitive Science*, New York 2010, 217–234.
Sheets-Johnstone, Maxine, *The Primacy of Movement*, Amsterdam 1999.
Sherrington, Charles S., *The Integrative Action of the Nervous System*, New Haven 1906.
Shusterman, Richard, *Body Consciousness: A Philosophy of Mindfulness and Somaesthetics*, Cambridge 2008.
Shusterman, Richard, "Body Consciousness and Performance: Somaesthetics East and West," in: *The Journal of Aesthetics and Art Criticism* 67.2 (2009): 133–145.
Wacquant, Loïc, *Body & Soul: Notebooks of an Apprentice Boxer*, Oxford 2004.
Weber, Ernst Heinrich, *The Sense of Touch*, New York 1978.
Weber, Ernst Heinrich, *Tastsinn und Gemeingefühl*, ed. Ewald Hering, Leipzig 1905.

Hilge Landweer

Choreographies With and Without a Choreographer

Intuitive and Intentional Corporeal Interactions[1]

I Phenomena and Questions

What do free jazz, classical ballet and a 'successful' football match have in common with pedestrians passing each other by in the street? What distinguishes these forms of interaction? In this essay, I want to look at different ways in which movements are aligned with other movements and at ways in which people corporeally react to each other, both in everyday situations and in specific contexts and settings. I will discuss which structure of the 'felt body'[2] enables us to do this and how these phenomena of 'aligning ourselves with each other' can be described and classified.

In classical ballet and in a football match (if played successfully), movements form a sort of higher-level entity, a 'comprehensive body' as it were. It appears as if its parts, the individual bodies, were not moving of their own accord but under the guidance of a higher entity. And although the individual bodies can be active in one instant and passive in the next of their own will, their movements are aligned in a way that they no longer appear to primarily result from the intention of the individuals. In free jazz (the improvised play of several musicians that does not rest upon a given melody or other predefined elements), the ways in which the individual voices respond to each other can similarly be described as movements that relate to each other and seemingly originate from one single entity. In this case, however, we are not looking at physical but at acoustic movements, even if they are produced by the body.

---

1 I would like to thank Maria Wirth, Ingrid Kasten, Thorsten Streubel and Roswitha Breckner for their detailed comments and suggestions.

2 In this essay, I will distinguish the notions of 'felt body' (German *Leib*) and 'material body' (German *Körper*). The adjectives 'corporeal' (relating to *Leib*) and 'bodily' or 'physical' (relating to *Körper*) will be used accordingly. I will explain below (section 2) how these notions are defined and used in the philosophy of Hermann Schmitz that I wish to discuss. The use of English terminology for Schmitz' theory largely draws on an English translation of a text by Schmitz, published by Rudolf Owen Müllan and Jan Slaby in: "Emotions Outside the Box: The New Phenomenology of Feeling and Corporeality", in: *Phenomenology and the Cognitive Sciences* 10.2 (2011): 241–259.

Although higher-level entities can emerge in all the cases we have mentioned, these entities occur in different ways. Differently from a football match and a free jazz performance, classical ballet bases itself on predefined movements. Leaving no space for spontaneous modifications or surprising variations, it operates with a fixed choreography that the individual bodies must comply with. But even in ballet, the performance of two or more dancers on stage can suggest the existence of only one body that the individual bodies are merely (not quite independent) parts of. Within the range of phenomena I have chosen as examples of corporeal interaction, the movements of pedestrians passing each other by in the street constitutes the opposite extreme. While pedestrians coordinate their movements only intuitively, they do so surprisingly smoothly. By getting out of each other's way, two bodies relate to each other for a short moment in time. Yet this moment, implying a new entity that comprises both bodies, quickly 'dissolves' and disappears in the crowd.

Team sports and the arts develop and cultivate highly complex forms of corporeal interaction, but when pedestrians get out of each other's way, they do so intuitively. Whether such movements are intentional or unintentional, in both cases we lack a precise terminology that could describe what exactly happens with the felt body and what we perceive respectively. The phenomenologist Hermann Schmitz refers to all these cases as examples of 'corporeal communication' and speaks of 'encorporation' as one possible form thereof. According to Schmitz, there are two modes of encorporation which he calls 'antagonistic' and 'solidary' respectively. In this paper, I will closely examine the concepts proposed by Schmitz. I shall discuss how useful they are and whether there are alternative terms that could be used instead. Subsequently, I will analyze how a corporeally bound 'us' emerges and which role joint intentions play in that process.

In order to do so, I will introduce the basic ideas of Schmitz' phenomenology in the following section (2), followed by a critical discussion (by means of examples) of the notions of 'antagonistic' and 'solidary' encorporation (3). The fourth section of my essay (4) will be devoted to the phenomenon of corporeal resonance, which can best be illustrated by looking at practices of artistic performance. This intentional form of corporeal interaction will be contrasted with the flight behaviour of a herd of animals as an example of intuitive solidary encorporation (5). The last section of my paper (6) will discuss the relationship between intentional and intuitive forms of corporeal interaction. Is the formation of joint intentions bound to a co-presence and thus to corporeal interaction? And if so, are all forms of corporeal interaction equally suited for this purpose?

II Phenomenology – Contraction and Expansion

The examples we have mentioned of cases in which movements are aligned with one another can generally be described in at least two different ways. We can either observe a sequence of movements from an outside perspective (as we do when watching a video recording), or we consider the process of movement from an internal perspective, looking at what the individuals feel while they move.

These two perspectives of observation and description ('external' point of view vs. perception of one's own movement) can essentially be regarded as a 'third-person perspective' and a 'first-person perspective' respectively. If we relate these perspectives to the body, we arrive at the phenomenological distinction made in German between *Körper* (the body as a physical object, i.e. 'material body') and *Leib* ('felt body' as experienced by the self). The 'felt body' refers to the body as it is experienced, perceived and described in the first-person perspective, while the term 'material body' (or simply 'body') denotes the body as it is perceived and observed from an external, a third-person perspective.[3] The material body is the objectified felt body, deprived of its subjectivity, for example when we look at what is quantifiable about pain. 'Material body' comprises externally observable behaviour and gestural expression as well as medical or kinetic data, in brief, all that can be perceived optically, tactually or acoustically about the self or the other and measured with the help of instruments.

The distinction between 'material' and 'felt body' can already be found in Edmund Husserl.[4] It was taken up by Helmuth Plessner who transformed it into the opposition of 'having a (material) body' and 'being a (felt) body'.[5] Hermann Schmitz' new phenomenology, however, proposes an even clearer distinction. When dealing with the difference between material and felt body in the context of interactions between different bodies, Schmitz' philosophy provides a detailed terminology which enables us to describe the above-mentioned situations, for instance a football match or a ballet performance, not only externally, but also from the internal perspective of those involved.[6] From the perspective of the audience

---

3 When we speak about a personal experience, we certainly go beyond the experience, because we use language, yet this is not equivalent to a third-person perspective.

4 Husserl, Edmund, *Ideen* II: *Hua* IV, The Hague 1952, 144 ff., *Krisis Hua* VI, The Hague 1976, 109.

5 'Having' a body here indicates the acts of objectification and appropriation, while 'being' a felt body refers to the existence itself (cf. Plessner, Helmuth, *Die Stufen des Organischen und der Mensch*, Berlin 1975).

6 Cf. Schmitz, Hermann, *System der Philosophie*, vol. V, Bonn 1980, §288; Schmitz, Hermann, "Leibliche Kommunikation mit und ohne Wort", in: Manfred Bauschulte/Volkhard Krech/Hilge Landweer (ed.), *Wege – Bilder – Spiele: Festschrift zum 60. Geburtstag von Jürgen Frese*, Bielefeld 1999, 251–260.

observing particularly well-performed processes of corporeal coordination, we must, above all, explain the impression that there is one single comprehensive body performing, consisting of the individual agents as its parts. How must the individual felt body be structured for such an entity to emerge?

The impression of observing one single entity is based on the perception of one's own felt body, both by those observed and by those observing the performance. According to Schmitz' analysis of the felt body, its condition constantly oscillates between tendencies of contraction and expansion. When we breathe in, for instance, we perceive a certain expansion until we reach a moment of obstruction which, in turn, is dissolved by breathing out. This new process of expansion is again, at some point, obstructed, forcing us to breathe in again. Shock is perceived as a sudden contraction, while pride is corporeally connected to expansion (see the expression 'to feel a swell of pride'), and fear can choke us (there are numerous other expressions that illustrate that fear is perceived corporeally as an intense form of contraction). The etymology of the German word *Angst* speaks for itself: the Old High German *angust* literally means 'Enge' (narrowness). Pain is always a phenomenon of corporeal contraction. By means of his 'Alphabet of the Felt Body' ('Alphabet der Leiblichkeit'), which cannot be described in detail in this essay, Schmitz distinguishes between different modes of contraction that he describes in a highly differentiated manner.[7]

The insight that contraction and expansion are always bound to each other in human experience is of particular importance for the description of corporeal interactions. Together, they form the economy of the felt body which can be characterized with the help of further categories.[8] From the interaction of the two main corporeal tendencies of contraction and expansion emerges the 'dialogical structure of the felt body'. Showing itself in the fact that the felt body perceptibly oscillates between tendencies of contraction and expansion, this dialogical structure enables it to be affected externally in the first place. Through certain 'channels' of corporeal contact (e.g., the gaze or the taking-up of a given rhythm), it is possible to corporeally 'respond' to gestalt behaviour that is connatural to the felt body,

---

7 Cf. Schmitz, *System der Philosophie*, vol. II.1, chap. 2, §§ 48–55.

8 These categories comprise tension and swelling as well as direction, intensity, rhythm, protopathic and epicritic tendencies (cf. Schmitz, *System der Philosophie*). The nature of corporeal economy can best be studied in boundary phenomena that occur when the balance between contraction and expansion is radically altered. When, corporeally, contraction completely disconnects itself from expansion, we lose consciousness (e.g. when pain becomes too constricting that corporeal expansion is no longer possible). Similarly, when the tendency of expansion becomes independent, waking consciousness is no longer possible (e.g. when we fall asleep or in a state of intoxication).

movements and suggestions of movement by modifying and adapting one's own tendencies of contraction and expansion. I will discuss in section 4 of this essay if this 'echo' can be understood as a phenomenon of resonance.

For the way in which the felt body deals with things and facts it perceives, Schmitz coined the notion of 'corporeal communication'. Corporeal communication occurs every time we are sensibly affected by something or when something forces us to take notice of it. Schmitz does not restrict this notion to interactions between individuals. He describes the subjectivity of perception as a corporeal process between a sentient being and his or her counterpart, whether that is a person or an object. Strictly speaking, it is not the subject that consciously communicates in this instance, but it is the corporeal directions of contraction and expansion that interact with the respective directions of movement, the corporeal tendencies of their counterpart.[9]

In this context, it may be misleading to use the term 'communication' which is strongly associated with human agents (I communicate with someone) and intentionality (I only communicate when I want to). However, these aspects Schmitz specifically does *not* have in mind. He is concerned with a process that often takes place below the threshold of consciousness: the process in which the corporeal directions of contraction and expansion are modified by the contact with somebody or something, so that a 'higher' entity emerges. For this reason, I suggest that we speak of 'corporeal *interactions*', a term that appears more adequate particularly with regard to situations of contact with inanimate objects (i.e. to 'normal' perception). As we have seen, Schmitz considers these to be examples of corporeal communication, too, and he uses his categories of contraction and expansion to describe the individual's common perception. In this essay, however, I want to focus on interactions between subjects. The notion of 'subject', as I use it here, comprises highly developed life forms whose felt bodies are, like ours, structured by a dialogue of contraction and expansion (e.g., all mammals). Before I proceed, let me make a few general remarks on perception.

According to Schmitz, perception works along the same principles of alignment as the contact with other individuals. He thus rejects the information-theoretical (and often causally interpreted) assumption that perception merely consists of the intake and processing of signals. Discussing the example of seeing (the 'channel' of this corporeal interaction) a threatening object that is drawing closer, Schmitz demonstrates the narrowness of such a summative concept. Un-

9 Schmitz speaks of 'suggestions of movement' originating from animate and inanimate things. We cannot discuss this in greater detail (cf. Schmitz, Hermann, *Der unerschöpfliche Gegenstand: Grundzüge der Philosophie*, Bonn 1990, esp. 140–151 and 282–284).

less one is hindered by someone or something, one is normally, and without deliberation, capable of moving out of the way. Obviously, the reason for this is not that we see both our own body and the approaching object and, based on this information, draw conclusions about the positions of and distance between them – in that case, reactions would often not be quick and precise enough. Besides, we cannot normally see the exact position of our entire body and sometimes cannot see it at all. Yet we do perceive our body in situations like this, not by seeing, but by feeling it, while our eyes remain fixed on the threatening object. The focusing sight is an activity of the felt body and not a merely passive intake of quickly succeeding stimuli. It allows for an active *co-acting* with the approaching object *without reaction time* (in the terms of information theory).

As the felt body is *in itself* dialogically structured, the corporeal directions of contraction and expansion can be 'allocated' spatially, without changing their structure, to partners or poles which pit these tendencies against each other. Like in a wrestling match, sometimes one part dominates in its tendency of expansion, and sometimes the other. The tendency of contraction is in the meantime taken on by the other part – until it gains the upper hand once again. In processes of this kind, the corporeal directions of those involved spontaneously form higher-level, quasi-corporeal entities structured just like the felt body itself. This, of course, affects the corporeal economy of the individual. The alignment of corporeal communication is of fundamental importance in everyday life. One example of this is the exchanging of glances (even if it is brief) between pedestrians who manage to coordinate their movements in an unplanned, highly intuitive manner and yet so smoothly that they very rarely bump into each other, not even in bigger crowds of people. Such co-acting *without reaction time* can be observed particularly well when craftspeople work and musicians play together, as well as in team sports.

III 'Antagonistic' and 'Solidary Encorporation'

Among the different phenomena of corporal interaction, Hermann Schmitz distinguishes *Einleibung* ('encorporation') and *Ausleibung* ('excorporation'). They correspond to the 'weak points' or poles where the corporeal economy can be radically altered through intervention. While encorporation forms part of the contraction pole of corporeal economy that can, in extreme cases, lead to unconsciousness (e.g. due to overwhelming pain or shock), excorporation, bound to the expansion pole, can generate ecstasy (intoxication) or make a person fall asleep.[10] In this case, corporeal communication goes hand in hand with a feeling of depth through outpouring and sinking, like when, on a moving train, we look out of the window and get lost in the scenery passing by. This form of corporeal communication, however, is not relevant for the questions I wish to discuss in this essay.

In everyday life, encorporation is a much more common case. According to Schmitz, it is always motivated by the concentration on someone or something (i.e. a counterpart). Corporeally, concentration is always associated with contraction. This aspect can be applied to all phenomena of corporeal interaction that we have discussed so far, and for this reason we can classify them as examples of 'encorporation'. This is not true, however, of the scenery example mentioned above which must be attributed to the opposing expansion pole and thus stands for excorporation. Through its connotations (intake of food, seizure), the term 'encorporation' implies the appropriation of world (and perhaps deliberately so). In processes of encorporation, the objects, animals and people perceived are related to the narrowness of one's own felt body. The term 'excorporation', on the other hand, seems to suggest that one's own felt body completely 'disappears' in the interaction with the object of perception (although this does not refer to the material body).

Considering it to be the most important form of corporeal communication, Schmitz distinguishes two different modes of encorporation: 'antagonistic' and 'solidary encorporation'. In situations of *antagonistic* encorporation, there are always two instigators or sources determining the corporeal directions (e.g. two individuals who corporeally relate to each other). If movements are concerned, they are either intuitively (e.g. pedestrians) or intentionally (e.g. in a fight) aligned with one another, potentially in a competitive manner (in order to attack effectively, for example). In processes of *solidary* encorporation, however, the impulse comes from an object or a theme that determines the corporeal alignment and movements of those involved. As examples of solidary encorporation, let us name the corporeal cooperation of craftspeople working together (e.g. cutting up logs), sports competitions between teams or pairs of athletes, music played by several people together, as well as fleeing herds of animals or groups of people, all relating to a threat or something else that gives them an impulse. At first glance, this form of encorporation seems to be more strongly associated with a sense of community than instances of antagonistic encorporation.

In Schmitz' characterization, antagonistic encorporation is based on the dominance of one of the two poles involved that give corporeal impulses. This domi-

---

10 See above, footnote 8.

nance can either be stable (Schmitz speaks of 'unilateral antagonistic encorporation') or unstable, in which case it alternates and oscillates (Schmitz speaks of 'reciprocal antagonistic encorporation'). He uses the examples of hypnosis, fascination and suggestion to illustrate his concept of unilateral antagonistic encorporation. It differs from forms of reciprocal antagonistic encorporation which I will mainly be looking at in this paper (specifically in contrast to instances of solidary encorporation).[11] If we look at these phenomena from an external perspective only, we see two bodies at play in each case, interacting as sources for movement impulses and corporeal directions. With the help of the dialogical structure of the felt body, they either align themselves with each other or are already aligned with each other. Examples of this are, among others, a handshake, pedestrians avoiding physical contact with each other, a conversation, a boxing match or the caresses exchanged between a mother and her baby. There may be certain rules for the coordination of reciprocal antagonistic encorporation (e.g. in a conversation), but corporeal interaction does not happen for the sake of rules, and the specific characteristics of the interaction are not determined externally (not even by rules). The oscillation of the dominating role, as described above, may well be competitive or show itself in a fight (particularly obvious in a boxing match). Because the dominant role alternates in this form of encorporation, it is possible that both partners increasingly incite each other as in some of the above-mentioned examples of corporeal communication or in the interaction between rider and horse, which can be described as a case of reciprocal antagonistic encorporation. An increase of energies, as it can similarly be perceived in solidary encorporation, seems even more typical of the solidary form of encorporation that often appears to specifically aim for such an increase.

The way Schmitz phrases his concept and theories, the use of the terms 'antagonistic' and 'solidary' as well as the emphasis he places on questions of dominance (in cases of reciprocal and unilateral antagonistic encorporation) suggest that corporeal interactions can be associated with processes of power, violence and domination. A rider will, to use a rather 'innocent' example, always aim to hold the reins tightly: it is he/she who directs the horse. Whoever gets involved in a wrestling match will try to dominate the fight in order to win it. The more playful an interaction is, the less those involved will aim for dominance.

---

11 In what follows, I will not use the specifying adjective 'reciprocal', because all examples of antagonistic encorporation that I will discuss refer to the antagonistic mode (as opposed to solidary encorporation).

However, contraction and expansion are constantly in competition with each other within the corporeal entity spanning the two bodies.[12]

What about solidary encorporation then, is it free from power and hierarchies, as Schmitz' designation seems to suggest? No, it is not, and if we were to assume that it was, we would clearly misconceive this form of corporeal interaction. It is true that Schmitz characterizes it as being free from dominance and subjection, but this only refers to the way in which two (or more) partners jointly relate, in solidary encorporation, to something else (e.g. a task or a theme). It has, however, no implication at all for the relationship between the partners beyond their joint corporeal relation to the task or theme. Social relations between individuals who are associated with each other in solidary encorporation can be characterized by specific roles which often have a hierarchical dimension: the foreman may do sawing work with his apprentice, football teams are divided into strikers and defenders, and an orchestra consists of complex hierarchies that are mostly predefined by the written composition (e.g. the roles of conductor, lead violinist and soloists who can set the pace and influence the performance in different ways). While individual instruments or groups of instruments can take the lead, all musicians keep concentrated, in a 'solidary' manner, on the composition that they can only play together and that predetermines the alternating impulses (e.g. in terms of dynamics or voice leading). Personal intentions or spontaneous decisions of the individual musicians, which can make themselves heard in the course of the performance, are secondary to this joint instigator.[13]

---

12 For Schmitz, looking at each other as a case of antagonistic encorporation can be described as a 'Ringkampf der Blicke' (wrestling match of gazes): 'Man kann niemals einem anderen ins Auge schauen, ohne dass sich ein "Ringkampf der Blicke" mit ganz derselben Struktur [wie bei einem Ringkampf; HL] einspielt. Das liegt nicht an irgendeiner herrschsüchtigen Absicht, sondern an der Struktur des Leibes, die den sich begegnenden Blicken als unteilbar ausgedehnten leiblichen Richtungen [...] eingeprägt ist. Blicke, die sich begegnen, sind wie Speere im Turnier; sie greifen tief in das leibliche Befinden beider Partner ein, die sich dadurch bedeutsame Signale geben, sind schwer auszuhalten, werden demgemäß – allerdings nicht überall [...] – gern vermieden [...] und entfalten sich [...] zu einem vielfältigen Wechselspiel.' [Whenever we look someone else in the eyes, a 'wrestling match of gazes' with the very same structure (as a real wrestling match) takes place. This is not due to some domineering intention, but to the very structure of the felt body which characterizes the gazes (as inseparable corporeal directions) when the eyes of two people meet. Mutual gazes behave like javelins in a joust: they deeply affect the corporeal condition of both partners who give each other signals full of meaning; they are difficult to bear; one often – but not always – tries to avoid them; and they unfold a multifaceted interplay.] (Schmitz, *Der unerschöpfliche Gegenstand*, 136; cf. Schmitz, *System der Philosophie*, vol. III.2, 378–383, and vol. 5, 31–33).

13 In this respect, free jazz constitutes an interesting boundary phenomenon (see below, section 4).

Solidary encorporation lacks dominance and subjection, because it essentially consists in the joint concentration on a task or theme. All the felt bodies involved focus on the same joint goal, and through their concentration, they become one higher-level entity. Rowing in one boat can be considered an example of solidary encorporation only if all athletes aim for speed and therefore pull the oars simultaneously. Otherwise a joint rhythm will not be accomplished.[14] Groups of people singing and playing music together are further examples, as are crowds of people who follow a spontaneous impulse or run from a threat. In this case, the joint 'theme' as a prerequisite for solidary corporeal communication consists in whatever triggers the impulse (e.g. the onstage performance of a star 'sparking' the people in the audience and making them interact in solidary encorporation).

I believe that what we have said clearly shows that Schmitz' distinction of antagonistic and solidary encorporation does not coincide with the distinction of combative and peaceful interaction. On the one hand, solidary encorporation (e.g. rhythmic rowing) can indeed be a part of a sporting competition like any other form of team sport. On the other hand, the exchange of caresses between a mother and her baby, despite being an example of antagonistic corporeal communication, bears no agonal traits at all. Obviously, the terms 'competition' and 'playful interaction' do not denote the phenomena that Schmitz wants to distinguish and do not generally relate to different types of genuine corporeal interacting. One could say that they refer to the contextual frame of the situation that is shared by all those involved, but not to specific features of the corporeal contact itself.

If we wanted to avoid the misleading connotations of the terms 'antagonistic' and 'solidary' encorporation with power and controversy, we could replace them with the terms 'bi-impulsive' and 'uni-impulsive' or rather 'bipolar' and 'unipolar'. At first glance, one might even be tempted to call forms of solidary encorporation 'multipolar', for they bear a more complex structure than forms of antagonistic encorporation, but this would be problematic because there is still only one instigator affecting more than one person. The 'poles' of encorporation must not be understood to be felt bodies, but as entities triggering the decisive impulse (or, in the case of antagonistic encorporation: the alternating impulses). This impulse (or these impulses) is (are) not necessarily bound to a felt body. The number of corporeal individuals involved does not permit any conclusions on the type of corporeal interaction: in a busy pedestrian area, a great number of people can be connected in antagonistic corporeal interaction, and two musicians playing together form a case of solidary encorporation without their instigator, the composition, being a subject. In what follows, I will use

14 Cf. Schmitz, *Der unerschöpfliche Gegenstand*, 152.

both terminological distinctions, 'antagonistic' and 'solidary encorporation' on the one hand, and 'bipolar' and 'unipolar encorporation' on the other.

It seems that a large part of the phenomena that Schmitz refers to as 'encorporation' can be described as cases of joint intentional action. This is true of all instances of solidary encorporation (with the exception of the fleeing herd or crowd that we will discuss below), as well as of numerous examples of antagonistic encorporation. A regulated boxing match, for example, requires certain joint intentions, for example the joint intention to accept the setting and comply with the rules, while pedestrians who get out of each other's way (another case of antagonistic encorporation) may do this intentionally or intuitively (possibly without even noticing it themselves).

Are intentions equally well suited to describe the phenomena we have been looking at as the corporeal aspect of interaction? Are they really the same phenomena? The joint intention to perform a regulated boxing match does not appear to be the same thing as the strife for dominance in the alternation of contraction and expansion during the match. Such processes of corporeal interaction may happen mostly intentionally (at least in a more or less professional boxing match), but the underlying *joint* intention seems to be restricted to the athletes' decision to follow the rules. It does not include the corporeal interaction which, per definition, consists in a fight that both opponents want to win and thus dominate physically and corporeally, which means that the individual intentions of the athletes involved are mutually exclusive.

Is it, therefore, possible that there are not and that there cannot be any joint intentions at the basis of antagonistic encorporation, for this type of corporeal interaction is characterized by an irregular constant alternation of dominance whose dynamics only result from interaction? When there are two instigators for corporeal communication, they are usually unsynchronized – otherwise there would have to be a specific demand set by a third pole as instigator. This source of corporeal impulses can, as we have seen, consist in a joint task that requires certain joint intentions. Two poles synchronized in such a way would not be able to give any impulses of interaction of their own, independent of the impulse for synchronization, because they would have to assume a subordinate role to the 'third' (virtually the only) instigator, in order to be able to reach a joint corporeal alignment. This, however, would correspond to Schmitz' description of solidary encorporation.

The assumption that all corporeal interactions would have to be *either* antagonistic *or* solidary is certainly incorrect. Schmitz does suggest, however, that these two forms of encorporation can change into each other. If we take into account the procedural character of corporeal interactions, we can say that at least initially, forms of antagonistic encorporation cannot be based on joint intentions. Although in most cases all those intentions will be shared that are required by the situation and provide the setting for corporeal interaction,[15] the *corporeal* side of the interaction and its dynamics of contraction and expansion is explicitly *not* guided by joint intentions – even if the interaction may give way to joint intentions (e.g. the plan to carry out another fight).

Can we deduce from these observations that the antagonistic form of encorporation is essentially characterized by a lack of joint intentions (referring to the corporeal movement), while joint intentions constitute a necessary prerequisite for any form of solidary encorporation? It would be too early to draw such a conclusion as long as we cannot determine with certainty that all potential cases of solidary corporeal interaction require the existence of joint intentions.

In the following section, I want to discuss the concept of 'corporeal resonance' which shares some features with phenomena of solidary (and possibly also with antagonistic) encorporation. Leading to a clearer differentiation of the phenomena we have been looking at, it opens up new possibilities of describing one of the two extremes of solidary encorporation, namely professionally concerted processes of artistic performance. In focusing on this aspect, I will first consider phenomena that require the existence of joint intentions.

IV Corporeal Resonance

In physics, 'resonance' refers to processes in which an oscillatory system ('resonator') with its own frequency is activated by an input of energy ('exciter'). It can oscillate at amplitudes much greater than the amplitude of the exciter. Such phenomena can be observed in hydromechanics (wave resonance), electrical engineering, in atomic and nuclear physics as well as in mechanic (swing) and acoustic phenomena (e.g. in the generation of sound using string or wind instruments and when a string that has not been touched resonates as soon as an instrument that is tuned in the same way is caused to sound) of everyday life.[16]

---

15 If we wanted to take the theory of corporeal interaction and situation perception further, it would be interesting to combine Schmitz' New Phenomenology with Erving Goffman's Frame Analysis (cf. Goffman, Erving, *Frame Analysis: An Essay on the Organization of Experience*, New York, London 1974).

16 Cf. 'Resonance', in: *Wikipedia*, http://en.wikipedia.org/wiki/Resonance, (13.7.2013).

It goes without saying that the different uses and transformations of the term 'resonance' offer a great potential.[17] However, there do seem to be certain restrictions: the term refers to one direction of interaction, i.e. the effect of the movement of one 'system' on the other.[18] When 'resonance' refers to the contact between subjects, it describes the fact that one is influenced by others. In the phenomenology of the felt body, it relates to phenomena in which corporeally felt directions of the 'resonator' intuitively repeat the corporeal directions or the 'exciter' and coincide with them. The opposite direction of taking influence (of ego on alter, or of the resonator on the exciter) is not explicitly expressed by this term, at least not as a consistent phenomenon. Thus, 'resonance' would denote only one side of what we call 'corporeal interaction', namely the reactive side. Schmitz' model of contraction and expansion, however, is neutral with regard to activity and passivity because it is not primarily concerned with movement but with perception, and perception is not synonymous with passivity or inaction. Strictly speaking, this kind of neutrality is inconsistent with a model that bases itself on action and reaction.[19]

At the same time, the fact that the term 'resonance' is restricted to only one 'side' or direction of the interaction opens up a great opportunity. With the help of this concept, it is possible to consider the increase of the amplitude (here: the energy of the 'resonator') enabled by the resonating of the exciter. A number of phenomena of corporeal interaction are characterized by certain asymmetries between 'exciter' and 'resonator'. This could be true, for example, of the corporeal interaction between artists on stage and their audience, as the atmosphere is in most cases determined rather by the former than the latter. It is mainly pro-

---

17 Cf. Lichau, Karsten/Tkaczyk, Viktoria/Wolf, Rebecca, "Anregungen", in: id. (ed.), *Resonanz: Potentiale einer akustischen Figur*, Munich 2009, 11–32, as well as the other contributions to this volume.

18 Fischer-Lichte uses the term 'resonance' to denote the echo and influence of older dramatic plays and theatre productions on current performances (cf. Fischer-Lichte, Erika, "Theater als Resonanz-Raum", in: Lichau/Tkaczyk/Wolf (ed.), *Resonanz*, 237–248).

19 This does not exclude the possibility to describe activity and passivity by means of the dialogical structure of the felt body. Shame, for example, can corporeally be understood as a passivization deriving from a massive contraction and the unrealizable impulse to get away from the shameful situation. Rage, on the other hand, is a form of activization (cf. Schmitz, *System der Philosophie*, vol. III.3, §172d, 44–47; Landweer, Hilge, *Scham und Macht: Phänomenologische Untersuchungen zur Sozialität eines Gefühls*, Stuttgart 1999, 40 ff. and 43 f.; the comparison of 'Nemesis' and 'Aidos' in Williams, Bernard, *Shame and Necessity*, Berkeley, Los Angeles, London 1993, 80 ff.).

duced by the performance of the artists,[20] while resonance mainly happens among the audience. On the other hand, many artists say that when their audience appears reserved and barely approachable, they are strongly aware of this and feel restrained in their performance on stage. Even if the term 'resonance' does not seem to grasp the reciprocity of this relationship adequately, it may be suited to capture first the resonance of the audience and then the resonance the artists feel in reaction to it.

In other words, when we say that an artist received a good response (resonance) to his/her performance, this describes his/her competence as an 'exciter' very well, while the audience acts as a 'resonator' in the physical sense. In these cases, the initiative lies on the side of the artist. If, in turn, the artist reacts to the response of the audience, this would have to be regarded as a second step in the overall process which could also be described as 'resonance', with the audience taking the role of the 'exciter'. It would thus be adequate to use the concept of resonance when describing the relationship between artist and audience, as well as all other cases in which we can easily distinguish between 'exciter' and 'resonator'.

In this sense, Undine Eberlein emphasizes that corporeal resonance 'happens' to us and that it is not only an echo of what we perceive but also a transformation of our own corporeal feeling.[21] She discusses dance performances during which the audience not only perceives, but corporeally 're-enacts' the individual figures and sequences of movement (in this context, she refers to research studies on mirror neurons). This 're-enactment' of movements takes influence on the audience's

20 Cf. Schmitz, Hermann, "Über das Machen von Atmosphären", in: Anna Blume (ed.), *Zur Phänomenologie der ästhetischen Erfahrung*, Freiburg 2005, 26–43.

21 On the one hand, Eberlein uses the term 'resonance' to describe a broad range of different phenomena: the corporeal interaction of dancers; the corporeal reaction of the audience to the dance performance; the perception of the corporeal situation of somebody else and one's own corporeal reaction to it; mass phenomena like 'contagion' and the corporeal feeling of being deeply moved by an atmosphere (cf. Eberlein, Undine, "Leibliche Resonanz: Phänomenologische und andere Annäherungen", in: Kerstin Andermann/ead. (ed.), *Gefühle als Atmosphären: Neue Phänomenologie und philosophische Emotionstheorie*, Berlin 2011, 141–152, here 144f.). Her use of the term thus comprises all phenomena that Schmitz describes as 'encorporation', which might suggest that it is just an alternative name for the same thing. On the other hand, however, Eberlein is concerned with more specific phenomena of resonance (mainly in the audience) which we will be looking at in this paper. Several passages in the text suggest that Eberlein considers 'encorporation' and 'corporeal resonance' to be two different phenomena, e.g. '[...] die wechselseitige Einleibung [ist] [...] für das Phänomen der leiblichen Resonanz von zentraler Bedeutung' [mutual encorporation is of central importance for the phenomenon of corporeal resonance] (144); later she speaks of 'Situationen faszinierter Einleibung und Resonanz' [situations of fascinated encorporation and resonance] (146).

own corporeal economy.[22] From the start, the movements on stage are perceived (or better: felt) as parts of one single entity. Because of this impression – of there being a comprehensive entity or 'wholeness' – a non-professional audience 'knows' almost as well as an experienced critic whether a performance is possibly perfect in technical terms, yet terribly boring, or whether its intensity and presence succeed in casting a nearly magical spell on the spectator.[23] We may thus assume that neither the experience of a 'wholeness' nor the insight that a performance does not succeed in making a deeper impression (because it does not create a concentrated or exhilarating atmosphere) requires a professional qualification.

There are, however, other prerequisites for corporeal resonance to happen. Among those, the dramaturgic advisor Lenz Prütting named, in a theoretical essay of 1995, a specific setting like stage and auditorium, but mainly the willingness of the spectators to be open and attentive, concentrate on and get themselves involved in the performance, and finally the 'resonating' of the audience.[24] Beatrix

---

22 'In der mit- und nachvollziehenden Einleibung der Bewegung der Tänzer durch das Publikum (und andere Tänzer) wird deren eigenleibliche Wahrnehmung durchle(i)bt, und zwar eben nicht durch die bewusste Apperzeption einzelner Bewegungselemente und Figuren, sondern als eine [...] weitgehend "widerfahrende" Ganzheit des Geschehens' [When they corporeally follow, reproduce and encorporate the movement of the dancers, the audience (and the other dancers) corporeally experience the dancers' own corporeal perception, not through the conscious apperception of individual elements or figures of movement, but as an entity of events largely 'happening' to them] (Eberlein, "Leibliche Resonanz", 146).

23 Cf. Eberlein, "Leibliche Resonanz", 142.

24 'Die eben zitierte Auflistung [gemeint ist: aufmerken, sich ausrichten im Sinne von sich zuwenden, 'aufmachen', einleiben und mitgehen; H.L.] impliziert [...] eine Tendenz vom eher aktiven zum eher passive Verhalten, ohne daß jedoch am Ende dieser Reihe das Mitgehen als reines Widerfahrnis stünde. Es ist vielmehr so, daß in allen angeführten Verhältnisweisen sich Aktivität und Passivität auf eine ganz eigene Art durchdringen und verschränken. Und deshalb ist der mitgehende Zuschauer auch durchaus und beileibe kein "passiver" Zuschauer [...]' [This enumeration of phenomena, i.e. taking notice, turning towards and opening up to something/someone, encorporating and getting emotionally involved, implies a tendency of moving from active to passive modes of behaviour, which does not mean, however, that the process of getting emotionally involved merely 'happens' to us in the end. In all of these modes of behaviour, activity and passivity pervade each other in their own specific ways, which is why the spectator who gets emotionally involved is not a 'passive' spectator at all.] (Prütting, Lenz, "Über das Mitgehen: Einige Anmerkungen zum Phänomen transorchestraler Einleibung", in: Michael Großheim (ed.), *Leib und Gefühl: Beiträge zur Anthropologie*, Berlin 1995, 141–152, 147). Similarly, Eberlein emphasizes that the spectators are 'nicht bloß passiv Überwältigte' [not just passively overwhelmed] and that they can, at any moment in time, '[sich] aus dem "Gesamtleib" ausklinken' [disengage themselves from the 'comprehensive felt body' (Eberlein, "Leibliche Resonanz", 146 f.).

Hauser speaks of 'somatic vigilance' as a precondition for 'somatic resonance'.[25] These observations specify Schmitz' characterization of solidary encorporation as a joint concentration on a job, work or task in view of what happens between the stage and the audience: for there to be corporeal resonance during a performance, a special kind of receptiveness is required.[26] As the quality and intensity of the corporeal 're-enactment' of the observed movement (e.g. in dance) is conditioned by the spectator's own experience and skilfulness, this aspect may depend on the degree of professionalism: a dancer watching a dance performance from the auditorium will normally, says Eberlein, 'die leibliche Resonanz in der Regel viel deutlicher und differenzierter erleben, als dies einem "normalen" Zuschauer möglich ist' [experience the corporeal resonance more distinctly and in a more differentiated manner than a 'normal' spectator ever could].[27]

These descriptions of artistic performances illustrate at least two different phenomena of resonance.[28] On the one hand, the audience reacts to the overall

---

25 Hauser, Beatrix, "Zur somatischen Erfahrbarkeit von Aufführungen", in: Sonderforschungsbereich 626, Ästhetische Erfahrung (ed.), *Gegenstände, Konzepte, Geschichtlichkeit*, Berlin 2006, 9 (qtd. in Eberlein, "Leibliche Resonanz", 146).

26 The American dance critic John J. Martin, adopting findings of neurophysiological and psychological research on kinaesthetics for the discourse on dance, regards the spectator him/herself to be a type of resonance chamber. In his view 'reagiert der Körper auf jegliche inneren wie äußeren Reize [...] mit Eigenbewegungen und einer Regulierung der Lage- und Spannungsveränderungen in seinen Gelenken, Organen, u.Ä., einer Regulierung von Balance, Muskelspannung, Orientierung und Haltung, was seinen gesamten Wahrnehmungsapparat, seine Emotionen und Erinnerungen bewusst oder unbewusst kinästhetisch strukturiert. Die Übertragung von Tanzbewegungen vom Tänzer auf den Zuschauer fasst Martin in vergleichbarer Weise sympathetisch auf. Der Zuschauer absorbiert quasi die Bewegungen mit ihren emotional gestimmen Spannungsmomenten in seinen Körper, reagiert also im höchsten Maß emphatisch, das heißt, ihm zeigt sich der Tanz nicht primär visuell, sondern er wird von ihm eigenkörperlich berührt' [The body reacts to all interior and exterior stimuli with its own movements and through a regulation of the changes of position and tension of its joints, organs, etc., a regulation of balance, muscle tension, orientation and posture, structuring its entire apparatus of perception, its emotions and memories kinaesthetically – and this may happen knowingly or unknowingly. Martin believes that the transmission of dance movements from the dancer to the spectator happens in similar, sympathetic ways. The spectator 'absorbs' the movements and their moments of emotional tension into his/her body, thus showing emphatic reactions to the highest degree. This means that dance does not mainly present itself visually, but the spectator's felt body is touched by it.] (Huschka, Sabine, *Moderner Tanz: Konzepte, Stile, Utopien*, Reinbek 2002, 78f., qtd. in Eberlein, "Leibliche Resonanz", 148).

27 Eberlein, "Leibliche Resonanz", 151.

28 The fact that (reciprocal bipolar or unipolar) encorporation on stage (i.e. among the artists themselves) can also be described as resonance could, if we followed Eberlein's wide use of the term (cf. above, note 21), be regarded as a third type of it.

impression that can be described as the specific atmosphere of the individual performance. On the other hand, the spectators' own corporal re-enactments[29] of individual movements performed by the artists can be classified as 'resonance'.

This interpretation raises several new questions. Is it really true that the dancers' movements on stage (as contrary to and entangled with each other as they can be) are 're-enacted' by the spectators (in different degrees of intensity depending on the spectator's previous corporeal experience)? This question will certainly have to be answered in the affirmative, because all movements that the spectators perceive can only affect them through corporeal interaction, mainly through the 'channels' of gaze and the rhythm of movements and music. But would it not be more adequate to understand the corporeal resonance felt by the spectator and connecting him/her affectively with the events on stage as a response to the aesthetically staged atmosphere altogether rather than to its individual elements? Even if this atmosphere is certainly not independent of the artists' movements on stage, we would be looking at different types of resonance: resonance reacting to individual movements or resonance reacting to the situation as a whole. If we follow the analysis proposed by Schmitz, both aspects must be understood on the basis of the assumption that the dialogical structure of the felt body creates the contact between the events on stage and the audience. And this is true both of the atmosphere of the situation as a whole and of the individual movements. The individual felt body, however, cannot be multiplied, even if it is dialogically connected to the object of its perception (or in this case to numerous objects, movements, actors etc.) through the principles of contraction and expansion.

It must be assumed that corporeal resonance is, like all other forms of encorporation, a complex process guided by attention. Depending on what a spectator focuses on in the course of a performance, the corporeal interaction may be dominated by individual aspects or by the situation as a whole. The spectator's attention and focus, then, depend on his/her corporeal disposition[30] as well as on the aspects mentioned by Prütting and Eberlein. As the spectator can focus on different things either at the same time or consecutively, the same must be true of corporeal interaction. The problem of the unity of the two different types of resonance could be solved only if the different corporeal 'responses'

---

29 In common terminology, we could speak of a 'mental' following of the movements, given that the spectators do not move themselves. However, this would mean ignoring the corporeal aspect of this re-enactment, i.e. the changes in the economy of contraction and expansion.

30 For the term of corporeal disposition see Schmitz, *Der unerschöpfliche Gegenstand*, 127–130 and 143–147.

could overlay in complex ways.[31] Although intuitively, this seems plausible, it certainly requires further empirical-phenomenological analysis. We have to bear in mind, however, that both corporeal resonance and corporeal interaction constitute processes in time. While the perception of the individual movements on stage is short and often fleeting (and not necessarily accompanied by affective reactions), the atmosphere will be established only in the course of the performance (very rarely will it emerge instantly). As a matter of fact, the perception of individual movements on stage shapes the overall impression of one entity that will be validated or modified by the movements that follow.

Is corporeal resonance generally a unipolar form of encorporation? This question is not easy to answer. When somebody watches a rowing regatta (by himself or in the company of others), his corporeal resonance is likely to be different from what it would be if he himself were rowing the race. When I listen to a melody at a concert, my corporeal feeling will be different from what it would be if I were singing along. And listening to something *in a group*, for instance at a concert, might again be a different phenomenon. At first sight, the comparatively passive process of listening or watching[32] seems to be structured like the perception of things, as a form of bipolar encorporation: there are impulses coming both from the felt body and from the things it perceives. If, on the other hand, two or more people concentrate on the same thing (a performance or a task), which requires that they are aware of each other, we are dealing with a form of unipolar encorporation, the impulse coming from the performance or the event perceived and experienced together. Even if corporeal impulses will certainly come from the individual performers, they do not result from their own intentions, but will, figuratively, be authorized by the piece acting as the superior instigator.

In dealing with these questions, we may gain further valuable insight by looking at the boundary phenomenon of free jazz. In free jazz, no conventional elements are predetermined. Several musicians improvise at the same time, and this process of playing together shapes a structured entity. Using the terms of corporeal phenomenology, we might describe this process by saying that the initial form of bi- or multipolar encorporation is gradually transformed into a form of unipolar interaction, at least as long as every musician involved makes his/her contribution to developing a new entity based on the improved play. To that end, a specific type of 'floating' concentration is required, and it is necessary that

---

31 I would like to thank Maria Wirth for calling my attention to the important role of interferences in corporeal interaction.

32 Following Prütting and Eberlein, we have already mentioned that the role of the spectator is not a merely passive one. When compared to acts of properly taking part, however, it must be described as relatively passive (see note 24 above).

everybody listens very carefully to the evolving structure and to the other musicians involved. Of course one might argue that the mere intention of playing music together constitutes a task and that this process can be classified as a variant of unipolar encorporation from the start.

In this instance, I would like to leave the question how to classify corporeal interaction in free jazz unanswered. I believe that it has been made clear that the form of encorporation depends, among other things, on the intentions of those involved, on whether they accept a joint frame or setting, and which one specifically. It can be assumed that in all performances spatially characterized by a setting comprising a stage and an auditorium, the performers are connected in solidary encorporation with the work or piece that is being staged, or, in other words, with the task that they have set themselves (e.g. in cabaret). The spectators, on the other hand, are put in a state of unipolar encorporation by the events on stage, as long as they give the impulse for joint corporeal concentration (the phenomenon of new forms of performance practices shown in the public sphere without a stage – when the spectators need time to realise that an art performance is taking place – will have to be discussed in a different paper).[33] Corporeally, the interaction between stage and auditorium is also characterized by the spectators' resonance to the atmosphere that is being created artistically. The events on stage, however, are not in all cases identical with the performed work or piece, and they are both not necessarily identical with the atmosphere that is being created. In successful cases, the atmosphere is developed in the course of the performance in such a way that it leads to the joint encorporation of artists and spectators through the piece performed on stage. Once again it is evident that further, particularly empirical research is required in order to reassess the applicability of the categories of 'bipolar' and 'unipolar encorporation' in view of the experience of the artists and spectators involved (especially for new forms of street performances).

The phenomenon of corporeal resonance between the artists on stage and their audience constitutes a special case not only in terms of the spatial-corporeal aspect of interaction, but mainly in terms of its character as an artistic production (which distinguishes it from other forms of solidary encorporation). For every artistic production happens intentionally, and the suggestions of movement it gives are professionally rehearsed, staged and controlled. Mostly, the spectators following these suggestions of movement are not accidentally fascinated, but because they have decided to go see the performance and concentrate on what happens

33 This is reflected in the spatial alignment of the chairs facing the stage.

on stage.[34] Would it make sense to distinguish intentional forms of solidary enactment from intuitive forms (e.g. the flight of a herd or a group of people)? The philosophical debate on 'shared intentionality' and joint intentions provides useful suggestions which might help us answer this question.

V Joint Intentions and Simultaneous Intuitive Behaviour

Scholars discussing shared, joint or collective intentionality disagree on how these concepts can be reconstructed and which of these alternative terms is most adequate in describing the aspect in question.[35] They all agree in their understanding of the term 'intention' which they do not use in Husserl's sense.[36] Secondly, most scholars believe that joint intentions cannot be reduced to the individual intentions of the people involved. For example, when two people go for a walk together, we cannot adequately describe this by saying that person A wants to go for a walk and take route xyz and that person B intends to do the same thing. In that case, person A and person B could walk alongside one another without taking a walk *together.* How can we then reconstruct the joint or shared intention of taking a walk together as an act (or as acts) of consciousness of the individuals involved, given that there is only the individual consciousness of person A and person B respectively, but no collective subject and no superior joint consciousness? In this highly differentiated debate, this remains a controversial question. For my purposes, it is not necessary to come to a

34 As I said above, I do not wish to deal here with new forms of performances in public places, i.e. performances that are not immediately recognizable as such. They constitute hybrid forms because they involve observers (who often just happen to be there) in antagonistic encorporation.

35 Schmid, Hans Bernhard/Schulte-Ostermann, Katinka/Psarros, Nikos (ed.), *Concepts of Sharedness: Essays on Collective Intentionality*, Heusenstamm 2008; Schmid, Hans Bernhard, *Wir-Intentionalität: Kritik des ontologischen Individualismus und Rekonstruktion der Gemeinschaft*, Freiburg, Munich 2005. An outline of the complex debate is given in Schmid, Hans Bernhard/Schweikard, David P., "Einleitung: Kollektive Intentionalität, Begriff, Geschichte, Probleme", in: id. (ed.), *Kollektive Intentionaliät: Eine Debatte über die Grundlagen des Sozialen*, Frankfurt 2009, 11–65. This volume presents texts from the 1990s as well as more recent contributions.

36 According to Husserl, the term 'intention' generally describes the relatedness of a consciousness to objects. It can thus also be used for the relatedness of perception to different objects and is not restricted to intentions in the usual sense (cf. Husserl, Edmund, *Logische Untersuchungen: Gesammelte Schriften*, vol. 3, ed. Elisabeth Ströker, Hamburg 1992 [1900/01]).

final terminological decision; it will be enough to assume that there is something we may call 'joint intentions' and thus 'shared intentionality'.

The philosophical debates on shared intentionality rarely take into account the corporeal co-presence of the agents, and they do not usually relate the development of joint intentions to specific forms of corporeal contact. Yet all examples of solidary encorporation that we have discussed so far can only take place on the basis of joint intentions. Artistic performances that lead to corporeal resonance, as we have discussed in the previous section, can well be used to illustrate this aspect. However, there is an exception that I have mentioned before: groups of animals or people fleeing in the same direction. This phenomenon is of particular interest because it seems to be contrary to the assumption that unipolar encorporation requires a certain degree of coordination and thus joint intentions.

When groups of animals or people flee together, encorporation is triggered by a threat or danger which acts as the contraction pole and instigator of the interaction. All 'parts' of the group, the individuals, are corporeally and affectively moved by fear (i.e. contraction) which makes them 'go on the run' together and 'expand' again in this way. Triggering the impulse, the threat makes the individuals move in the same direction. But does this necessarily involve a joint intention? Even if everybody wants the same thing at the same time, for instance run away from a fire through a narrow passageway, in many cases they cannot all want this *together*, for the door and passageway might be too narrow to allow everybody to pass through quickly enough. And mainly, there is simply no time to develop and communicate joint intentions when danger is imminent.

But what is 'solidary' about this form of corporeal interaction if not the intentions? Does the element of 'solidarity' consist in the fact that everybody moves away from the threat in the same direction? This is not quite clear. There are definitely transitional forms between unipolar and bipolar interaction in which the individuals do not move in the same direction, for example, two people working together with a saw. In this case the joint goal (e.g. to saw through a branch or plank) determines how the two individuals must corporeally interact with each other. Yet they do not move in the same but work in opposite directions: when person A applies pressure, the movement of person B will slow down, and the moment of reversing the movement must be well-coordinated for the work to succeed. This is true of musical performances involving different parts and of many forms of European couple dance as well.[37] It is, in fact, not the direction of movement that makes Schmitz speak of 'solidary' corporeal com-

37 Schmitz classifies these forms as transitional (see Schmitz, *System der Philosophie*, vol. II.5, 96f.).

munication, but the (single) impulse that affects at least two felt bodies in the same way and 'aligns' them. If we follow Schmitz' line of thought, we could say that this impulse forces them to coordinate their movements in a certain, but not necessarily in a parallel way.

Elias Canetti's description of flight crowds is a compelling example of solidary encorporation taking place without joint intentions (particularly because this phenomenon is not restricted to human corporeal interaction). According to Canetti, 'crowd-fear' counts among the collective experiences 'of all animals who live together in herds and whose joint safety depends on their speed'.[38] The impetus necessary for the escape is increased by the 'simultaneity of identical movements'.[39] In analogy to flight movements of animals, a human flight crowd can be characterized by 'the exaltation of common movement'[40] resulting from the excitement and energy of the group that stays close together. 'People flee together because it is best to flee that way. They feel the same excitement and the energy of some increases the energy of others [...]'.[41] Canetti emphasizes that as long as they stay together, they feel that fear and danger appear to be distributed,[42] while the direction and goal keep the group together.

Although Canetti's description was not meant as a contribution to corporeal phenomenology, it does mention three aspects characteristic of this special form of corporeal communication: the increase of energy through the synchronization

---

38 Canetti, Elias, *Crowds and Power*, New York 1962, 26 (http://cva.ap.buffalo.edu/courses/f07/arc588/files/f07/arc588/readings/canetti.pdf).

39 Canetti, *Crowds and Power*, 26 (see also Kuhnau, Petra, *Masse und Macht in der Geschichte: Zur Konzeption anthropologischer Konstanten in Elias Canettis Werk 'Masse und Macht'*, Würzburg 1996, 107).

40 Canetti, *Crowds and Power*, 54.

41 Canetti, *Crowds and Power*, 53.

42 Canetti, *Crowds and Power*, 53. In this context, it is important to note Canetti's distinction between mass flight and panic: 'The impetus of the flight continues to multiply so long as everyone recognizes that there are others fleeing with him. He may press them forward, but he must not push them aside.' (ibid.). When the direction of the flight has been repeatedly impeded, the crowd 'grows confused about its direction and thus loses its coherence. The danger which, till then, had united its members and given them wings, now sets each man up as an enemy of the next. Everyone is intent only on saving himself. The mass flight, on the other hand, contrary to the panic, derives its energy from its coherence. As long as it remains one powerful and undivided river and does not allow itself to be dispersed and split, so long does the fear by which it is driven remain bearable.' (Canetti, *Crowds and Power*, 54). – Differently from mass flight, which is a clear example of unipolar encorporation, panic would have to be described, in Schmitz' terms, as a phenomenon between bipolar and unipolar encorporation – bipolar due to the individuals' uncontrolled movement against each other, and unipolar due to their joint focus on the threat.

of direction and movement; the reduction of fear through distribution; and the fact that the crowd is (in a phenomenological perspective: corporeally and physically) kept together by their direction and objective. Each one of these aspects describes specific processes of corporeal alignment, without having to refer to the existence of explicit joint intentions. This is true of human flight crowds as well as for fleeing herds of animals.

It is possible, however, to detect joint intentions shared by all those involved: everybody wishes to increase his/her speed of flight (which can best be done when being part of a crowd), everybody wants to reduce his/her fear, and everybody wants to achieve the same objective. Linguistically, we can attribute an intention to the herd as a whole. We might say, for example, that the herd of antelopes intends to elude the lion in a northerly direction. The behaviour of the herd seems to suggest that thinking of a joint intention does make some sense. We do not assume, however, that beyond the process of corporeal coordination animals within a herd would communicate and agree on joint intentions. And this is not simply because animals do not possess language. It is rather because, as far as we know, animals cannot reflect on nor communicate about their (individual or collective) intentions.[43] As I will try to show, the abilities to distance oneself from one's own immediate needs and to hypostatize them as a (medium- or long-term) goal constitute two preconditions for the development of truly joint intentions (as opposed to joint goals that we follow intuitively).[44]

VI Intentions and Unipolar Encorporation

In this context, recent (evolutionary) psychological research on a 'theory of mind' (TOM) can prove to be particularly insightful. A TOM implies having assumptions about the processes of consciousness of one's conspecifics and recog-

---

43 Plessner describes this by means of his terms of 'centric' and 'ex-centric positionality'. The term 'positionality' refers to the fact that all living beings are placed into and positioned in their environment. All living beings are determined by their boundaries to their specific environment, differently from inorganic bodies which can be ascribed a rim but no boundary because they have no relation to their environment. While animals are 'centrically' positioned, human beings are characterized by 'ex-centric' positionality. Animals have a drive, i.e. a centre from which derive their will and impulses, yet they cannot themselves relate to their centre and live, in Plessner's words, 'from their centre'. Human beings, on the other hand, can relate to their centre and distance themselves from their spontaneous impulses. Animals lose themselves in their experience. Human beings can do that, too, yet they can also relate to their experience (cf. Plessner, *Die Stufen des Organischen*).

44 For an overview see Schmid/Schulte-Ostermann/Psarros, *Concepts of Sharedness*.

nizing them in oneself, in short, the ability to ascribe feelings, needs, ideas, intentions, expectations and opinions to oneself and to others. When discussing the difference between intuitive and intentional solidary interaction, it seems important to bear in mind that even highly developed animals possess no or, at best, only a rudimentary theory of mind.[45] In fact, some primates are able to ask for things (using ritualized signals), understand intentions and perceptions and draw practical conclusions.[46] But primates are barely capable of using the complex type of communication that we can already observe in young human children: they are not able to communicate in order to help others or share opinions with them. This, however, forms the basis for a next stage in the development of communication: the recursive recognition of intentions, the ability to feel empathy and to change perspectives, leading to fully developed forms of cooperative communication that can only be observed in humans,

> turning helping and sharing into mutual expectations or even norms of cooperation; turning the understanding of goals and intentions into joint goals and Gricean communicative intentions;[47] turning the understanding of attention into joint attention and common ground; turning practical reasoning into cooperative reasoning; and turning imitated signals into bidirectional, shared conventions.[48]

The ability to adopt reciprocal roles of interaction, presupposing the ability to ascribe intentions,[49] must be considered a precondition for the development of joint

45 For a theory of mind in animals, see Tomasello, Michael/Call, J./Hare, B., "Chimpanzees Understand Psychological States – The Question is Which Ones and to What Extent", in: *Trends in Cognitive Sciences* 7.4 (2003): 153–156.

46 Cf. Tomasello, Michael, *Origins of Human Communication*, Cambridge (MA), London 2008, esp. chap. 2: "Primate Intentional Communication", 13–55.

47 Grice's basic assumption is that an action can be called communicative when it aims to accomplish its purpose by making the recipient understand that the communicator wants to accomplish this purpose (cf. Grice, H. Paul, "Intendieren, Meinen, Bedeuten", in: Georg Meggle (ed.), *Handlung, Kommunikation, Bedeutung*, Frankfurt 1993, 16–51).

48 Tomasello, *Origins of Human Communication*, 106 [online, Google Books].

49 In what I have said, I have implicitly distinguished between two types of intentions: 1) Simple intentions that are ascribed only externally; living beings with this kind of intentionality do not have to be able to refer to their intentions. They cannot develop a longer-term will because in order to do so, they would have to be able to defer sudden impulses for the idea of a superior intention. 2) Intentions capable of reflection. Living beings with this kind of intentionality can position themselves to their own wishes and intentions, e.g. speak about and inform others of them. I would like to thank Justus Schollmeyer for his suggestions. Brandom operates with a similar distinction (cf. Brandom, Robert B., *Expressive Vernunft: Begründung, Repräsentation und diskursive Festlegung*, Darmstadt 2000, 112f.). It is important to bear in mind,

intentions. In my opinion, the term implies a form of communicative understanding (mostly among individuals who are co-present) but not necessarily language. It is possible to come to an agreement on how to act and what to do together just by exchanging gazes (an intentional form of corporeal interaction). In doing so, everybody involved must assume that the others perceive and understand the situation in a similar way and that they all share a common ground. Additionally, everybody must understand his/her own actions as contributing to the joint action and expect a similar contribution from the others. These intentions must be made explicit, at least through gazes or gestures, and the individuals must be able to accept at least minimal obligations for the realization of these 'contributing actions'. Having a joint intention means seriously wanting to realize it (as is the case with individual intentions). The realization of one's own intentions does not depend on other people, and therefore they can more easily be modified if we change our minds. Joint intentions cannot be modified quite as easily, because they require a certain degree of normativity in the sense that every individual involved feels obliged to make his or her contribution to the joint action.[50]

I want to argue that the development of joint intentions rests on the ability of solidary encorporation which, in phenomenological terms, we share with all gregarious animals. It is *grounded* in this form of encorporation. In terms of developmental psychology, more complex intentional forms of unipolar encorporation[51] (when compared to bipolar encorporation) require competences that other primates do not possess or possess only rudimentarily: the ability to ascribe intentions to other individuals and the awareness that they, in turn, will develop an idea of our intentions (as well as of the fact that we know what they are doing and vice versa), the ability to develop well-coordinated forms of imitation as a prerequisite for adopting reciprocal roles of interaction and for practising specific physical-corporeal techniques requiring a certain degree of self-discipline and the ability to refrain from immediate needs and corporeal impulses. Even an action as simple as waiting in line until it is our turn to be served at the food counter requires the ability of solidary encorporation (everybody is focused on a fair distribution of food as a joint theme), a certain degree of self-discipline, a commitment to joint intentions and a common ground (e.g. agreeing that it is fair that who has come first will be served first). We could say that *because* this behaviour requires us to subordinate our own corporeal im-

---

however, that living beings who are, in theory, capable of reflection will not only show reflected forms of intentionality.

50 Cf. Gilbert, Margret, "Walking Together", in: *Midwest Studies in Philosophy* 15 (1990): 1–14.

51 Flight crowds as a simple form of solidary encorporation are not subject to these conditions.

pulses to the joint objective of being treated fairly, it can be classified as a form of solidary encorporation.

In my opinion it is significant that small children learn most forms of unipolar corporeal interaction by practising them repeatedly;[52] and this is true even of seemingly simple forms such as singing together in a group. Particularly complex forms of cooperation (joint handwork, team sports or playing music in a group) require a high degree of commitment, self-control and physical training and can only be accomplished in processes of *intentional* solidary encorporation. Probably the most highly developed form of solidary encorporation can be found in complex artistic stage performances comprising the action of dancers, singers, choirs, actors and an orchestra as well as the response and resonance of their audience. They are 'highly developed' in the sense that the complexity of necessary corporeal coordination requires a high level of concentration on the joint, flawless action as well as a perfect sense of the details of encorporation and outstanding technical skills.

Intuitive solidary encorporation (patterns of behaviour in groups and crowds that do not need to be learned and practised) is, however, probably an important prerequisite for the development of more complex forms based on explicit joint intentions. In all likelihood, the sense of community, which develops in situations of intuitive unipolar encorporation, constitutes a significant motif for the intentional creation of such situations: the feeling of being jointly affected by a specific atmosphere has a strong bonding force. This specific attachment also frames a further, intrinsically human form of solidary encorporation, the resonance of the audience in the course of a stage performance. It seems that a positive form of attachment is at the basis of negative experiences as well, for example crowd-fear that, as Canetti puts it, can lead to 'exaltation'. Certain narratives of war, describing the sense of companionship in the trenches, seem to betray, in most ambiguous ways, a positive emotional experience, as opposed to the experience of joint shock, existential fear, suffering and misery. In *intentional* forms of unipolar encorporation, this exaltation is purposefully created under freely chosen circumstances and, in a certain fashion, 'cultivated'. The purpose of the – often complex – enactment of this form of encorporation seems to be the creation of such exaltation, particularly in the context of athletic and artistic performances. Additionally, solidary encorporation serves to increase the strength and/or mental capacities of the individuals. Without intentional solidary encorporation, joint forms of handwork would not be possible.

52 For learning by imitation, Schmitz coins the notion of 'spielerische Identifizierung' (playful identification) (cf. Schmitz, *Der unerschöpfliche Gegenstand*, 175).

Earlier in this paper, I have characterized the phenomenological concept of the 'felt body', as opposed to the concept of the (material) body, by referring to the difference between the first- and third-person perspectives. In view of the different phenomena of encorporation that we have looked at, this distinction can be complemented by two more dimensions. The antagonistic form of encorporation (with its poles competing for expansion) can be reconstructed as a me–you relation or as a second-person perspective: my own felt body feels the other one, not as a thing, but as another felt body that I am dialogically associated with through contraction and expansion. Solidary encorporation, on the other hand, permits the connection of first- and second-person perspective forming an 'us', a joint plural perspective that could not be accomplished without the experience of corporeal interaction between co-present individuals.

But aren't there joint intentions without corporeal interaction? Thanks to modern communication technology, it is certainly possible to medially transmit and communicate about joint intentions. We would have to ask, however, whether and how we can reach synchronized, joint decisions without any form of co-presence at all (e.g. vocal encorporation in a teleconference). Even if that was possible, two arguments help us defend the hypothesis that unipolar encorporation seems to be a precondition for the development of joint intentions. Firstly, such abstract forms of developing joint intentions derive from direct corporeal interaction. Without it, the ability to communicate in more abstract ways could never have evolved, because it would be impossible to make trust-based commitments. Trust, on the other hand, can only grow with co-presence for it has a corporeal basis; we must be able to exchange glances or at least have acoustic encorporation through voice.[53] Secondly, we must not forget that making joint decisions *without* co-presence often requires additional affirmation in processes of encorporation. Despite the possibilities of modern technology, for example, even busy, internationally operating top managers cannot do without occasional get-togethers, and scholars do not only communicate by Skype (a form of vocal encorporation, like old-fashioned phone conversations) or in publications, but prefer to do their networking in situations of co-presence (e.g. at academic conferences).

I would thus like to propose the following thesis: What allows us to enter into complex relations of cooperation and thus increase the performance of

53 Cf. Demmerling, Christoph/Landweer, Hilge, *Philosophie der Gefühle: Von Achtung bis Zorn*, Stuttgart 2007 (chap. "Liebe" [love]), 127–165.

the individual, is grounded in the ability to develop those forms of solidary encorporation which permits the development of joint intentions.

Translated by Christine Henschel

References

Brandom, Robert B., *Expressive Vernunft: Begründung, Repräsentation und diskursive Festlegung*, Darmstadt 2000.

Canetti, Elias, *Crowds and Power*, New York 1962 (http://cva.ap.buffalo.edu/courses/f07/arc588/files/f07/arc588/readings/canetti.pdf).

Demmerling, Christoph/Landweer, Hilge, *Philosophie der Gefühle: Von Achtung bis Zorn*, Stuttgart 2007.

Eberlein, Undine, "Leibliche Resonanz: Phänomenologische und andere Annäherungen", in: Kerstin Andermann/ead. (ed.), *Gefühle als Atmosphären: Neue Phänomenologie und philosophische Emotionstheorie*, Berlin 2011, 141–152.

Fischer-Lichte, Erika, "Theater als Resonanz-Raum", in: Lichau/Tkaczyk/Wolf (ed.), *Resonanz*, 237–248.

Gilbert, Margret, "Walking Together", in: *Midwest Studies in Philosophy* 15 (1990): 1–14.

Goffman, Erving, *Frame Analysis: An Essay on the Organization of Experience*, New York, London 1974.

Grice, H. Paul, "Intendieren, Meinen, Bedeuten", in: Georg Meggle (ed.), *Handlung, Kommunikation, Bedeutung*, Frankfurt 1993, 16–51.

Husserl, Edmund, *Ideen* II: *Hua* IV, The Hague 1952.

Husserl, Edmund, *Krisis Hua* VI, The Hague 1976.

Husserl, Edmund, *Logische Untersuchungen: Gesammelte Schriften*, vol. 3, ed. Elisabeth Ströker, Hamburg 1992 [1900/01].

Kuhnau, Petra, *Masse und Macht in der Geschichte: Zur Konzeption anthropologischer Konstanten in Elias Canettis Werk 'Masse und Macht'*, Würzburg 1996.

Landweer, Hilge, *Scham und Macht: Phänomenologische Untersuchungen zur Sozialität eines Gefühls*, Stuttgart 1999.

Lichau, Karsten/Tkaczyk, Viktoria/Wolf, Rebecca (ed.), *Resonanz: Potentiale einer akustischen Figur*, Munich 2009.

Lichau, Karsten/Tkaczyk, Viktoria/Wolf, Rebecca, "Anregungen", in: id. (ed.), *Resonanz*, Munich 2009, 11–32.

Müllan, Rudolf Owen/Slaby, Jan, "Emotions Outside the Box: The New Phenomenology of Feeling and Corporeality", in: *Phenomenology and the Cognitive Sciences* 10.2 (2011): 241–259 (transl. of a text by Hermann Schmitz).

Plessner, Helmuth, *Die Stufen des Organischen und der Mensch*, Berlin 1975.

Prütting, Lenz, "Über das Mitgehen: Einige Anmerkungen zum Phänomen transorchestraler Einleibung", in: Michael Großheim (ed.), *Leib und Gefühl: Beiträge zur Anthropologie*, Berlin 1995, 141–152.

Schmid, Hans Bernhard, *Wir-Intentionalität: Kritik des ontologischen Individualismus und Rekonstruktion der Gemeinschaft*, Freiburg, Munich 2005.

Schmid, Hans Bernhard/Schulte-Ostermann, Katinka/Psarros, Nikos (ed.), *Concepts of Sharedness: Essays on Collective Intentionality*, Heusenstamm 2008.
Schmid, Hans Bernhard/Schweikard, David P., "Einleitung: Kollektive Intentionalität, Begriff, Geschichte, Probleme", in: id. (ed.), *Kollektive Intentionaliät: Eine Debatte über die Grundlagen des Sozialen*, Frankfurt 2009, 11–65.
Schmitz, Hermann, *System der Philosophie*, vol. V, Bonn 1980.
Schmitz, Hermann, *Der unerschöpfliche Gegenstand: Grundzüge der Philosophie*, Bonn 1990.
Schmitz, Hermann, "Leibliche Kommunikation mit und ohne Wort", in: Manfred Bauschulte/Volkhard Krech/Hilge Landweer (ed.), *Wege – Bilder – Spiele: Festschrift zum 60. Geburtstag von Jürgen Frese*, Bielefeld 1999, 251–260.
Schmitz, Hermann, "Über das Machen von Atmosphären", in: Anna Blume (ed.), *Zur Phänomenologie der ästhetischen Erfahrung*, Freiburg 2005, 26–43.
Tomasello, Michael, *Origins of Human Communication*, Cambridge (MA), London 2008.
Tomasello, Michael/Call, J./Hare, B., "Chimpanzees Understand Psychological States – The Question is Which Ones and to What Extent", in: *Trends in Cognitive Sciences* 7.4 (2003): 153–156.
Williams, Bernard, *Shame and Necessity*, Berkeley, Los Angeles, London 1993.

Gabriele Brandstetter

"Listening"

Kinesthetic Awareness in Contemporary Dance

"Listen!"

This appeal is generally intended to attract the attention of another person. "LISTEN" was the title given to a series of "lecture demonstrations" by the American musician and performer Max Neuhaus in the 1960s:

"What interested me at first," he wrote,[1] "was the imperative expressed by the word 'listen.'" He goes on to tell a private anecdote about how he hit upon this theme: "My girlfriend at that time – who was of French-Bulgarian origin – used to shout this word when she got into a rage before proceeding to throw objects at me." His first work as a free-lance artist was to take people who wanted to attend a "concert" of his on a "walk" through Manhattan's Lower East Side. He took a rubber stamp and stamped the word "listen" on the hands of those who had come to the concert and walked with them through certain streets and districts. Concentrating entirely on hearing, he "just set off"[2] and remained silent during this guided tour of the everyday environment. "At first people found it somewhat embarrassing," Neuhaus continues, "but my concentration was usually infectious." After this attentive walk through these everyday sonic landscapes, "many of them had found a new kind of hearing for themselves."[3]

So much for Max Neuhaus. But who could attend this "Listen" lecture demonstration without being reminded of John Cage's performance of *4'33*, that silent piece in which the solo instrument, the piano, was *not* heard? What happened instead was that a space of attention was opened for the numerous noises and sounds heard inside and outside the concert hall.

Listen: The reduction of sensory attention to an "act of hearing" was used in yet another way by Xavier le Roy in his solo performance *Self Unfinished* (1998). Le Roy begins the piece by entering a vacant white space and going towards a tape deck and pressing a button, as though starting the music for a (dance) piece. However, not a sound is heard, neither music nor noise. Le Roy makes the *gesture* of opening a sonic dimension and thus brings about an *act* of "listening." The drawing of the

---

1 Neuhaus, Max, "LISTEN," in: *Welt auf tönernen Füßen. Die Töne und das Hören*, ed. by Kunst- und Ausstellungshalle der Bundesrepublik Deutschland (Bonn), Göttingen 1994, 125–127.

2 Ibid., 126.

3 Ibid.

audience's attention to a "*possible*" hearing event changes the hearing *and* the attention. This focusing of an act of "listening" facilitates – synesthetically speaking – a *different* perception of space and movement. "*Listen!*": listening to the movement, hearing the body-space movement, means that the seeing receives a different, an additional sensory (kinesthetic) quality. As Paul Valéry, for whom the ear was the preferred sense for conveying attention, said, the ear "keeps watch, so to speak, at the frontier beyond which the eye does not see."[4]

"Listening" is a term belonging to the basic vocabulary of contact improvisation. The following remarks are intended to investigate the use of the concept and see how it ties in with the discussions and practices of "kinesthetic awareness." Cheryl Pallant remarks in her introduction to "contact improvisation" as a "dance form"[5] that "listening," "listening to motion," is a term

> regularly used in contact improvisation. Listening, according to contact improvisation's metaphorical use of the word, refers to paying attention to all sensory occurrences arising from touch, from the play of weight as partners move through space, and from the event of one body encountering the presence of another. Listening refers to noticing stimuli not only within oneself but also from another.[6]

The range of meanings covered by the word "listening" refers to one of those – open – scenarios of the metaphorical which Lakoff called "metaphors we live by."[7] Thus the image of a summons to an act of "listening" refers to a field of perception of the sensory that is not just limited to acoustics: it is a syn-aesthetic network of experiences of the body, of its internal and external states at rest and in movement. It involves awareness, which in contact improvisation is exercised and refined in a multitude of ways in and through synesthetic-kinesthetic addressings of perception. A selection of sentences that act as a guide to such (synesthetic) perception may illustrate the range covered by "listening" (quoted from Cheryl Pallant):

- Listen to the click of cartilage, the slap of skin, of the whisper of your will typically silenced by a shout.
- Notice a part of your body for which you have no name, no history, no awareness [...].
- Feel weight push into your stubbornness, your expectations, against your habit of always yielding to aggression or constantly fighting it. [...]

4 Valéry, Paul, *Cahiers*, vol. II, Paris 1974, 934 [translated by G.B.], cf. Waldenfels, Bernhard, *Phänomenologie der Aufmerksamkeit*, Frankfurt am Main 2004, 198.

5 Pallant, Cheryl, *Contact Improvisation: An Introduction to a Vitalizing Dance Form*, North Carolina, London 2006.

6 Ibid., 31 f.

7 Lakoff, George/Johnson, Mark, *Philosophy in the Flesh*, New York 1999; Johnson, Mark, *The Body in the Mind: The Bodily Basis of Meaning, Imagination and Reason*, Chicago 1987.

- Sniff the circumstances, the leg extending into view, the hand urging direction [...].
- Watch time dissolve [...].
- Follow the sound into the garden past the bench in the corner on [...].
- Tend your body as if it were the body of a lover [...].
- Drink the elixir of expansion, the release within repose.
- Find the edge between comfort and discomfort, the familiar and the unknown. Balance there, however precariously.
- Devolve into protozoa. [...]
- Let your body call you back into yourself [...].[8]

From this list of addressings of a sensory awareness that is important for the preparation and setting *in* contact improvisation it can be seen that “listening” refers not so much to *hearing* as a sensory form of registering acoustic events (although this is part of it), but to a very broad and open state of sensuous/sensory perception. It also includes the sensing, the tactility of touching. Thus “listen” does not refer primarily to a hearing event. It refers rather to an intersection of action and event (*Ereignis*) – which in German breaks down into the terms *zuhören*, *hören auf*, *horchen*, and *lauschen* – thus implying the reference to the self as well as the reference to the other and to space.[9] In contact improvisation “listening,” to “listen” to motion, refers to synesthetic and kinesthetic forms of *awareness* which embrace both conscious and unconscious “subliminal” perceptions.

I Contact Improvisation and Kinesthesia

Contact improvisation is a dance practice in which two (or more) moving partners, always in contact, explore their possibilities of movement. Curt Siddall, an “early exponent of contact improvisation,” defines this dance form as “a combination of kinesthetic forces: Contact improvisation is a movement form, improvisational in nature, involving two bodies in contact. Impulses, weight, and momentum are communicated through a point of physical contact that continually rolls across and around the bodies of the dancers.”[10] Historically this movement *praxis* goes back to Steve Paxton’s movement and improvisation explorations at Oberlin College in 1972. Since then this dance form has established itself internationally and

8 Lakoff/Johnson, *Philosophy in the Flesh*, 7 f.

9 Cf. Nancy, Jean-Luc, *Zum Gehör*, Berlin, Zürich 2010, 15 f. and 38 f. Nancy argues that “listening” relates to the whole register of the senses, being touched, and within a difference of the interior and the exterior.

10 Quoted after Albright, Ann Cooper/Gere, David (ed.), *Taken by Surprise: A Dance Improvisation Reader*, Middletown 2003, 206.

assumed different forms, whether as performance improvisation or social dance in the shape of regular jams, or as a means of exercising in combination with various release techniques, which in turn determine the esthetics of contemporary dance. "The physical training of contact improvisation emphasizes the release of the body's weight into the floor or onto a partner's body [...]," writes Ann Cooper Albright. "The experience of internal sensations and the flow of the movement of two bodies is more important than specific shapes or formal positions."[11] In an early article in *The Drama Review* (1975), Steve Paxton identifies the six main elements of contact improvisation: "attitude, sensing time, orientation to space, orientation to partner, expanding peripheral vision, and muscular development, which includes centring, stretching, taking weight, and increasing joint action."[12]

The emphasizing of the motor aspects of movement – such as working with "momentum," "gravity," "mass"/"weight," "chaos," "inertia," the attention to highly differentiated states of muscle tone between release/inertia and contraction, and finally the shifting of spatial perception between the focus on the interior of the body and the exterior of space make clear that an accent of the overall concept of contact improvisation lies on the conscious work with the "sixth sense," kinesthesia.

This is where two fields of the kinesthetic mesh together: kinesthetic proprioception and working with kinesthetic communication – contact and shifts of weight and spatial position, which are shifts of the dynamically interacting dancer-bodies.

"Both the gross motor awareness of kinesthesia and the less conscious sensory feedback mechanism of proprioception form the basis of the physical dialogue which is so pivotal to creating dance," writes Cheryl Pallant.[13] That is why experienced "contacters" like Nancy Stark Smith constantly stress the spatial orientation created not only by vision but by the entire physical perception, a condition which Stark Smith calls "telescoping awareness,"[14] a "shifting between narrow and wide views, from up-close sensation to perceptions of the wider world,"[15] accompanied by the "sensation of dropping through space," "the forces of gravity, momentum, and mass."[16]

---

11 Ibid. This is not the place for a more detailed examination of the historical development and esthetic structures of contact improvisation. Cooper Albright has pointed out that it is a "tricky business" to give a coherent description of contact improvisation: "the form has grown exponentially over time and has travelled through many countries and dance communities. Although it was developed in the seventies, contact improvisation has recognisable roots in the social and aesthetic revolutions of the sixties" (ibid., 205). On the history of contact improvisation see: Novack, Cynthia, *Sharing the Dance: Contact Improvisation and American Culture*, Madison 1990.
12 Steve Paxton, quoted after Pallant, *Contact Improvisation*, 12 f.

Who leads? Andy Wichorek and Kelley Lane

Ease in stillness: Susan Singer and Keith Winston

There has been a delay between the highly differentiated *praxis* with kinesthesia in dance and theoretical references to the results of physiological, phenomenological, and neuroscientific research into kinesthesia. This research has only recently attracted more attention in the field of dance studies.[17] The initial impetus for this discussion was provided in the 1980s by an article by Mary M. Smyth: *Kinesthetic Communication in Dance.*[18] The researches and discoveries connected with "kinesthesia" as the ability to feel movement of the limbs and body, from Sherrington (1906)[19] to Gibson (1966),[20] are examined by Smyth to see if they yield an answer to the question as to how "watching some-

13 Ibid., 32.
14 Cf. in Albright/Gere, *Taken by Surprise*, 153.
15 Ibid., 154.
16 Ibid., 157.
17 Cf. Montero, Barbara, "Proprioception as an Aesthetic Sense," in: *The Journal of Aesthetics and Art Criticism* 64.2 (2006): 231–242; and cf. Coleman, Jonathan/Montero, Barbara, "Affective Proprioception," in: *Janus Head* 9.2 (2007): 299–317.
18 Smyth, Mary M., "Kinesthetic Communication in Dance," in: *Dance Research Journal* 16.2 (1984): 19–22.
19 Sherrington, Charles, *The Integrative Action of the Nervous System*, New Haven 1906.
20 Gibson, James J., *The Senses Considered as Perceptual Systems*, Boston 1966.

one dance could link to the movement system of the observer."[21] For Smyth this question ultimately remains unanswerable.[22] Her work reviews several hypotheses, all of which arose before more recent research findings on the function of "mirror neurons" were discussed in dance and dance studies.[23] The subject of "kinesthesia," on the other hand, is a topical one in dance research, since interest has become focused on the meaning of "energy," "rhythm," "synchronization" of movements in Modern and Contemporary Dance. Thus Dee Reynolds has devoted her study of "Rhythmic Subjects"[24] to the "uses of energy" and the question of kinesthesia, not only in relation to bodily position, muscle tension, and movement, but also with regard to kinesthetically "embodied" cultural "imaginations" of and attitudes to the kinesthetic. Rudolf von Laban's concept of "effort" and Edmund Husserl's and Maurice Merleau-Ponty's phenomenological theories inform her approach. By using the concept of "kinesthetic imagination" – which refers not only to the subjective aspects of proprioception, but also to questions of cultural imprint and transfer of energy – she manages to detach the phenomenon of kinesthesia from the issue of self-perception in dancer *praxis* and open it up to questions of (syn-)esthetic perception by the observer.[25]

In the practice of various body techniques which are of relevance to contemporary dance (though not only to it), the subject of kinesthesia is of increasing importance, even if the term itself is not part of the vocabulary of the discourse. Thus in a newly held series of interviews not few representatives of body techniques – such as Feldenkrais and proponents of the Alexander technique and Body-Mind Centering – admitted to having worked with the basic principles of kinesthesia long before they became aware of the term and the research associated with it. The dancer Julyen Hamilton, for example, stated that his work was very spatially oriented. "This spatial sense is highly informed through the kines-

---

21 Smyth, *Kinesthetic Communication in Dance*, 19.

22 Mary M. Smyth notes that "Somehow remains as a gap in the process. Even if dancers were happy that such a process could in any way relate to the experiences which they called 'kinesthetic communication', we still do not know how it is effected [...]. We do not yet know how seen movement can do this." (Smyth, *Kinesthetic Communication in Dance*, 22).

23 Rizzolatti, Giacomo et al., *Mirrors in the Brain: How our Minds Share Actions, Emotions, and Experience*, Oxford 2008.

24 Reynolds, Dee, *Rhythmic Subjects: Uses of Energy in the Dances of Mary Wigman, Martha Graham, and Merce Cunningham*, Hampshire 2007.

25 Cf. Foster, Susan, *Movements Contagion: The kinesthetic impact of performance*, 2008, online publication of: University of California, International Performance and Culture Multicampus Research Group, http://uc-ipc.com/wp-content/uploads/2008/06/movementscontagion-11.pdf (1.6.2008). Regarding kinesthesia and empathy cf. Foster, Susan, *Choreographing Empathy: Kinesthesia in Performance*, London 2010.

Establishing a Contact Point: Corrine Mickler and Brandon Crouder

thetic sensing of the inner body."[26] Furthermore, the "choices," which are always movement decisions, are influenced by the fact that they are not made from outside – "via an outside eye," but they are "choices made from the proprioceptive abilities within the body as it senses itself and its environment." The result is a "radically spatial event permitted by the public and performers sharing of space." Susan Klein, the founder of the "Klein Technique™," stresses the extraordinary importance of "kinesthetics," both as a "tool that allows us to understand the body" and as an aid to artistic work.

> For me the beauty and excitement in kinesthetics is bringing a body-felt understanding of movement to consciousness. It is fine-tuning our ability to feel, on subtle levels. [...] Kines-

26 Cf. Corpus, *Kinesthetics: Four Questions*, 2010, http://www.corpusweb.net/kinesthetics-four-questions.html (28.9.2010), 2.

Following impulses: Brandson Crouder and Corinne Mickler

> thetics is our tool to bring the body into a deep state of balance, to its optimal state of movement potential.[27]

The associated "process of discovery," which is able to trigger a periphrasis of fixed blockades, of postures of muscles, bones, and tissue, leads to a kinesthetically informed "internal knowing."[28] The aim is – as in most concepts of body techniques which operate with Body-Mind Centering, "ideokinesis" or "functional integration" (such as the Feldenkrais method) – "body alignment, ease of movement and overall body harmony in dance," as Linda Rabin puts it.[29] Here, as in all works of kinesthetically oriented practices, it is not a "beautiful" bodily form resulting from a course of training dictated by an esthetic style or movement code that is the guiding principle of the idea of dance and choreography, but the question posed by Linda Rabin: "What would dance performance be like if dancers drew from this essential source?"[30]

II Attention: Kinesthetic Awareness

A key concept that plays a pivotal role in nearly all texts and discourses of the above-mentioned body techniques and contact improvisation is that of attention – in the double sense of attention and awareness, of directed attention (perception) and "noticing," a distinction made by the philosopher Bernhard Waldenfels on the basis of the phenomenological theory of Husserl and Merleau-Ponty.[31]

"It all starts with paying attention," says Linda Rabin.[32] "Kinesthetics, the sense that tells us where and how we exist in our internal environment and how we connect and relate to our external environment,"[33] can lead to a conscious perception. Of particular interest in this connection is the *division* of attention. According to Susan Klein: "It requires a split level of consciousness: one level is doing while the other level is observing what is done. Kinesthetic aware-

27 Ibid., 3.
28 Ibid.
29 Ibid., 4. Linda Rabin worked with Lulu Sweigard on "ideokinesis," learned the Alexander technique (through Rika Cohen), practiced "Body-Mind Centering®" (founded by Bonnie Bainbridge Cohen), and finally became a teacher of the "Continuum Movement" founded by Emile Conrad.
30 Cf. ibid.
31 Cf. Waldenfels, *Phänomenologie der Aufmerksamkeit.*
32 Corpus, *Kinesthetics*, 5.
33 Klein, Corpus, *Kinesthetics*, 3.

ness allows us to keep track of what we are doing with our bodies as well as how we are doing it."[34]

Attention – as attention and awareness – embraces the entire sensory and action scenario which is addressed, for example, in the movement explorations of contact improvisation. Attention implies attentiveness – both to the processes of one's own physicality and to the experience of contact with the Other. Attention thus opens up the entire range of the concept as it might be described in an anthropological-phenomenological specification. Waldenfels points out with recourse to Immanuel Kant that the boundaries between the deliberate direction of attention (*attentio, abstractio, distentio*), noticing (*animadvertere*), and observing (*observare*), are fluid.[35] To this must be added the mode of self-affection, i.e., "the affection of the inner sense by ourselves with an 'Actus of attention.'"[36] It is this doubling of only partially controllable awareness and self-affection by images in the (inner) perception that marks the potential of kinesthesia and "kinesthetic imagination" (Dee Reynolds). This is where the key formula "*listening*" opens the synesthetic-kinesthetic spectrum of possible modes of attention: perception and awareness. A small episode may serve to illustrate the shifts and transfers between movement and (observer) perception:

In summer 2010 I drove with two colleagues through the Brandenburg landscape to a village where there was a kind of "Dance Land" farm called "Ponderosa," where a workshop on contact improvisation was being held by Nancy Stark Smith, one of the best-known personalities in this field. We had announced our arrival and had permission to observe the workshop as a small research team. The workshop was taking place in a large, somewhat dilapidated barn set amidst an overgrown, elderberry-scented "Paradise" – like a relic of the hippy 1970s: in a large, light-filled room supported by wooden beams, whose atmosphere had put all the "participants" into the right mood, thanks to its spaciousness, openness, conduciveness to concentration, calm, *and* dynamic character, and the rhythmical division of the space by windows and beams. The nineteen participants of the workshop – and we as observers – distributed ourselves about this space. What was striking was how much this space and its divisions actually helped to promote the whole process of the workshop, with its various action centers of movement and "contact." Here *participation* was not just about "sharing" the place, but also and equally a constituting of "space" in (inter-)action, in motion, and in watching–listening.

34 Ibid.
35 Waldenfels, *Phänomenologie der Aufmerksamkeit*, 230 f.
36 Ibid., 231.

The question of the relationship between movers (workshop participants) and observers was in the air throughout the entire process – and it changed! Relationships transform perception.

The workshop was intended "to pay special attention to the delicate transition from intimate, private authenticity to making art intended to be viewed by the public. [...] Are the subtle experiences of perception and action inside improvised dance visible to the watcher?"[37] These *experiences* and these *questions* were shared. At the end we were asked: "What did you see?" – with regard to a process of movement involving constant changes between "dancing, watching, listening, and being watched." It is remarkable that all themes and processes that occurred in the workshop were linked with the question of "attention": attention as a sensory-kinesthetic mode of participation.

In reply to the question that all participants put to us in the closing interview: "What did you see?," one participant of our research group said that the entering into (and sharing of) this workshop on *space* and *framework* had been a striking experience. Why? Because of the difference between a stress situation (one's normal job, university life, an arduous journey) and this space offering opportunities of meeting people, lots of peace and quiet, freedom to organize one's own affairs, and release from perfection constraints. This feedback was very well received by the workshop participants. It was clear that the experience of release from tensions in everyday life, scope for selective contacts, and the remoteness of output-oriented tasks, unconsciously responded to the concept of the workshops and contact improvisation. In addition, however, it was the experience of difference itself in this case which caused the outside observer to be accepted as a participant – in the sense of "sharing" – in the group of "movers."

Our questions lead us to consider whether and in what way those parameters which constitute the *kinesthetic sharing* could be reconsidered. In this case this would also mean, for example, that the mutual "responsiveness" between workshop-movers and workshop-observers was not the "answer" to the question of participation. But it does raise a question concerning context-dependent changes. The question is: What does this tell us about forms of kinesthetic and synesthetic empathy, if the relaxing of tension or a change in breathing is seen as an emergent effect of such a transference? And in what way are different dimensions of experience and knowledge addressed in such processes? Was this, in the case of our example, addressing a "tacit knowledge" (Michael Polanyi) of a liminal attention? It is hard to describe a state in which one is re-

37 Nancy Stark Smith in the Program note.

ceptive to signals that one is not – intentionally – focused on and that are received in a distracted, casual manner.[38]

III "Listening": "Small Dance"

The questions which were to be illuminated by this episode of a "contact" between dancers and observers at a contact improvisation workshop are complex. Neither in a neuroscientific nor in an esthetic-theoretical sense is the multiplicity of aspects easy to solve. There is, for example, the question of *how* the intricate and microscopic kinesthetic processes which take place during an hour of Susan Klein technique or in a sequence of contact improvisation are perceptible to an observer.

In the context of contact improvisation discourse it is "listening" – as a quality of attention *and* awareness – in which "voluntary" and "involuntary" movement processes are open both to *mover* and *observer*: "Remaining present and listening go hand in hand."[39]

As the phenomenological studies by Husserl, Merleau-Ponty, and Waldenfels show, kinesthesia is a physical embodied space–time experience. As such it assumes a particular shape within a hear-sound space.[40] This implies an "acoustic epoché," i.e., a breach (in) the resonance. "Listening" thus means (following Waldenfels)[41] that a "different kind of hearing" (ein "Andershören") is needed to break through the hearing order and reach the synesthetic-kinesthetic quality of that "movere" (in the sense of a sensory and emotional "being moved") which is in-

38 Our questions, on the other hand, focus on *modes of participation* with such formulas and criteria that are inadequate to describe: for example, in what way do I belong or am "included" in a process of actions, exercises, movements – as in that contact improvisation workshop in "Dance Land" Ponderosa – even if I as "audience," observer, spectator am "there" and yet remain outside, a situation that is not accurately reflected in the words "exclusion," or "not belonging." Does this not show how potentially interesting it might be to rethink *participation* on the basis of experience and performance theory? Should we not recognize that the attributions of active and passive and the semantic range of the concepts of action and performance cannot ultimately be determined, and that the shifting nature of relationships (e.g., between performers and spectators) and changes in the background against which they play out, give rise to all sorts of temporary possibilities of participation? Cf. Rancière, Jacques, *The Emancipated Spectator*, London 2009.

39 Pallant, *Contact Improvisation*, 34.

40 Cf. Waldenfels, *Phänomenologie der Aufmerksamkeit*, 199; cf. Taylor, Carman, "The Body in Husserl and Merleau-Ponty," in: *Philosophical Topics* 27 (1999): 205–286.

41 Waldenfels, *Phänomenologie der Aufmerksamkeit*, 194.

volved in the "contact," the touch play of contact improvisation. Thus the concept of "*Kinästhese*," as Edmund Husserl introduces the term, is not

> to be understood as a sensation of movement that is only distinguished from other sensations by a special sensibility, but 'Kinästhese', which the ego ascribes to itself, means a moving sensation before a sensing movement, the chiastic formulation indicating that kinesis and aisthesis are not fully congruent either in a phenomenal or a neuronal sense.[42]

This hesitation, this delay – that kinesthetic *epoché* that keeps open a gap in the relationship between movement and (self-)perception[43] – is constantly registered in the descriptions of kinesthetics and contact improvisation. Linda Rabin poses the question: "What is movement within the movement?"[44] Nancy Stark Smith constantly emphasizes the elementary significance of the kinesthetic experience of "disorientation" for contact improvisation and of a "gap" which interrupts the control of movement orientation. At this point the – long since obsolete – question may be raised again as to where the boundary work with kinesthetic perception between the *praxis* of "social dance" and artistic performance runs, shifts, becomes diffuse. The orientation to "flow," to "coordinating falling, following momentum, blending with 'partner's movement'" turns – according to Stark Smith – into a game "against":

> making myself heavy instead of light when a lift starts, [...] insisting instead of yielding, adding no to yes. [...] I've been in the harmony business a long time now. [...] As much as I love running around, I think I'm going to try running *into* things more often, or at least against them.[45]

This refraction – on the part of a counter force, a kinesthetic resistance – reflects the pendulum, the balancing between motion and stillness, in which the potentiality/reflexivity of the kinesthetic is articulated: the extreme reduction of movement as a "deep inner dance" – and the question of what happens "by reducing the outer movement to a minimum, and by slowing down the speed to a degree, I continued to explore the *inner* world of the dance," as Linda Rabin relates.[46] She was eager to learn "what the audience would perceive when all extraneous movement was removed, if the simplicity of a dancer's walk, sitting or raising

42 Cf. Waldenfels, Bernhard, *Sinnesschwellen*, Frankfurt am Main 1999, 68 ff.
43 Waldenfels, *Phänomenologie der Aufmerksamkeit*, 221.
44 In: *Corpus*, *Kinesthetics*, 5.
45 Stark Smith in an article in: *Contact Quarterly* (1984), quoted in: Albright/Gere, *Taken by Surprise*, 162.
46 In: *Corpus*, *Kinesthetics*, 5.

an arm, could communicate the intriguing world of sensations and feelings coursing through the performer."[47] Even if this inner journey, this "inner dance," cannot be comprehensible in every detail, there are links in kinesthetic perception – a "sharing" between "mover" and "observer" – which is a vital aspect of a syn- and kinesthetic (empathic) movement synchronization in contact improvisation and other body practices mentioned here.

This reduction, the attention to the microscopic "inner dance," opens a specific field of (kin-)esthetic movement experience bordering on "standstill"; a standing still that is not a standstill, but a scenario full of risky inner movement: "Even standing, we execute a continuous fall."[48] Where would we find beginning and end, rest and movement in a dance which consists of nothing but a standing still? Nancy Stark Smith describes that "dance" that Steve Paxton invented in the 1970s and which he called "*small dance*."[49]

A dance that consists of nothing but "standing"? She comments on her experience as follows:

> Relaxing erect, the intelligence of the body is revealed as it fires the appropriate muscles just enough to keep the body mass hovering within the range of its vertical supports. The micromovements that occur to keep me balanced are so tiny and yet so magnified, and arise from such a deep feeling of stillness and space, that I get giddy, tickled by the impossible magnitude of such subtle sensations. The disorientation in the stand comes from the feeling that inside the apparent solidity and stillness of standing, there is nothing but movement and space![50]

Disorientation, the intensity of movement – a tumult in the heart of standing still – these moments of the kinesthetic experience of an act of "listening" are what open and transmit the potential for inventing movement in contemporary dance.

Translated by Iain Taylor

---

47 Ibid.
48 Ann Woodhall formulated this central paradox, quoted in: Albright/Gere, *Taken by Surprise*, 157.
49 On Steve Paxton's instructions on "small dance" see the "reconstruction" by Nora Heilmann (2006): in *Rupture in Space* (http://www.ruptures.wordpress.com); cf. Erin Manning, who comments on "A Mover's Guide to Standing Still," referring to Steve Paxton's "Small Dance" (Manning, Erin: *Relationscapes: Movement, Art, Philosophy*, Cambridge/Massachusetts, London 2009, 43–49).
50 Quoted in: Albright/Gere, *Taken by Surprise*, 162f.

References

Coleman, Jonathan/Montero, Barbara, "Affective Proprioception," in: *Janus Head* 9.2 (2007): 299–317.

Albright, Ann Cooper/Gere, David (ed.), *Taken by Surprise: A Dance Improvisation Reader*, Middletown 2003.

Corpus, *Kinesthetics: Four Questions*, 2010, http://www.corpusweb.net/kinesthetics-four-questions.html (28.9.2010).

Foster, Susan, *Choreographing Empathy: Kinesthesia in Performance*, London 2010.

Foster, Susan, *Movements Contagion: The kinesthetic impact of performance*, 2008, online publication of University of California, International Performance and Culture Multicampus Research Group, http://uc-ipc.com/wp-content/uploads/2008/06/movementscontagion-11.pdf (1.6.2008).

Gibson, James J., *The Senses Considered as Perceptual Systems*, Boston 1966.

Manning, Erin, *Relationscapes: Movement, Art, Philosophy*, Cambridge/Massachusetts, London 2009.

Johnson, Mark, *The Body in the Mind: The Bodily Basis of Meaning, Imagination and Reason*, Chicago 1987.

Lakoff, George/Johnson, Mark, *Philosophy in the Flesh*, New York 1999.

Montero, Barbara, "Proprioception as an Aesthetic Sense," in: *The Journal of Aesthetics and Art Criticism* 64.2 (2006): 231–242.

Nancy, Jean-Luc, *Zum Gehör*, Berlin, Zürich 2010.

Neuhaus, Max, "LISTEN," in: *Welt auf tönernen Füßen. Die Töne und das Hören*, ed. by Kunst- und Ausstellungshalle der Bundesrepublik Deutschland (Bonn), Göttingen 1994, 125–127.

Novack, Cynthia, *Sharing the Dance: Contact Improvisation and American Culture*, Madison 1990.

Pallant, Cheryl, *Contact Improvisation: An Introduction to a Vitalizing Dance Form*, North Carolina, London 2006.

Rancière, Jacques, *The Emancipated Spectator*, London 2009.

Reynolds, Dee, *Rhythmic Subjects: Uses of Energy in the Dances of Mary Wigman, Martha Graham, and Merce Cunningham*, Hampshire 2007.

Rizzolatti, Giacomo et al., *Mirrors in the Brain: How our Minds Share Actions, Emotions, and Experience*, Oxford 2008.

Sherrington, Charles, *The Integrative Action of the Nervous System*, New Haven 1906.

Smyth, Mary M., "Kinesthetic Communication in Dance," in: *Dance Research Journal* 16.2 (1984): 19–22.

Taylor, Carman, "The Body in Husserl and Merleau-Ponty," in: *Philosophical Topics* 27 (1999): 205–286.

Valéry, Paul, *Cahiers*, vol. II, Paris 1974.

Waldenfels, Bernhard, *Phänomenologie der Aufmerksamkeit*, Frankfurt am Main 2004.

Waldenfels, Bernhard, *Sinnesschwellen*, Frankfurt am Main 1999.

This text was first published in: Sabine Flach/Jan Söffner/Jörg Fingerhut (ed.), *Habitus in Habitat III: Synaesthesia and Kinaesthetics*, Bern 2011.

Corinne Jola

Do You Feel the Same Way Too?

This book chapter describes an unusual research project in which the cognitive neuroscientist, myself, participated not only as a subject in the experiments but also in the daily rehearsals of the dance company Emio Greco | PC as part of the research. Hence, I called it 'embodied neuroscience'. As both subject-matter and framework of this research project are unusual, I also chose an atypical format for presenting my research: a virtual interview. I collected questions that I have been asked by different people in regard to this project or my research approach in general and provide answers that I have either given or would give in case my response was still outstanding. Some of the questions presented in the interview were raised in advance of the project and some at conference presentations after the project. My aim is to address issues in a way that a conventional paper would not. I would like to provide the reader with the background and specific aims of the research project and the particular experiments conducted, as well as concede the atmosphere of the project and a feeling for some of the issues involved.

The Aims

The project had two aims. First, to investigate the neuronal underpinnings of sensory dance experience in order to better understand the shared experience of spectators when they watch dance. Second, to inform scientific research through the experiences gained in participating in the company's rehearsals.

At some point in our lives, dancing is or has been important for all of us and we all have danced. The reasons for dancing may have been different, but for most people, they were related to joyful moments. Nowadays, many may enjoy dance passively, by observing others. In the UK, for example, the BBC programme *Strictly Come Dancing* is very popular.[1] Clearly, dance uses rhythmic body movements to communicate something – in other words, dance is a means of exchanging information[2] which is an important part of everyday life. An increasing number of scientific studies have thus investigated the brain's responses to watching dance for the purposes of understanding action observation and other perceptual, cognitive and

---

1 http://www.bbc.co.uk/strictlycomedancing/ (13.7.2013).

2 Cf. Stevens, C./McKechnie, S., "Thinking in action: thought made visible in contemporary dance.", in: *Cognitive Processing* 6.4 (2005): 243–252.

memory-related processes involved in watching dance.[3] It is generally found that spectators internally mirror the dance actions they observe on stage according to their own motor or visual expertise. Trained in dance and choreography myself, I know that each dancer has a personal preference for certain moves – probably as much as non-dancers have their preferred moves in the disco. Hence, how could it be possible that spectators mirror movements in a similar manner amongst each other? Furthermore, I know that movements can be experienced differently depending on the momentary fitness level. Therefore, can the perception of a movement really be dependent on the neuronal state alone and independent on physiological parameters? To better understand the link between the body, the subjective experience and the brain when watching dance, I thus studied spectators' neuronal responses in the brain using modern imaging technologies while at the same time acknowledging and measuring subjective experiences and physiological changes in the body when watching dance before and after training.

The Project

The project was financially supported by a grant from the Faculty of Arts and Human Science at the University of Surrey, where I had started working as a Research Fellow in Psychology shortly before. The project evolved into two separate studies: a single-case study (on myself) which I called *Phenomenal Experience and Neuronal Changes*, and a more conventional scientific experiment that tested neuronal changes in first-year dance students of the University of the Arts in Amsterdam (AHK), entitled *Shared Motor Experience*. The latter study on shared experience was run in collaboration with and supported by Professor Christian Keysers and Dr Valeria Gazzola at the Netherlands Institute of Neuroscience (NIN)/Spinoza Centre. Both studies were hugely exciting and, with the combined support from all the scientists, dancers and dance students involved, they ran more smoothly than any other study I have conducted so far. This is surprising, considering that the *Shared Motor Experience* experiment took place abroad, outside my everyday research environment, which meant working with new scientists, other technologies and within another culture. Moreover, using the experimenter (myself) as an experimental subject as I did in *Phenomenal Experience and Neural Changes*, is very unusual in my field of research, yet it has, at numerous conferences, been received surprisingly well.

3 Bläsing, B./Calvo-Merino, B./Cross, E.S./Jola, C./Honisch, J./Stevens, C.J., "Neurocognitive Control in Dance Perception and Performance", in: *Acta Psychologica* 139.2 (2012): 300–308; Sevdalis, V./Keller, P.E., "Captured by motion: Dance, action understanding, and social cognition", in: *Brain Cognition* 77 (2011): 231–236.

The Background

When my previous contract with the Watching Dance project at the University of Glasgow[4] was coming to an end, I decided at first not to worry about my next position and instead to start thinking of what I wanted to do next. Dance, of course! So I proposed a research project to Bertha Bermúdez Pascual from the Dance Company Emio Greco | PC in Amsterdam[5] which would not only require me to measure all sorts of data but also to rehearse with the company. I have collaborated on the company's research projects since 2006.[6] This has provided me with a deep and experiential understanding of the company's (research) agenda and how their research interests match with mine, which was – I believe – one of the reasons for why I was in such a favourable position to join them. For two months (February/ March 2011) I actively joined the training of the internationally acclaimed dance company Emio Greco | PC in Amsterdam. The training in general involved a warm-up of about 2 hours in the morning. On most days, this training was led by Barbara Meneses Gutierrez, the company's repetitor, and consisted of the Double Skin | Double Mind workshop (DS|DM). This workshop was initially developed by Emio Greco and Peter Scholten for teaching purposes, to transmit the process of creating a performance that best relates to their work. The DS|DM prepares the body for the creative process, and its main modes are jumping, breathing, expanding, reducing.[7] These modes do not have a particular order, but what characterises this work, compared to other classical and contemporary dance warm-ups, is that they are of uninterrupted extended duration and thus physically strenuous and cardiovascularly challenging. After the workshop I was taught specific phrases of their repertoire and copied the dancers' movements from one of the studios' corners during their rehearsals.

The Questions

One of the overarching questions of dance is what it is that makes it so joyful and beautiful. The entrainment to music and/or rhythm is one of the aspects often

4 http://www.watchingdance.org.

5 http://www.ickamsterdam.com.

6 Jola, Corinne, "Movement intention: dialectic of internal and external movements – Reflections from cognitive neuroscience", in: Scott deLahunta (ed.), *Capturing intention*, Amsterdam 2007, 62–67.

7 DeLahunta, Scott, "The moment to question ... Double Skin/Double Mind", in: Scott deLahunta (ed.), *Capturing Intention*, Amsterdam 2007, 20–21.

mentioned. Importantly, dancing has beneficial effects: previous studies found evidence for dance-specific cognitive enhancements and increased emotional well-being.[8] Is it possible that dance training-related changes in the body and the brain further increase the pleasure of dance? Could dance simply be enjoyed because of its beneficial effects on fitness, or because of increased emotional balance and acquired coping strategies, potentially gained through sharing emotions and sensations with others via movements?

Numerous conferences attended by both scientists and artists have led to debates on dance perception, cognition and the health of dancers. Moreover, how best to study these questions is an often discussed topic. Specific for the projects described here is their interdisciplinary character, as they were both strongly influenced by the collaboration with the arts. The complete dataset of both studies are currently being analysed. Nevertheless, I describe the primarily 'practice-as-research' parts of the projects as the format of these research endeavours allowed me to make some interesting observations which I would like to share here.

The Experiments

In both studies –*Phenomenal Experience and Neural Changes* and *Shared Motor Experience*– I combined recently established methods to assess changes in the body and the brain. The studies were longitudinal, comparing before and after, and practically no risks were involved for the subjects – of which I was one. The second study on shared motor expertise followed a more classic factorial design, testing a group of students, and its set-up and results will be presented elsewhere. In this chapter, I am therefore mostly referring to the former study, which was on myself. It was less conventional, had more types of measures involved, and comments on more classical models of action observation – which makes it much more exciting in this context. In this study, I measured the anatomy and functional processes of my brain before and after the dance training by means of a Magnet Resonance Imaging (MRI) scanner. MRI is a non-invasive means of measuring the activity and the structure of the human brain. It was developed in the late 1970s and routinely applied in clinical diagnostics since the 1980s. To study the activity in the

8 Kattenstroth, J.-C./Kolankowska, I./Kalisch, T./Dinse, H.R., "Superior sensory, motor, and cognitive performance in elderly individuals with multi-year dancing activities", in: *Frontiers in Aging Neuroscience* 2.31 (2010): 1–9; Sze, J.A./Gyurak, A./Yuan, J.W./Levenson, R.W., "Coherence between emotional experience and physiology: does body awareness training have an impact?", in: *Emotion* 10.6 (2010): 803–814.

brain related to specific functional tasks, such as the perception of dance movements, I conducted a functional Magnet Resonance Imaging (fMRI). Furthermore, to visualise anatomical structures of white matter (i.e. fibre tracts), I employed Diffusion Tensor Imaging (DTI). These are two different means of looking inside the brain using an MRI scanner. These days, MRI scanners can be found in many hospitals and universities, as they are used for clinical assessment as well as research. Some participants feel a little uncomfortable during an MRI scan due to the restricted space and the noise created by the magnetic pulses of the machine (in particular the DTI), but it is deemed safe if the guidelines are followed; thus it has become a very popular non-invasive method of visualising neural structures and their activity. The visualisation of neural activity is done by measuring changes in blood flow while the participant lies inside the scanner, a huge tube-like construction with a strong, static magnetic field. By changing the magnetic field gradients, the direction of the naturally-occurring spinning motion of protons in the fluid parts (such as blood) is manipulated and measured. In blood with high levels of oxygen, there are more protons that align with the magnetic gradient. Hence, the measured signal is 'stronger' in 'fresh' blood. The assumption is that oxygen is required in areas where the brain's activity is increased. This type of measurement is called Blood Oxygenated Level Dependent (BOLD) contrast, and it provides information on different activity levels in the brain. The researcher compares the different levels of BOLD for each condition the participant is exposed to, as for example while watching dance phrases that are familiar vs. phrases that are unfamiliar. The different activity levels are then indexed in images of the brain by different colouring. In my case, the different conditions were simply watching a set of dance videos from the company before and after training with them. One trailer consisted of movement sequences I had learned during my stay with the dance company,[9] whereas the other contained movement sequences I had seen but never physically practised. Different areas of the brain are predicted to respond to the two videos, depending on whether I have previously embodied the sequences shown or not. However, due to the low resolution of functional scans, the data has first to be pre-processed and then assigned to higher resolution, so-called anatomical scans. For a picture of me and an anatomical structure of my brain of this type, see Figures 1 and 2.

9 See: http://tinyurl.com/c5ndf2y for the learnt sequences.

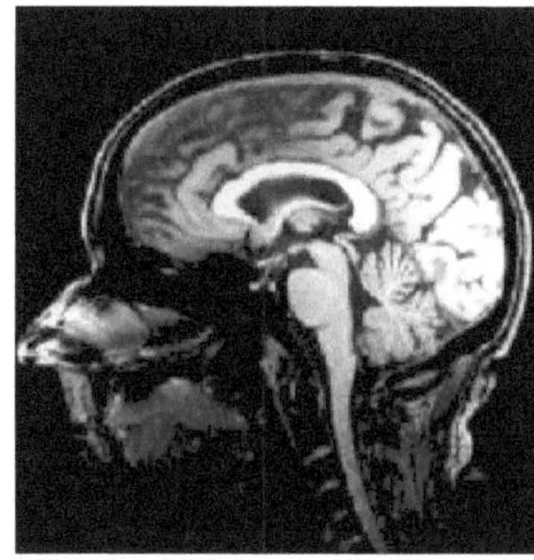

Figure 1 and 2: Picture of my head while dancing (Figure 1, left side, copyright: Venture Photography) and of my brain while watching dance (Figure 2, right side) This is an anatomical scan that does not reveal functional activity.

In addition to the functional and anatomical scans, I also measured my fitness by means of a cardiopulmonary exercise test (CPET), which is a non-invasive method used to assess the performance of the heart and lungs at rest and during exercise, which was cycling in this case. CPET allows us to analyse the heart, lung and metabolism capacity during incremental, maximal exercise in order to determine fitness levels, develop personalised training programmes (including weight loss) and aid in rehabilitation. Finally, I kept a daily journal during the rehearsal period.

In the other study on shared motor experiences, I measured functional changes in the brain of a group of dance students while they watched a video in the MRI scanner showing different dance phrases performed by one of the dancers from Emio Greco | PC. The students were tested twice, once before and once after learning one of the phrases. To make sure they have only learnt one of two phrases, I was always present in the dance classes. As this study relied on a larger number of volunteers, it had to be shorter than the study on myself. I therefore focused on one measurement, the functional brain activity (besides qualitative assessment). In contrast, a large number of different types of measures is one of the demands of contemporary single-subjects studies. Furthermore, the more explorative character of self-experiments can also be seen as one of its benefits.

The History of Self-Experimentation

Figure 3: Caricature example of a laughing gas party among experimenters and their friends. Laughing gas turned out to be a useful anaesthetic gas, 1820 (copyright: Wellcome Library, London).

For reasons of objectivity, research investigators are generally not supposed to study their own brains and behaviours. However, this has not always been the case. In order to study the functioning of the human body, a process that could sometimes be very dangerous and harmful,[10] scientists and doctors have often experimented on themselves. Two well-known examples are Isaac Newton who examined distortions of vision by poking his eye with needles,[11] and Dr Barry J. Marshall who swallowed heliobacter pylory to provide evidence that this bacterium causes ulcers.[12] Finally, another less harmful example of self-ex-

10 See: Altman, Lawrence K., *Who Goes First? The Story of Self-experimentation in Medicine*, Los Angeles 1998.

11 See for example his statement "I once made upon my self with ye hazzard of my eyes" in a letter to John Locke, 1691 (cited in: Iliffe, Rob, "'That puzleing Problem': Isaac Newton and the Political Physiology of Self", in: *Medical History* 39.4 [1995]: 433–458, here: 440).

12 See: Roberts, Seth, "Self-experimentation as a source of new ideas: Ten examples about sleep, mood, health, and weight", in: *Behavioral And Brain Sciences* 27 (2004): 227–288.

Figure 4: Early self-experimentation in the development of transcranial magnet stimulation (TMS). The figure shows five coils placed on top of each other in an attempt to produce a large, rapidly-changing electromagnetic field that elucidates visual phenomena. Reprinted with permission of the publisher (Taylor & Francis Ltd, http://www.tandf.co.uk/journals).

perimentation are the 'extensive' studies of the effects of laughing gas that became very popular amongst experimenters and their friends (see Figure 3).

Contemporary neuroscientific measurements of brain activity have also been partly developed on the basis of self-experimentation. For example, the first successful transcranial magnetic stimulation (TMS), nowadays extensively used to probe the motor cortex excitability or to inhibit specific functional processes of the brain, involved many self-stimulations by Dr Merton. For a rather spooky looking example of early TMS explorations, see Figure 4 and also Walsh and Cowey

(1998). Fortunately, due to safety regulations and technical improvements, serious incidences of contemporary TMS applications are however extremely rare.[13]

The Rationale

Self-experimentation in my research project was a good way to save other people from strenuous activity. In general, it allows scientists to generate new research ideas if specific effects are found.[14] Unlike the very early self-experiments, the potential harm of my studies was very small. The danger I was most aware of was the potential of injury during the dance training.[15]

The aim was to recognise specific aspects of dance training by participation-based experience (myself and dance students) which I would otherwise not have noticed. Furthermore, since I was not a novice in dance, I could potentially reveal aspects that another less expert person may not notice. Last but not least, I had a strong personal motivation because I enjoy dancing.[16]

Contrary to my concerns and the criticism of others regarding self-experimentation, it has been a great pleasure to present the study on myself at two conferences[17] where it was well-received. I could only compare two time-points (before and after), but I explored related changes in several measurements (fMRI, DTI, fitness). Admittedly, it is an unconventional study with very limited results if evaluated from a scientific perspective. Nevertheless, the observations are of relevance for my scientific practice and I choose a virtual interview to make some of the insights available. The interview can be seen as the equivalent of conceiving this kind of single-case self-study project – both forms are unusual, or even unorthodox, in 21st-century psychological research. Studies in the field of

---

13 Rossi, S./Hallett, M./Rossini, P.M./Pascual-Leone, A./The Safety of TMS Consensus Group, "Safety, ethical considerations, and application guidelines for the use of transcranial magnetic stimulation in clinical practice and research", in: *Clinical Neurophysiology* 120.12 (2009): 2008–2039.

14 See: Roberts, "Self-experimentation as a source of new ideas".

15 See also interview below.

16 See also: Roberts, Seth, "The unreasonable effectiveness of my self-experimentation", in: *Medical Hypotheses* 75.6 (2010): 482–489; and interview questions below.

17 Jola, Corinne, "Preparing the Body. Talk at the 1st International Congress of Dance in Health", *From motivation to movement: towards an empirical understanding of the role of dance in health*, May, 24th–25th 2009, Bedfordshire, Bedford; Jola, Corinne, "Do You Feel the Same Way Too?", Invited presenter at the Interdisciplinary Symposium *Touching and to Be Touched. Kinesthesia and Empathy in Dance*, July, 7th–9th 2011, Cluster of Excellence, Languages of Emotion. Deutsche Forschungsgemeinschaft e.V., Freie Universität Berlin, Charité Berlin and others, at Radialsystem V, Berlin.

cognitive neuroscience have related neuronal functions to behaviour by means of reaction times or accuracy, whereas I was interested in linking the phenomenal experience of dancing with changes in the brain. Interestingly, the single-case study seems to be an optimal means to deviate from a common scientific format towards a more performative event when presenting. Hence, in line with these changes, I have decided to create an artificial interview to communicate some of my experiences. This has taken a complete different format from how I would normally write up my research. Importantly, all the people who have asked the questions have agreed to be named in this chapter.

So, how do you feel now? Do you feel a change after two months of rehearsing with us? Suzan Tunca, Dancer Emio Greco I PC

Thank you for this question! I feel great and yes, I do feel different, but not in the sense I was expecting. I don't feel tense; also, I do not have any injuries or strains, which I was so worried about because of previous experiences. For instance, when I danced only at recreational level, I sometimes suffered hip strain. This occured occasionally since I had developed a hip injury during two months of intense dance training at IWANSON School of Contemporary Dance in Munich. This injury used to cause me pain whenever I over-used certain muscles in training. But not now. After two months of strenuous and intense dancing with the Emio Greco | PC dance company, I feel soft on the outside and strong on the inside. It is a great sensation and slightly different from how I remember feeling when dancing extensively in other forms of dance, such as ballet, jazz or contemporary dance. I am relieved that I feel so good. Clearly, I was worried about getting injured once more, which – strangely enough – only happened before the training in a bicycle accident! Even this did not manifest in any further pain nor did my hip injury re-occur. I have a bodily sensation that makes me believe that I am flexible, responsive and at ease with the environment, and at the same time I feel resilient, incredibly grounded and self-confident.

Has your fitness changed? Dr Emma Redding, Head of Dance Science, Trinity Laban Conservatoire of Music and Dance

Yes – but not as much as I was expecting. At the beginning of the project, I was disheartened to find out how low my fitness levels were before training. I was even more surprised to hear that in spite of the training and my physical sensation of inner strength, that some fitness values were even lower in the second compared to the first measurement. The first 20-watt-per-minute ramped bicycle test on January 28th, 2011 lasted 12 minutes and 28 seconds. The oxygen intake was measured in the expired air by gas analysers and used as an indication of maximum performance when the maximum oxygen uptake (i.e. VO2max) was reached. The maximum is defined at the point where higher workload does not lead to more oxygen intake showing a 'plateau' of oxygen intake. In other words, VO2max indicates that the maximum performance is reached of transporting oxygen through breathing to the muscles. Some researchers suggest using the VO2max itself as an index of endurance performance or a measure of maximal cardiopulmonary function in healthy populations.[18] In this test, I had reached VO2max at 34.3 ml/kg/min. As a comparison, university class athletes' VO2 max in 1968 ranged from 30.6 to 69.1 ml/kg/min.[19] My VO2max was at the lower end despite considerate preparation: before I began working with Emio Greco | PC, I followed a fitness regime outlined by Sara Beck, a fitness instructor and dance scientist at Laban Trinity College, in order to prepare my body for the strenuous training with the company. Nevertheless, my fitness was classified as 'fair' on a scale that ranged from 'poor' and 'fair', to 'average', 'good', and 'high'. I was aware that during the previous two years, I hadn't been moving as much as I used to. During those two years, I had my first proper post-doctoral research position studying audiences' responses to watching dance at University of Glasgow, and I was working many hours a day. As for so many of us, this means sitting in front of a computer screen. However, besides the immobile computer work of about 10 hours a day, I went jogging, swimming, and played badminton occasionally, besides doing other non-cardiovascular activities such as yoga or Pilates whenever possible. Hence, I did not expect my fitness to be so low! Regardless of my aims to be active, for two years I could probably be best described as a sedentary desk worker. By contrast, in Amsterdam, I trained on average 6 hours a day and spent only about 5 hours a day on the computer. After the training with Emio Greco | PC, I underwent the same cardiovascular fitness test. This time, my fitness was even classified as 'low'. What went wrong?

One explanation is that I had over-trained, leading to low values. Another explanation is that conventional dance training does not enhance cardiovascular

---

18 Astorino, T.A./Willey, J./Kinnahan, J./Larsson, S.M./Welch, H./Dalleck, L.C., "Elucidating determinants of the plateau in oxygen consumption at VO2max", in: *British Journal for Sports Medicine* 39.9 (2005): 655–660.

19 Shepard, R. J./Allen, C./Benade, A.J.S./Davies, C.T.M./Di Prampero, P.E./Hedman, R./Merriman, J.E./Myhre, K./Simmons, R., "The Maximum Oxygen Intake", in: *Bulletin de l'Organisation mondiale de lasanté* 38 (1968): 757–764.

fitness.[20] However, the training I underwent with the company could be thought of as cardiovascular: it involves extended periods of jumping, some movement sequences at high speed for long durations and a continuous warm-up. Hence, it is possible that the CEPT is measuring some specific fitness which cannot be transferred to optimal fitness in dance. Indeed, the CEPT has been criticised as not being the appropriate measure for fitness in dance and thus specific training and measurement tools for dancers have been developed.[21] Interestingly, however, I could not find any longitudinal studies on changes in fitness with the CEPT on dancers.

Finally, it is possible that a measurement is flawed and as with single-subject research, any errors or variances will not be averaged out or identified as outliers as they normally would. There are a number of points at which any measurement is prone to potential errors. Of the above-mentioned values, some measures are based on automatic detection, such as the anaerobic threshold. Frequently, the anaerobic treshold is located where relatively more carbon dioxide is eliminated compared to the intake of oxygen. These points are detected automatically and as can be seen in Figure 5, the anaerobic thresholds are set at a very early point of cross-over of VO2 and VCO2. In addition to the automatic detection of the values, visual inspection and potential manual threshold adjustment is deemed to be crucial to amend for momentary individual changes. Indeed, my data required some visual inspection and hand-made adjustments: the uptake curve in the test after training shows more of a peak than a plateau. This could be indicative of a sub-maximal performance, as during maximal performances a plateau of O2 is usually seen (see Figure 5, left graph). Secondly, the plateau mentioned above can be seen in the VO2 and may have happened due to a misunderstanding between the sport exercise scientist (i.e. the experimenter) and myself. I interpreted her signs erroneously as telling me to stop as the maximum had been reached. Hence, I started talking to her which led to a leak in the oxygen mask and I may have stopped too early which unfortunately cannot be corrected or replaced by another subject. For a single subject, the measures would have had to be repeated a number of times to enhance validity. For in-

---

20 Wyon M., "Preparing to perform: periodization and dance", in: *Journal of Dance Medicine and Science* 14.2 (2010): 67–72; Rafferty, S., "Considerations for integrating fitness into dance training", in: *Journal of Dance Medicine and Science* 14.2 (2010): 45–49.

21 Twitchett, E./Nevill, A./Angioi, M./Koutedakis, Y./Wyon, M., "Development, validity, and reliability of a ballet-specific aerobic fitness test", in: *Journal of Dance Medicine and Science* 15.3 (2011): 123–127; Wyon, M./Redding, E./Abt, G./Head, A./Craig, N./Sharp, C., "Development, Reliability, and Validity of a Multistage Dance Specific Aerobic Fitness Test (DAFT)", in: *Journal of Dance Medicine and Science* 7.3 (2003): 80–84.

stance, the anaerobic treshold was initially located at 8 min 40 s, leading to lower values (see Table 1–IIa). After manual correction of the anaerobic threshold to 10 min 56 s, the measures indicated slight improvements (see Table 1–IIb). In particular, my maximum load was higher after training than before without a significant increase in heart rate. These changes show that my values were indeed better after training in the sense that I was stronger and could carry a higher maximum load. However, even after these changes, my fitness was still classified as 'low'. If one looks at the additional measures, two relevant changes should be highlighted due to potential effects of the dance training: (1) my lung function improved: both the amount of air I could forcefully exhale and the amount I could exhale in one second increased; and (2) my BMI went down from 25.2 to 25.0.

Importantly, my reduction in BMI was small and due to a change in measured height rather than a change in weight. Moreover, BMI has been criticised as a reliable measure for body fat.[22] Hence, the manual height measure could be based on a measurement error (it was not the same person measuring my height before and after training), or a general stretch of the body from being more upright during the training. Furthermore, since I was stronger after training, it is likely that muscle mass increased and body fat decreased which not necessarily results in lower weight. I like to believe that my height increased in response to the specific training of Emio Greco | PC which consists of several modes, one of which is 'expanding'. In the expanding mode, dancers have to stretch their bodies beyond their imagined boundaries and to continue expanding for a prolonged period of time. Hence, the training includes a lot of body stretches and stamina training to keep the elongations. To verify if increased height is a specific effect of this particular training, a study comparing at least two groups undergoing different dance training forms should be conducted.

Based on this insight into my low fitness before as well as after the training, I was motivated to enhance my cardiovascular fitness and train on a bike on average 2 to 3 times a week for 1 hour starting at 150 Watt and increasing it to 200 Watt on average over the time course of a year. I already started this project with an intrinsic motivation; but it would definitely be attractive, if such studies motivated more people to increase their levels of fitness. I do experience that my lung volume capacity has increased. However, it would be interesting to do another CEPT to validate the assumption. Moreover, it would be an interesting hypothesis to test if enhanced

---

22 Nevill, A.M./Stavropoulos-Kalinoglou, A./Metsios, G.S./Koutedakis, Y./Holder, R.L./Kitas, G.D./Mohammed, M.A., "Inverted BMI rather than BMI is a better proxy for percentage of body fat", in: *Annals of Human Biology* 38.6 (2011): 681–684.

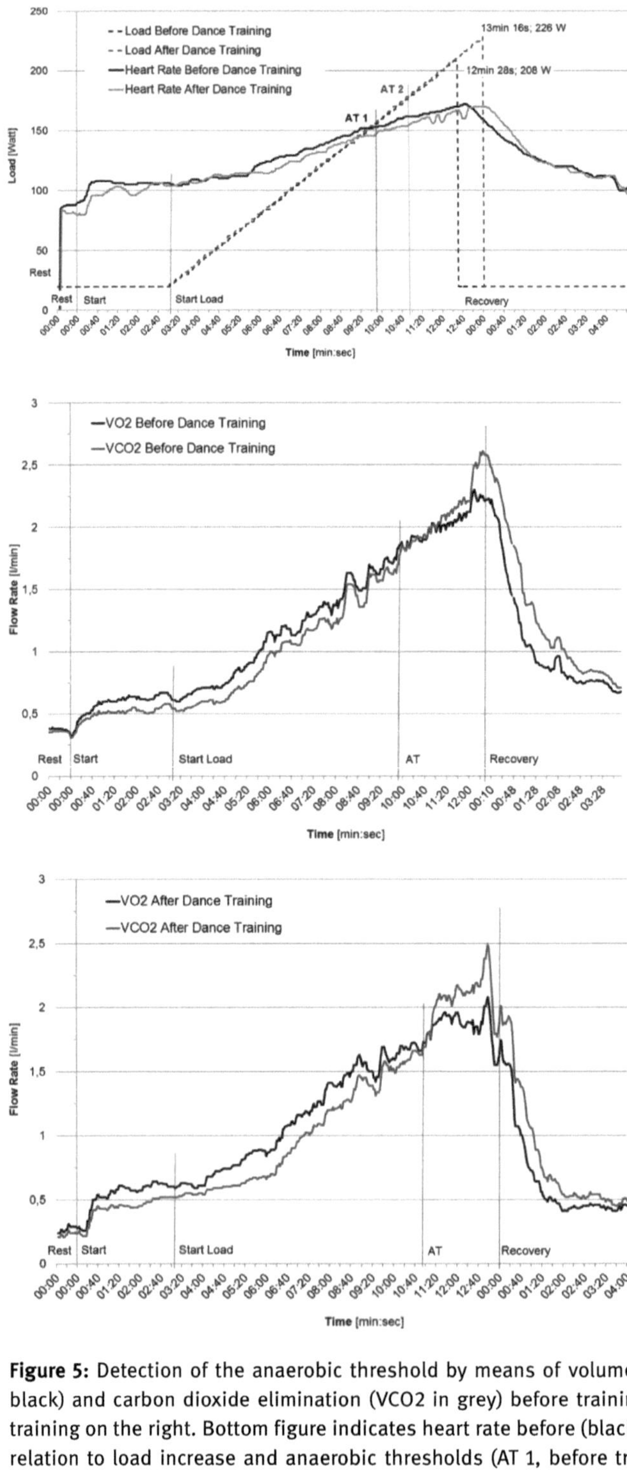

Figure 5: Detection of the anaerobic threshold by means of volume of oxygen intake (VO2 in black) and carbon dioxide elimination (VCO2 in grey) before training on the left and after training on the right. Bottom figure indicates heart rate before (black) and after training (grey) in relation to load increase and anaerobic thresholds (AT 1, before training: AT 2, after training).

	Before Training	After Training	
	I	IIa	IIb
BMI	25.2	25.0	
FVC *(l)*	3.83	4.11	
FEV1 *(l)*	3.31	3.55	
Maximum HR *(bmp)*	172	170	
Maximum Load *(watts)*	208	226	
HR @ Max Load *(bmp)*	170	170	
AT @ timepoint *(min:s)*	**9:51**	8:40	**10:56**
HR @ AT *(bmp)*	**153**	140	**156**
HR @ $VO_{2\ max}$ *(bmp)*	**168**	160	**169**
Load @ AT *(watts)*	**156**	134	**180**
Load @ $VO_{2\ max}$ *(watts)*	**204**	200	**222**
VO_2 @ AT *(l/min)*	1.83	1.49	1.73
VO_2 @ Max Load *(l/min)*	2.23	1.93	1.55
VO_2 max *(ml/min/kg)*	34.3	29.0	31.0

Table 1: Values of CEPT before (left column, I) and after (right columns, II) 9 weeks of dance training with Emio Greco | PC. Scores without (IIa) and with (IIb) manual correction of anaerobid threshold. BMI=Body Mass Index (weight in relation to height; FVC= Forced Vital Capacity (total volume of air that can be exhaled from the lungs during a forced expiration following a maximal inspiration); FEV1=Forced Expiratory Volume at 1st second; HR=Heart Rate; VO2=Volume of Oxygen; AT=Anaerobic Threshold (point at which the energy pathway begins to burn fuel without oxygen with a rapid increase in blood lactate concentration).

functional response in sensorimotor areas when watching dance is related to increased oxygen intake. For instance, trained dancers have increased VO2max, hence, their cardio-respiratory fitness is higher which may lead to different responses even when passively watching dance compared to sedentary desktop workers. This has never been tested, and research into the relationship between breathing and watching dance has just started independently of this observation.[23]

23 It has been shown that watching dance with music synchronises more brain areas between spectators than without (Jola, C./McAleer, Ph./Grosbras, M-H./Love, S.A./Morison, G./Pollick, F.E., "Uni- and multisensory brain areas are synchronised across spectators when watching unedited dance", i-Perception 4 (2013): 265 – 284. Further, hearing the breathing of dancers leads to very individual responses (Jola, Corinne, "Some like it hot. The Personality factor 'Openness' is correlated with liking to see effort and hear the dancers' breath", in preparation; Reason, M./Pollick, F.E./Jola, C./Grosbras, M.-H./Kay, R./Reynolds, D., "Researching audience responses to sound and movement in dance performance", submitted).

How did you measure your happiness? Dr Isabel Dziobek, Head Junior Research Group 'Understanding Interaffectivity', Freie Universität Berlin

I kept a daily journal which can be analysed for reoccurring themes. Overall, I remember being very happy throughout the project – however, this may not fully correspond to how I felt at the time. Namely, by flicking through the journal and looking at the photographs taken, I can clearly indicate a number of moments of frustration. I was aching, tired, exhausted, and had to fight a cold 6 weeks into training. Hence, my memory of happiness does not necessarily match the actual daily experiences. Ideally, I would have measured levels of happiness randomly and quantitatively. In the future, I would consider conducting pen-and-paper questionnaires on happiness/well-being as well as a measurement tool that can record responses in a more controlled manner during the day. At this point in time, I would like to answer the question back and say – how would you measure happiness?

Did you find any brain activity in touch-related areas? Joa Hug, dancer/performer, PhD student, Theatre Academy in Helsinki/Berlin

The theme of this book is 'being touched' by dance; hence, your question is vital. We use this analogy to describe the cocktail of somatosensory and emotional experiences we can have and often seek when we watch dance or dance ourselves. Indeed, there seems to be a partial overlap of the sensation of touch and watching dance: the somatosensory cortex and the limbic brain areas are activated through active touch and through being touched. Interestingly, these regions are also activated when subjects observe somebody else dancing. Below, I will describe the functional responsiveness of the somatosensory cortex and the limbic areas in more detail.

The somatosensory cortex has recently been recognised as an important part of the extended network that is activated when we passively observe actions performed by others. Action observation (such as watching dance) has been found to activate areas in the ventral premotor and inferior-parietal regions,[24] where

24 For a review on the topic of action observation in the context of dance, see: Sevdalis, V./Keller, P.E., "Captured by motion: Dance, action understanding, and social cognition", in: *Brain Cognition* 77 (2011): 231–236; Bläsing, B./Calvo-Merino, B./Cross, E.S./Jola, C./Honisch, J./Stevens, C.J., "Neurocognitive control in dance perception and performance", in: *Acta Psychologica* 139.2 (2012): 300–308.

the so-called mirror neurons were initially detected in the macaque monkey.[25] Mirror neurons are neurons that fire both during action observation and action execution. Hence the spectator internally 'mirrors' or resonates with the observed actions. In humans, the activation during action observation is normally seen in several areas; and is thus often called a 'mirror-neuron network'. One of these areas is in the somatosensory cortex, which has frequently been found to be activated during action observation,[26] during the observation of dance,[27] and in studies on tactile perception. Touch – independent of whether it is active (self-touch) or passive touch (being touched) – is reflected in activity in somatosensory areas.[28] Accordingly, it has been suggested that the somatosensory cortex plays an important role in understanding others' actions, namely by giving the spectator the sensation of the action: 'what it could feel like to act as seen'.[29] In both action observation and touch, it is predominately the left hemisphere in the somatosensory cortex that is activated.[30] In my study on the students' shared experiences, I found some preliminary evidence for more similar activity in somatosensory areas across students for movements they had learnt.

The limbic structure is the other brain region where touch perception is located, in particular in its orbitofrontal parts and posterior insula which are responsive to

---

25 Iacoboni, M., "Imitation, empathy, and mirror neurons", in: *Annual Review of Psychology* 60 (2009): 653–670.

26 Van den Stock, J./Tamietto, M./Sorger, B./Pichon, S./Grèzes, J./de Gelder, B., "Cortico-subcortical visual, somatosensory, and motor activations for perceiving dynamic whole-body emotional expressions with and without striate cortex (V1)", in: *Proceedings of the National Academy of Sciences of the United States of America* 108.39 (2011): 16188–16193.

27 E.g. Calvo-Merino, B./Jola, C./Glaser, D.E./Haggard, P., "Towards a sensorimotor aesthetics of performing art", in: *Consciousness and Cognition* 17 (2008): 911–922; Jang, S.H./Pollick, F., "Experience influences brain mechanisms of watching dance", in: *Dance Research* 29.2 (2011): 350–375; Cross, E.S./Hamilton, A.F./Grafton, S.T., "Building a motor simulation de novo: observation of dance by dancers", in: *NeuroImage* 31.3 (2006): 1257–1267.

28 E.g. Ackerley, R./Hassan, E./Curran, A./Wessberg, J./Olausson, H./McGlone, F., "An fMRI study on cortical responses during active self-touch and passive touch from others", in: *Frontiers in Behavioral Neuroscience* 6 (2012): 51; Keysers, C./Wicker, B./Gazzola, V./Anton, J.L./Fogassi, L./Gallese, V., "A touching sight: SII/PV activation during the observation and experience of touch", in: *Neuron* 42 (2012): 335–346; Grèzes, J./Armony, J.L./Rowe, J./Passingham, R.E., "Activations related to 'mirror' and 'canonical' neurones in the human brain: an fMRI study", in: *Neuroimage* 18.4 (2003): 928–937.

29 Gazzola, V./Keysers, Chr., "The observation and execution of actions share motor and somatosensory voxels in all tested subjects: single-subject analyses of unsmoothed fMRI data", in: *Cerebral Cortex* 19 (2009): 1239–1255.

30 Except: Cross/Hamilton/Grafton, "Building a motor simulation de novo".

pleasant touch on hairy rather than glabrous skin.[31] Of these, the insula is often found to be activated during general action and dance movement observation,[32] but in particular in relation to the experience of taste and distaste.[33]

Notably, in most studies, the observations of actions are not independent of observations of touch. For instance, grasping an object not only involves the grasping action, but also the touch of the object. Johnson-Frey et al. (2003) attempted to dissociate actions from hand–object interactions and found that observing touch alone, without an action, does lead to less activity in the inferior-frontal area than the action combined with the hand–object interaction. However, the authors looked only at those regions of interest and the paper does not report differences in activity in other areas, such as somatosensory cortex or the insula. On the contrary, Keysers et al. (2004) found that being touched and observing videos of someone else being touched commonly activated the somatosensory cortex – importantly, though, action and touch were again combined. Finally, the insula is enhanced when subjects watch actions in which pain is inflicted on others.[34] In all these examples, both actions as well as touch are employed. Using one set of stimuli with emotional facial expressions, the authors found evidence for a dissociation of the insular activity in response to mirroring of the emotional content of the facial expressions and the activity in somatosensory areas to be related to motor mirroring of the actual facial move-

31 McGlone, F./Olausson, H./Boyle, J.A./Jones-Gotman, M./Dancer, C./Guest, S./Essick, G., "Touching and feeling: differences in pleasant touch processing between glabrous and hairy skin in humans", in: *European Journal of Neuroscience* 35.11 (2012):1782–1788.

32 Iacoboni, "Imitation, empathy, and mirror neurons"; Cross, E./Kirsch, L./Ticini, L.F./Schütz-Bosbach, S., "The impact of aesthetic evaluation and physical ability on dance perception", in: *Frontiers in Human Neuroscience* 5 (2011): 102.

33 Wicker, B./Keysers, C./Plailly, J./Royet, J.P./Gallese, V./Rizzolatti, G., "Both of us disgusted in My Insula: The common neural basis of seeing and feeling disgust", in: *Neuron* 40.3 (2003): 655–664. See also: Bastiaansen, J.A./Thioux, M./Keysers, C., "Evidence for mirror systems in emotions", in: *Philosophical Transactions of the Royal Society London B: Biological Sciences* 364 (2009): 2391–2404.

34 Jackson, P.L./Meltzoff, A.N./Decety, J., "How do we perceive the pain of others? A window into the neural processes involved in empathy", in: *NeuroImage* 24 (2005): 771–779. Some authors claim that somatosensory cortices are involved in the perception of the pain felt by others (Voisin, J.I./Marcoux, L.A./Canizales, D.L./Mercier, C./Jackson, P.L., "I am touched by your pain: limb-specific modulation of the cortical response to a tactile stimulation during pain observation", in: *Journal of Pain* 12.11 [2001]: 1182–1189) whereas others found no somatosensory activity for observation of pain (Singer, T./Seymour, B./O'Doherty, J./Kaube, H./Dolan, R./Frith, C.D., "Empathy for pain involves the affective but not sensory components of pain", in: *Science* 303 [2004]: 1157–1162).

ments that produced the expressions.[35] One may wonder whether a similar study using dance movements was possible, considering that most movements seem to evoke an emotional content. To my knowledge, no similar study has yet been conducted to disentangle the specific roles of the interior-frontal, insula and somatosensory cortex in response to a successful dissociation of sensation of touch and observation of action. Studies have emphasised the benefit of using dance as a means to study action observation without action–object integration. However, if we agree on the first note made here, that we sometimes feel touched when we watch dance, then even studies that employed dance to investigate action observation independent of objects do measure both the participants' response to actions and to touch – at least in a metaphorical manner. The feeling of being touched by actions may led to brain activity in areas related to touch as a response to this sensation – or vice versa, the sensation may be triggered by the activity in these brain areas. It seems that touch and action observation are intrinsically intertwined, either literally, or metaphorically.

Hence, for future studies on action observation, it may be a relevant but difficult task to disentangle the brain's activity as a function of mapping others' actions from the function of touch perception. Otherwise, we may have to change our views about touch being a separate entity from action. In other words, we may be able to emphasise with others through the mirroring of actions because actions and touch sensations are so closely related.[36]

If you design an experiment, don't you know exactly what stimuli you need at the very beginning? Barbara Meneses Gutierrez, Repetitor of Emio Greco | PC

In the optimal case, yes, we know exactly what the stimuli should look like! – And then, the scientists often 'create' them. To give some background to this question: Barbara helped me choose the stimuli for the study on students' shared motor experience when watching dance phrases before and after they learnt them. Hence, the phrases had to be from the repertoire of the Emio Greco | PC dance company, with whom they had a one-week workshop. What I remember Barbara saying was: 'I am getting frustrated with the many uncertainties we have in designing the experiment, in particular in regarding the choice of the stimuli – you seem to be adding more and more restrictions as we go along.' Normally, an experiment is based

35 Budell, L./Jackson, P./Rainville, P., "Brain responses to facial expressions of pain: emotional or motor mirroring?", in: *NeuroImage* 53.1 (2010): 355–363.

36 Carr, L./Iacoboni, M./Dubeau, M.C./Mazziotta, J.C./Lenzi, G.L., "Neural mechanisms of empathy in humans: a relay from neural systems for imitation to limbic areas", in: *Proceedings of the National Acadamy of Sciences U.S.A.* 100 (2003): 5497–5502.

on previous research findings, so you would know quite well which potential variables may interfere and affect your results – and it is important to control these variables when creating your own stimuli. However, whenever you are not conducting a strictly controlled experiment with custom-designed stimuli, but an ecologically more valid study that uses a selection of existing stimuli, excluding potential interfering variables can be more difficult. If you fail to control specific factors that are not present or absent equally across your conditions, it could distort your results, and you will make false conclusions (so-called 'confounding variables' that play a significant role in what you are measuring, but often are not of interest for a particular study). A simple but typical example of a confounding variable is if you correlate the size of shoes with gender and find a significant effect for male having larger feet, whereas in reality, shoe size is determined more strongly by height than gender. The problem is that experiments in general and including dance can be much more complicated, and that there are normally several relevant variables which you are unaware of until you find them. The more your stimuli resemble what you find in the real world, the more likely it is that you have a larger number of these variables. And, moreover, you cannot fully control them. Thus, when using existing stimuli in an ecologically valid study, as movement phrases from the EG|PC repertoire here, you need to conduct analyses of the stimuli prior to the experiment proper. These should preferably be of qualitative as well as quantitative character. Then, a number of so-called pilot studies ahead of the 'real' one can help to investigate potential effects and make sure that a) the stimuli actually do contrast in the variable you want to measure in a balanced manner (occur equally often); b) other potentially confounding variables are randomly distributed in the stimuli; and c) they have equal effects on valence and arousal.[37] Hence, if one wanted – as we did – to compare the spectators' responses to watching sections of dance performances that have been created by choreographers for artistic rather than scientific reasons, these sections first need to be inspected in-depth before we can identify potential factors we might need to control. Therefore, the selection of stimuli is not easy and must be seen as a process, during which – whenever we consider the inclusion of a particular

37 See for example: Jola, C./Abedian-Amiri, A./Kuppuswamy, A./Pollick, F./Grosbas, M.H., "Motor simulation without motor expertise: enhanced corticospinal excitability in visually experienced dance spectators", in: *PLoS ONE* 7.3 (2012): e33343; Jola/Clements/Christensen/Calvo-Merino (in preparation), "Moved by stills: vertical implied motion enhances kinesthetic sensation"; Jola, C./Grosbras, M.-H., "In the here and now. Enhanced motor corticospinal excitability in novices when watching live compared to video recorded dance", in: *Cognitive Neuroscience* 4.2 (2013): 90 – 98.

movement phrase – additional potential variables come up that needed to be controlled.

How important is the aesthetic appearance of experimental stimuli for you? Julia F. Christensen, PhD student, Laboratory of Human Systematics, University of the Balearic Islands, Spain

Very important. Do you feel touched by a bad dance performance? Aesthetic experiences are sensory experiences. And since we are investigating sensorimotor responses to watching dance, the aesthetic appearance of stimuli and their appreciation is an important element that needs consideration. In the preparation of one of our past studies,[38] we experienced that the quality of the movement performance can modify the spectators' sensorimotor responses. In the pilot study preceding the experiment proper, some of the performers were semi-professional and we did not find an enhancement of the spectators' motor simulation to their preferred dance style. Based on the audiences' feedback regarding the stimuli, we hired professional performers for the ballet and the acting control condition and found the expected audience responses. Hence, it is very important that the stimuli are of the highest possible aesthetic quality.

In addition, dance is the culmination of movement, music, expression, costume, and light, the so-called 'strands' of dance.[39] Therefore, my aim is always to ensure – although that is clearly not possible in all cases – that the dances used do not consist of basic movements only. For example, different types of costume are not only related to but demand different types of dance styles – because of practical applicability. In the 20th century, for example, dance experienced a trend called 'natural movement' where Isadora Duncan and other pioneers emphasised the female body and created dances that were less formal and consisted of more free-flowing forms of personal expression. The dancers wore Greek-style gowns and often performed barefoot. The movements involved everyday movements such as skipping and running, and the torso was used to initiate movements. Hence, not only the movements but also the costumes of Isadora Duncan needed to be in stark contrast to the classical female ballet outfit, consisting of a skirt (called 'tutu'), a tight-fitting body ('basque'), accessories (e.g. hair band), and point shoes. Before the 19th century, the tutus worn by female dancers were long and heavy, restricting free movement. Nowadays, the tutus

38 See: Jola/Abedian-Amiri/Kuppuswamy/Pollick/Grosbas, "Motor simulation without motor expertise".

39 Preston-Dunlop, Valerie, *Looking at Dances. A Choreological Perspective on Choreography*, London 1998.

are very short and show the technically challenging leg and foot point-work. I am using these examples to illustrate how different movement qualities (free-floating vs. upright) are related and can in some circumstances not be dissociated from clothing (free-floating vs. tight). Also, consider you are studying brain responses of spectators watching these two types of dances in the 'wrong' or neutral costumes – would you expect no effect of costume on movement perception when watched by spectators who frequently watch specific dance styles?

To summarise, it is important that the movements are rhythmic and/or fluent body limb moves embedded conceptually. To achieve this, they need to be of a certain duration and performed in the appropriate costumes, etc. Furthermore, the best way to experience dance is to watch a live performance. Because of this, we conducted an experiment in the theatre where four subjects watched a dress rehearsal of *Sleeping Beauty* while we probed their motor-corticospinal excitability using transcranial magnet stimulation (TMS). However, the uncontrolled environment led to high interference and made our data very noisy.[40] In another more controlled study, we compared a group of novices who watched dance live as opposed to dance on video. Importantly, it is much more evident in the video recording that the background was not completely neutral. Furthermore, in correspondence to a pilot study,[41] we found that it is absolutely crucial that the dance is performed professionally. Hence, the performance quality, the choreography and the recording need to be considered even if all the other relevant strands of dance are not.

A number of experimental studies, however, investigated perceptual, cognitive or sensorimotor processes using dance in a deconstructed manner in such a way that it may not be recognised as dance. Professional recording space, lighting and editing is expensive, and under laboratory conditions, these factors have not mattered for a long period of time. Also, some of the recordings we used had clear aesthetic limitations. Hence, it is still uncommon to invest in aesthetically perfect stimuli, especially if the research question does not concern aesthetic perception. Nevertheless, we know that the way we perceive and respond to our environment is affected by its aesthetics. Why would we not expect a different response to stimuli that are beautiful compared to those that aren't?

Does neuroscience mean anything to you? Dr Jason Frowley, PhD Psychology, Lecturer, University of St Andrews and Open University

40 Jola, C./Grosbras, M.H./Pollick, F.E., "Arousal decrease in 'Sleeping Beauty': audiences' neurophysiological correlates to watching a narrative dance performance of 2.5 hrs", in: *Dance Research Electronic* 29.2 (2011): 378–493.
41 Jola et al., "Motor simulation without motor expertise".

This is an abrupt question! (Saying this gave me some time to think before I could answer). Yes, it does. It actually means a lot to me. Hence, that is why I am keen to find ways in which methodological issues can be solved to get closer to the actual dance – with the aim not only to understand more about the culture of dance, but also to better understand the human brain. My argument is that cognitive neuroscience needs to study the human brain and behaviour in more complex environments. Indeed, a number of researchers have developed and still research novel methods to study the human brain in a more ecological manner.[42] As part of the watching dance project[43] we also have successfully increased the ecological validity of our studies by using live performances[44] as well as long dance movement segments.[45] However, these studies are still under-represented. Common studies take place in the laboratory, with artificially created stimuli that often don't resemble what we encounter in the real world.[46] From my point of view, the assumption of laboratory validity, objectivity and generalisation is not without its flaws. Also, I believe it is important to recognise the relevance of the subjective experience and how that can be related to the objective approach emphasised in scientific experiments. I would question objectivity per se; the manner in which we conduct studies greatly affects our data and interpretations. This is acknowledged by some researchers and/or discussed by others, but in neuroscience, there is a general tendency to gloss over it and continue research making strong assumptions: these are, firstly, that we are all the same (more or less), and, secondly, that we behave (our brain behaviours) the same under restricted conditions in the laboratory and outside the lab. Hence, it is assumed that 'our behaviour in the lab can be shuttled to the real world' easily. In other words, the brain can be studied by micro-questions and then these micro-findings can be pieced together. Clearly, I am not the first person

42 For example: Hasson, U./Nir, Y./Levy, I./Fuhrmann, G./Malach, R., "Intersubject synchronization of cortical activity during natural vision", in: *Science* 303 (2004): 1634–1640; Grosbras, M.H./Tan, H./Pollick, F., "Dance and emotion in posterior parietal cortex: a low-frequency rTMS study", in: *Brain Stimulation* 5.2 (2012): 130–136.

43 http://www.watchingdance.org.

44 Jola/Grosbras/Pollick, "Arousal decrease in 'Sleeping Beauty'"; Jola/Abedian-Amiri/Kuppuswamy/Pollick/Grosbas, "Motor simulation without motor expertise", Ida/Grosbras, "In the here and now".

45 Pollick, F./Noble, K./Darshane, N./Murphy, H./Glowinski, D./McAleer, P./Jola, C./Penfield, K./Camurri, A., "Using a novel motion index to study the neural basis of event segmentation", in: *i-Perception* 3.4 (2012): 225, Jola et al., "Uni- and multisensory brain areas".

46 See: Jola, Corinne, "Research and choreography – merging dance and cognitive neuroscience", in: Bettina Bläsing/Martin Puttke/Thomas Schack (ed.), *The Neurocognition of Dance. Mind, Movement and Motor Skills*, Hove 2010, 203–234.

to question whether reductionist psychological experiments are applicable in real life. Professor Ulrich Neisser (1928–2012), named by many the father of Cognitive Psychology, was a driving force in pushing for ecologically oriented research in psychology. Accordingly, I believe – and there is (yet) partial evidence – that ecologically valid research is needed as data from reductionist lab experiments do not always match those from real life;[47] more research is needed.

So what do you think is the solution? Gary Philippo, Mathematician and dancer

Recognising experiential factors. The experience of conducting experiments (experiencing the subjects) and doing the tasks (experiencing the object) is really important and should not be ignored. Self-experimentation leads to new observations, and together with the personal experience that provide a sensitivity for relevant factors it can enhance the validity of new studies. I would modulate the statement 'the best way to learn is to do'[48] to: the best way to find out is to do.

Combining dance (or art in general) and neuroscience has become really fashionable in the last couple of years and the number of studies in the area has increased quickly. What have you gained by doing this particular study? And what was the benefit for the dance company? Beatriz Calvo-Merino, Neuroscientist, London – Madrid

The research project had an immediate effect on me: I was in a good mood and I have grown (not only metaphorically). As for the contribution to the scientific world, this research project is one of the few recent attempts to emphasise the relevance of self-experimentation and combination of different disciplines. Normally, for scientific work to manifest itself, and to initiate changes in paradigms, much more time is needed. For the time being, however, I can say that:

- The study of cardiovascular fitness and dance – including watching dance – has gained in relevance. In particular, the breathing capacity and behaviour needed to be studied as an important factor when experiencing dance.
- Importance of issues regarding the ambivalence of health and well-being in dance training, including social components, weight loss and BMI index.

47 See for example: Risko, E.F./Laidlaw, K./Freeth, M./Foulsham, T./Kingstone, A., "Social attention with real versus reel stimuli: toward an empirical approach to concerns about ecological validity", in: *Frontiers in Human Neuroscience* 6 (2012): 143.

48 Halmos, P.R./Moise, E.E./Piranian, G., "The problem of learning to teach", in: *The American Mathematical Monthly* 82.5 (1975): 466–476, here: 466 (taken from Roberts, "Self-experimentation as a source of new ideas").

- Relevance of advances on perspective of warm-up training (e.g. warm-up to recalibrate a sensorimotor network that has become slightly miscalibrated rather than warming up the muscles, tendons and ligaments[49]). If a generic warm-up is not sufficient and the contemporary view of warm-up training is correct, then we should find neurophysiological evidence.

All of the above were indications for potential future research. Other ideas on how dancers resonate with audiences and how the dancers as subjects themselves communicate something to the audience have also emerged.

The contribution of this research to the artistic world is, however, on a number of levels, less clear. It is just as important for the dance company as it is for scientists to contribute to a better understanding of neuronal and physiological processes. Then, the book chapter I wrote on the last collaboration with Emio Greco | PC[50] has since been used and much appreciated as a teaching tool for dance students as well as professional dancers. I was keen to hear though what impact my stay with the company and the project eventually had on the company. Hence, at the end of this recent project I presented the research at ICK Amsterdam and asked the dancers whether they could see themselves in the research or whether they have any concerns, questions or other notions they would like to share. There was no response. This does not, however, mean that the research doesn't contribute at all. It may be more a matter of different means of communication. For instance, the research is often confirmatory: on a number of occasions, my approaches overlapped with current issues addressed in the company's performances. For example, in my presentation on how to measure movement, I started with the notion that produced energy in form of heat is a means to measure motion before I presented one of our studies that measured brain activity in response to motion intensity of dance movements. Interestingly, around the same time, the guest choreographer created a performance *Litium*, aimed at how dancers can create energy, how they can light up the theatre with their dancing. And finally, the collaboration has given them some media coverage beside the one they normally would have.

So all in all, interdisciplinary research is good, but it has its limitations and often it is not even very 'successful'.[51] Based on the idea of self-experimentation:

---

49 See Ajemian, R./D'Ausilio, A./Moorman, H./Bizzi, E., "Why professional athletes need a prolonged period of warm-up and other peculiarities of human motor learning", in: *Journal of Motor Behavior* 42.6 (2010): 381–388.

50 Jola, "Movement intention".

51 Strober, Myra H., *Interdisciplinary Conversations: Challenging Habits of Thought*, Stanford 2010.

would it be better to be interdisciplinary within oneself rather than between two people? Clearly, cross-disciplinary research requires experts of both fields. However, two experts alone don't guarantee interdisciplinary exchange. In order to conduct interdisciplinary research, one has to be or become a bit of an expert in all the disciplines involved. For example, I am not a professional dancer who earns her living from dancing or choreographing. Hence I may not be considered a professional. But with degrees in dance (Choreography MA, dance teaching diploma IWANSON, and Dance Culture postgraduate Diploma from the University of Berne) and some work experience in teaching, performing and choreographing, I am not a novice either. I had the rare but powerful position of the subject-matter knowledge of an insider, but the freedom of an outsider in both, the dance world and the neuroscience. The benefit of self-experimentation and doing several parts yourself is that one does continuous multi-factorial monitoring. We are all capable of continuously registering events and sensorial stimuli. And this for example is why I noticed the relevance of the novel theory on 'warm-up' as mentioned above.

Please keep me updated on your work! Scott deLahunta, Senior Research Fellow, Coventry University[52]

Sure. Most importantly, the outcome of this research has taken a slightly new angle on health and well-being which is very different from some of my previous research on spatial cognition and perception, for example. This was not intended at the start of the project. Post-hoc, though, it makes sense that the emphasis of the outcomes have taken on a new route: when conducting embodied neuroscience (what I have called this kind of experiential practice based research), one experiences things that are vital for the actual object of study.

Acknowledgements

Emio Greco | PC, Bertha Bermudez
FAHS University of Surrey
Labex Arts-H2H – Université Paris 8
Prof A Sterr, Dr Lisa Harris, Ari Lingeswaran (scanning UK)
Barbara Menses Gutierrez and dancers of the company
Dr M Raats, Dr M Cropley (general project support)
Dr Jason Frowley (for his comments en an earlier version of this manuscript)
Joe Wainwright (for CEPT test feedback and adjustments)

52 See: http://www.sdela.dds.nl.

References

Ackerley, R./Hassan, E./Curran, A./Wessberg, J./Olausson, H./McGlone, F., “An fMRI study on cortical responses during active self-touch and passive touch from others”, in: *Frontiers in Behavioral Neuroscience* 6 (2012): 51.

Altman, Lawrence K., *Who Goes First? The Story of Self-experimentation in Medicine*, Los Angeles 1998.

Ajemian, R./D’Ausilio, A./Moorman, H./Bizzi, E., “Why professional athletes need a prolonged period of warm-up and other peculiarities of human motor learning”, in: *Journal of Motor Behavior* 42.6 (2010): 381–388.

Astorino, T.A./Willey, J./Kinnahan, J./Larsson,s.M./Welch, H./Dalleck, L.C., “Elucidating determinants of the plateau in oxygen consumption at VO2max”, in: *British Journal for Sports Medicine* 39.9 (2005): 655–660.

Banzett, R.B./Lansing, R.W./Brown, R./Topulos, G.P./Yager, D./Steele, S.M./Londoño, B./Loring, S.H./Reid, M.B./Adams, L. et al., “‘Air hunger’ from increased PCO2 persists after complete neuromuscular block in humans”, in: *Respir Physiol* 81.1 (1990): 1–17.

Bastiaansen, J.A./Thioux, M./Keysers, C., “Evidence for mirror systems in emotions”, in: *Philosophical Transactions of the Royal Society London B: Biological Sciences* 364 (2009): 2391–2404.

Bläsing, B./Calvo-Merino, B./Cross, E.S./Jola, C./Honisch, J./Stevens, C.J., “Neurocognitive control in dance perception and performance”, in: *Acta Psychologica* 139.2 (2012): 300–308.

Budell, L./Jackson, P./Rainville, P., “Brain responses to facial expressions of pain: emotional or motor mirroring?”, in: *NeuroImage* 53.1 (2010): 355–363.

Calvo-Merino, B./Jola, C./Glaser, D.E./Haggard, P., “Towards a sensorimotor aesthetics of performing art”, in: *Consciousness and Cognition* 17 (2008): 911–922.

Carr, L./Iacoboni, M./Dubeau, M.C./Mazziotta, J.C./Lenzi, G.L., “Neural mechanisms of empathy in humans: a relay from neural systems for imitation to limbic areas”, in: *Proceedings of the National Acadamy of Sciences U.S.A.* 100 (2003): 5497–5502.

Cross, E./Kirsch, L./Ticini, L.F./Schütz-Bosbach, S., “The impact of aesthetic evaluation and physical ability on dance perception”, in: *Frontiers in Human Neuroscience* 5 (2011): 102.

Cross, E.S./Hamilton, A.F./Grafton, S.T., “Building a motor simulation de novo: observation of dance by dancers”, in: *NeuroImage* 31.3 (2006): 1257–1267.

De Lahunta, Scott, “The moment to question ... Double Skin/Double Mind”, in: Scott De Lahunta (ed.), *Capturing intention*, Amsterdam 2007, 20–21.

Dendy, Leslie/Boring, Mel, *Guinea Pig Scientists: Bold Self-Experimenters in Science and Medicine*, New York 2005.

Fiks, Arsenij S., *Self-Experimenters: Sources for Study*, edited by Paul A. Buelow, Westport/London 2003.

Franklin, Jon/Sutherland, John, *Guinea Pig Doctors: The Drama of Medical Research Through Self-Experimentation*, New York 1984.

Gazzola, V./Keysers, Chr., “The observation and execution of actions share motor and somatosensory voxels in all tested subjects: single-subject analyses of unsmoothed fMRI data”, in: *Cerebral Cortex* 19 (2009): 1239–1255.

Grèzes, J./Armony, J.L./Rowe, J./Passingham, R.E., "Activations related to 'mirror' and 'canonical' neurones in the human brain: an fMRI study", in: *NeuroImage* 18.4 (2003): 928–937.
Grosbras, M.H./Tan, H./Pollick, F., "Dance and emotion in posterior parietal cortex: a low-frequency rTMS study", in: *Brain Stimulation* 5.2 (2012): 130–136.
Hasson, U./Nir, Y./Levy, I./Fuhrmann, G./Malach, R., "Intersubject synchronization of cortical activity during natural vision", in: *Science* 303 (2004): 1634–1640.
Halmos, P.R./Moise, E.E./Piranian, G., "The problem of learning to teach", in: *The American Mathematical Monthly* 82.5 (1975): 466–476.
Iacoboni, M., "Imitation, empathy, and mirror neurons", in: *Annual Review of Psychology* 60 (2009): 653–670.
Iliffe, Rob, "'That puzleing problem': Isaac Newton and the political physiology of self", in: *Medical History* 39.4 (1995): 433–458.
Jackson, P.L./Meltzoff, A.N./Decety, J., "How do we perceive the pain of others? A window into the neural processes involved in empathy", in: *NeuroImage* 24 (2005): 771–779.
Jang, S.H./Pollick, F., "Experience Influences brain mechanisms of watching dance", in: *Dance Research* 29.2 (2011): 350–375
Johnson-Frey, S.H./Maloof, F.R./Newman-Norlund, R./Farrer, C./Inati, S./Grafton, S.T., "Actions or hand-object interactions? Human inferior frontal cortex and action observation", in: *Neuron* 39.6 (2003): 1053–1058.
Jola, Corinne, "Research and choreography – merging dance and cognitive neuroscience", in: Bettina Bläsing/Martin Puttke/Thomas Schack (ed.), *The Neurocognition of Dance. Mind, Movement and Motor Skills*, Hove 2010, 203–234.
Jola, Corinne, "Movement intention: dialectic of internal and external movements – Reflections from cognitive neuroscience", in: Scott deLahunta (ed.), *Capturing intention*, Amsterdam 2007, 62–67.
Jola, C./Abedian-Amiri, A./Kuppuswamy, A./Pollick, F./Grosbas, M.H., "Motor simulation without motor expertise: enhanced corticospinal excitability in visually experienced dance spectators", in: *PLoS ONE* 7.3 (2012).
Jola, C./McAleer, Ph./Grosbras, M-H./Love, S.A./Morison, G./Pollick, F.E., "Uni- and multisensory brain areas are synchronised across spectators when watching unedited dance", i-Perception 4 (2013): 265–284.
Jola, C./Grosbras, M-H., "In the here and now. Enhanced motor corticospinal excitability in novices when watching live compared to video recorded dance", in: *Cognitive Neuroscience*, 4.2 (2013): 90–98.
Kattenstroth, J.-C./Kolankowska, I./Kalisch, T./Dinse, H.R., "Superior sensory, motor, and cognitive performance in elderly individuals with multi-year dancing activities", in: *Frontiers in Aging Neuroscience* 2.31 (2010): 1–9.
Keysers, C./Wicker, B./Gazzola, V./Anton, J.L./Fogassi, L./Gallese, V., "A touching sight: SII/PV activation during the observation and experience of touch", in: *Neuron* 42 (2012): 335–346.
McGlone, F./Olausson, H./Boyle, J.A./Jones-Gotman, M./Dancer, C./Guest, S./Essick, G., "Touching and feeling: differences in pleasant touch processing between glabrous and hairy skin in humans", in: *European Journal of Neuroscience* 35.11 (2012): 1782–1788.
Nevill, A.M./Stavropoulos-Kalinoglou, A./Metsios, G.S./Koutedakis, Y./Holder, R.L./Kitas, G.D./Mohammed, M.A., "Inverted BMI rather than BMI is a better proxy for percentage of body fat", in: *Annals of Human Biology* 38.6 (2011): 681–684.

Pollick, F./Noble, K./Darshane, N./Murphy, H./Glowinski, D./McAleer, P./Jola, C./Penfield, K./Camurri, A., "Using a novel motion index to study the neural basis of event segmentation", in: *i-Perception* 3.4 (2012): 225.

Preston-Dunlop, Valerie, *Looking at Dances: A Choreological Perspective on Choreography*, London 1998.

Rafferty, S., "Considerations for integrating fitness into dance training", in: *Journal of Dance Medicine and Science* 14.2 (2010): 45–49.

Risko, E.F./Laidlaw, K./Freeth, M./Foulsham, T./Kingstone, A., "Social attention with real versus reel stimuli: toward an empirical approach to concerns about ecological validity", in: *Frontiers in Human Neuroscience* 6 (2012): 143.

Roberts, Seth, "The unreasonable effectiveness of my self-experimentation", in: *Medical Hypotheses* 75.6 (2010): 482–489.

Roberts, Seth, "Self-experimentation as a source of new ideas: Ten examples about sleep, mood, health, and weight", in: *Behavioral And Brain Sciences* 27 (2004): 227–288.

Rossi, S./Hallett, M./Rossini, P.M./Pascual-Leone, A./The Safety of TMS Consensus Group, "Safety, ethical considerations, and application guidelines for the use of transcranial magnetic stimulation in clinical practice and research", in: *Clinical Neurophysiology* 120.12 (2009): 2008–2039.

Sevdalis, V./Keller, P.E., "Captured by motion: Dance, action understanding, and social cognition", in: *Brain Cognition* 77 (2011): 231–236.

Shepard, R. J./Allen, C./Benade, A.J.S./Davies, C.T.M./Di Prampero, P.E./Hedman, R./Merriman, J. E./Myhre, K./Simmons, R., "The Maximum Oxygen Intake", in: *Bulletin de l'Organisation mondiale de lasanté* 38 (1968): 757–764.

Singer, T./Seymour, B./O'Doherty, J./Kaube, H./Dolan, R./Frith, C.D., "Empathy for pain involves the affective but not sensory components of pain", in: *Science* 303 (2004): 1157–1162.

Stevens, C./McKechnie, S., "Thinking in action: thought made visible in contemporary dance.", in: *Cognitive Processing* 6.4 (2005): 243–252.

Strober, Myra H., *Interdisciplinary Conversations: Challenging Habits of Thought*, Stanford 2010.

Sze, J.A./Gyurak, A./Yuan, J.W./Levenson, R.W., "Coherence between emotional experience and physiology: does body awareness training have an impact?", in: *Emotion* 10.6 (2010): 803–814.

Wyon M., "Preparing to perform: periodization and dance", in: *Journal of Dance Medicine and Science* 14.2 (2010): 67–72.

Twitchett, E./Nevill, A./Angioi, M./Koutedakis, Y./Wyon, M., "Development, validity, and reliability of a ballet-specific aerobic fitness test", in: *Journal of Dance Medicine and Science* 15.3 (2011): 123–127.

Van den Stock, J./Tamietto, M./Sorger, B./Pichon, S./Grèzes, J./de Gelder, B., "Cortico-subcortical visual, somatosensory, and motor activations for perceiving dynamic whole-body emotional expressions with and without striate cortex (V1)", in: *Proceedings of the National Academy of Sciences of the United States of America* 108.39 (2011): 16188–16193.

Voisin, J.I./Marcoux, L.A./Canizales, D.L./Mercier, C./Jackson, P.L., "I am touched by your pain: limb-specific modulation of the cortical response to a tactile stimulation during pain observation", in: *Journal of Pain* 12.11 (2001): 1182–1189

Walsh, V./Cowey, A., *Magnetic Stimulation Studies of Visual Cognition*, vol. 2, issue 3,1 (March 1998): 103–110.
Wicker, B./Keysers, C./Plailly, J./Royet, J.P./Gallese, V./Rizzolatti, G., "Both of us disgusted in My Insula: The common neural basis of seeing and feeling disgust", in: *Neuron* 40.3 (2003): 655–664.
Wyon, M./Redding, E./Abt, G./Head, A./Craig, N./Sharp, C., "Development, reliability, and validity of a multistage Dance Specific Aerobic Fitness Test (DAFT)", in: *Journal of Dance Medicine and Science* 7.3 (2003): 80–84.

Dee Reynolds

Empathy, Contagion and Affect

The Role of Kinesthesia in Watching Dance

This paper is informed by research carried out under the aegis of the project entitled *Watching Dance: Kinesthetic Empathy.*[1] The *Watching Dance* team used a range of studies to explore how spectators responded to watching different genres of dance, mainly performed live, but also encompassing a dance film directed by Rachel Davies and performed by Akram Khan.[2] Subsequently to the *Watching Dance* project I carried out audience research on a dance piece choreographed by Rosie Kay, *5 SOLDIERS: The Body is the Frontline*, which was performed on stage and also remade in a film version posted on the internet. While considering kinesthetic empathy in the context of debates in different disciplines alongside the responses of spectators explored in the audience research, I found that the complexity and variety of audience responses involved aspects of affect and contagion as well as empathy. The Davies and Kay films contrast in many respects and I shall explore how the differences in viewers' responses to these films might be linked to relative degrees of empathy, contagion or affect. One advantage of discussing film rather than live performance is that it is more easily accessible; in this case, both films are available online at the time of writing.

In its broadest definition kinesthesia can be understood to refer to sensations of movement and perception of the position of the parts of one's own body. Kinesthesia has a complex relationship with proprioception, which is articulated in diverse ways by scholars coming from different disciplinary and epistemological standpoints.[3] Maxine Sheets-Johnstone has recently argued that "kinesthetic experience is not a matter of sensations, but a matter precisely of

---

1 www.watchingdance.org.

2 Karen Wood, a PhD student on the project, completed a thesis on "Kinesthetic Empathy and Screendance Audiences".

3 See e.g. Gallagher, Shaun, *How the Body Shapes the Mind*, Oxford 2005; Gallagher, Shaun/Zahavi, Dan, *The Phenomenological Mind: An Introduction to Philosophy of Mind and Cognitive Science*, London, New York 2008; Gallagher, Shaun/Sasma, Michal, "Phenomenology and Neurophenomenology: An Interview with Shaun Gallagher", in: *Aluze: Revue proliteraturu, filozofii a jinè* 2 (2003): 92–102 (http://pegasus.cc.ucf.edu/~gallaghr/GallINTERVIEW.htm; 3.8.2012); Noland, Carrie, *Agency and Embodiment: Performing Gestures/Producing Culture*, Cambridge/Mass. 2009; Sheets-Johnstone, Maxine, *The Primacy of Movement*, Amsterdam 2011; Sheets-Johnstone, Maxine, "Movement and Mirror Neurons: A Challenging and Choice Conversation", in: *Phenomenology and the Cognitive Sciences* 11.3 (2012): 385–401.

dynamics. [...] When we move, we feel the dynamics of our movement kinesthetically; we feel the dynamics of an unfolding form."[4] Movement dynamics unfold in time and I shall argue that this temporal dimension is a key aspect of kinesthetic empathy.

There is an extensive literature on the subject of empathy, its history and implications.[5] Kinesthesia was a key aspect of the concept of empathy elaborated by German philosopher Theodor Lipps, who postulated that empathy ("Einfühlung") consisted in a process of "inner mimesis", where the subject internalized the movement of an observed object or another subject. "Einfühlung" was in fact a kind of performance, where contemplation of an external object led the spectator to enact an inner movement evoked by its forms.[6] For instance, a spectator might attribute virtual movement to a vertical line by perceiving it as "rising" or "sinking".[7] Empathy, then, was implicitly kinesthetic, in that it was based on a mode of perception in which visual objects were perceived in terms of movement dynamics.

More recent theories of empathy foreground its function of communicating emotions in interpersonal relations rather than its kinesthetic aspects. Cognitive judgement plays a role in empathically understanding the situation and intentions of others. "Empathy is the ability to perceive and understand other people's emotions and to react appropriately."[8] So-called "perspective taking", an empathic attitude which involves putting oneself in the shoes of another, relies on "a cognitive capacity to take the perspective of the other person".[9] For this reason I adopt the term 'kinesthetic *affect*' rather than 'empathy' when I want to place the emphasis on embodied responses which take the form of an "affective encounter" rather than an "emotional identification" with others.[10]

As with empathy, there is an extensive literature on affect.[11] Following Brian Massumi and others, I do not consider affect to be bound by the "received psy-

---

4 Sheets-Johnstone, "Movement and Mirror Neurons", 395.

5 For a recent example, see Coplan, Amy/Goldie, Peter, *Empathy: Philosophical and Psychological Perspectives*, Oxford 2011.

6 Lipps, Theodor, *Ästhetik: Psychologie des Schönen und der Kunst*, vol. 2: *Die ästhetische Betrachtung und die bildende Kunst* [1906], Leipzig 1920, 97 f.

7 Lipps, Theodor, *Ästhetik: Psychologie des Schönen und der Kunst*, vol. 1: *Grundlegung der Ästhetik* [1903], Leipzig 1923, 226.

8 Leiberg, Silke/Anders, Susanne, "The Multiple Facets of Empathy: A Survey of Theory and Evidence", in: *Progress in Brain Research* 156 (2006):419–440, 419.

9 Blair, Rhonda, "Cognitive Neuroscience and Acting: Imagination, Conceptual Blending, and Empathy", in: *The Drama Review* 53.4 (2009): 92–103, 98 f.

10 Bennett, Jill, *Empathic Vision: Affect, Trauma, and Contemporary Art*, Stanford 2005, 10.

11 See e.g. Blackman, Lisa/Couze, Venn, "Affect", in: *Body and Society* 16.1 (2010): 7–28.

chological categories"[12] which are often applied to emotion by cognitive psychologists.[13] I have elsewhere discussed the implications for empathy of the recent 'affective turn' in the arts and humanities.[14] Affect is pre-cognitive and refers to that point at which the body is activated, 'excited', in the process of responding, but this process has not yet reached consciousness. By contrast with empathic behaviour, to be 'affected' is to be moved in an embodied manner which is not dependent on emotion or cognition. Affect is felt in the body but not yet 'captured' in emotion; it is an embodied imaginative event in which we experience our own body differently through its felt connections across different bodies. In lived experience, affect and empathy are not always distinguishable, as affect may implicitly inform and amplify empathy by intensifying internalised simulation of another's behaviour and contributing to our sense that we are empathically 'entering into' the world of another. With affect, however, the focus is on materiality of sensation rather than interpersonal relations between individual subjects, although it can involve a sense of shared materiality across bodies and the environment.

Empathy is closely related to the phenomenon known as emotional contagion. However, unlike empathy, emotional contagion does not require preservation of the awareness of distinctions between self and other. Philosopher Edith Stein cites as an example of emotional contagion when "a child seeing another crying, cries too".[15] A similar example is used by theatre scholar Bruce McConachie in his book *Engaging Audiences:* "Put two babies in a room together, and if one of them begins crying the second will cry as well, in empathetic response to the first."[16] Stein argues that such behaviour does not constitute empathy, precisely because it collapses the difference between self and other. What happens in emotional contagion is that the other's feelings are transferred to us and we are "saturated" by them, rather than "turning toward or submerging ourselves in the foreign experience, which is the attitude characteristic of empathy".[17]

Rather than a clear-cut distinction, in many cases the difference between empathy and contagion is one of degree rather than kind. In contagion, however,

---

12 Massumi, Brian, *Parables for the Virtual*, London 2002, 28.

13 See e.g. Ortony, Andrew/Turner, Terence J., "What's Basic about Basic Emotions?", in: *Psychological Review* 97.3 (1990): 315–311.

14 Cf. Reynolds, Dee, "Kinesthetic Empathy and the Dance's Body: From Emotion to Affect", in: Dee Reynolds/Matthew Reason (ed.), *Kinesthetic Empathy in Creative and Cultural Practices*, Bristol 2012, 121–136.

15 Stein, Edith, *On the Problem of Empathy* [Halle 1917], The Hague 1970, 22.

16 McConachie, Bruce A., *Engaging Audiences: A Cognitive Approach to Spectating in the Theatre*, New York 2008, 67.

17 Stein, *On the Problem of Empathy*, 22.

the focus is on the embodied impact of the emotion, rather than on interpersonal understanding. Also, it is generally involuntary – we do not necessarily want to 'catch' emotion any more than we want to catch a cold – and it can take us unawares. Similarly to contagion, affect is highly infectious and does not respect individual boundaries. An example of how this works can be found in film scholar Vivian Sobchack's description of watching the opening scene of Jane Campion's *The Piano:*

> At the moment when Baines touches Ada's skin through her stocking, suddenly my skin is both mine and not my own: that is, the 'immediate tactile shock' opens me to the general erotic mattering and diffusion of my flesh, and I feel not only 'my own' body but also Baines's body, Ada's body, and what I have elsewhere called the 'film's body'.[18]

What takes place here is not the empathy of interpersonal relationships, but rather an affective encounter which brings about what Sobchack calls a "carnal subversion of fixed subject positions".[19] In this case, a visual experience becomes affective by being translated into the sense of touch, which is sometimes known as haptic visuality, where space becomes tactile, and "the eyes themselves function like organs of touch".[20] Watching another person or even objects being touched can activate "the brain regions responsible for our own experience of touch"[21] and make us feel as if we ourselves were being touched. The impact of touch can be very intense, and may be associated with pleasure (including erotic pleasure) and pain. Foregrounding of the haptic is often connected with disruption of the objectivising distance which provides favourable conditions for clear vision. When the boundaries between self and other are blurred through the "carnal subversion of fixed subject positions", this intensity can be experienced as generalized and not "belonging" to a single subject. Sobchack argues that this involves a "carnal identification with material subjectivity" which is "prepersonal and globally located".[22]

In the field of dance, kinesthetic empathy and related concepts came to the fore in the late 1920s and the 30s in the writings of John Martin, dance critic for the *New York Times.* Martin was responding to new dance forms being performed at that time, by Martha Graham and Mary Wigman, among others, which were

---

18 Sobchack, Vivian, *Carnal Thoughts: Embodiment and Moving Image Culture*, Berkeley 2004, 66.
19 Ibid., 67.
20 Marks, Laura U., *The Skin of the Film: Intercultural Cinema, Embodiment, and the Senses*, Durham 2000, 162.
21 Keysers, Christian, *The Empathic Brain*, 2011, 2114.
22 Sobchak, *Carnal Thoughts*, 65.

much less dependent on narrative than ballet had been, and therefore required new strategies of engagement from spectators. Martin argued that kinesthesia was a key element of spectator response to the new dance, and that "kinesthetic sympathy"[23] enabled dance spectators to "reproduce (movement) vicariously in our present muscular experience". Not only would this allow spectators to experience movement sensations without actually moving, but it would also awaken "such associational connotations as might have been ours if the original movement had been of our own making",[24] thereby enabling spectators to understand the intentions behind the dancer's movement. Martin also referred to this effect as contagious, arguing that "the inherent contagion of bodily movement [...] makes the onlooker feel sympathetically in his own musculature the exertions he sees in someone else's musculature".[25] Martin's approach, then, encompasses both automatic kinesthetic 'contagion' and interpersonal emotional communication.

The recent revival of interest in Martin's ideas has been influenced by the neuroscience of so-called mirror neurons. These neurons are activated both when one executes an action oneself and when one observes without moving an action being performed by others. They were first discovered in the brains of macaque monkeys observing goal-oriented actions,[26] and there is ongoing discussion about the extent to which this research is applicable to humans. Gallese and others have posited that understanding the intentions of others through observing their actions is grounded in embodied simulation through activation of our motor neurons.[27] This argument is eloquently summarised by philosopher Don Zahavi:

> When we encounter somebody, and observe their actions, or their displayed emotions or sensations, we don't just see them. In addition to the sensory information we receive from the other, internal representations of the body states associated with the other's ac-

23 Empathy and sympathy are sometimes used interchangeably but an accepted semantic distinction is that sympathy involves feelings and empathy understanding (see Oxford English Dictionary).

24 Martin, John, *America Dancing: The Background and Personalities of the Modern Dance* [1936], New York 1968, 121 f.

25 Cited in Foster, Susan, *Choreographing Empathy: Kinesthesia in Performance*, London 2011, 49.

26 Cf. Di Pellegrino, G./Fadiga, Luciano/Fogassi, Leonardo/Gallese, Vittorio/Rizzolatti, Giacomo, "Understanding Motor Events: A Neurophysiological Study", in: *Experimental Brain Research* 91 (1992): 176–180; Gallese, Vittorio/Fadiga, Luciano/Fogassi, Leonardo/Rizzolatti, Giacomo, "Action Recognition in the Premotor Cortex", in: *Brain* 119 (1996): 593–609.

27 Sheets-Johnstone turns this argument around by suggesting that mirror neurons are not foundational but are themselves "contingent on morphology and corporeal-kinetic tactile-kinesthetic experience". It will be very interesting to see how this discussion develops (Sheets-Johnstone, "Movement and Mirror Neurons", 385).

> tions, emotions and sensations are evoked in us, and it is 'as if' we were doing a similar action or experiencing a similar emotion or sensation. It is because of this automatic, non-predicative and non-inferential embodied simulation mechanism, it is because the activation of these neural mechanisms allows us to share actions, intentions, feelings and emotions with others, that we are able to understand others.[28]

There are close parallels between these sorts of claims regarding the function of mirror neurons and Martin's belief that watching dance could lead spectators to simulate dancers' movements by activating their neuromuscular system without actually leaving their seats, and that this simulation could in itself enable them to access the emotions being expressed by the dancer. However, Martin made his claim on the basis that the act of watching can activate neuromuscular memories and thereby "translate" what we see into our own embodied experience. Martin argued that "the sense organs which report movement and postural change are closely connected with that part of the nervous system where emotions are generated".[29] Therefore our movements, or our preparations for movement, can revive "memories of previous experiences over the same neuromuscular paths".[30] Martin has been accused of assuming that "inner mimicry secured the universal capacity of dance to mean the same thing to everyone".[31] Indeed, his use of 'we' is universalising and does not take account of cultural or other diversities. However, although he did not himself address its implications in terms of individual, social and cultural diversity, his argument that the connections between movements and emotions are dependent on past experiences itself indicates that these connections are not universal. Rather than straightforward mirroring, Martin's argument indicates that simulation also involves a process of translating observed movement into neuromuscular memories that are inevitably individually and culturally specific.

Neuroscience accounts of mirroring mechanisms and embodied simulation have also been accused of universalism. On one level this is indeed the case. Freedberg and Gallese, for instance, argue that "embodied mechanisms encompassing the simulation of actions, emotions and corporeal sensation [...] are universal". However, they do not see an opposition between the universality of these

28 Zahavi, Dan, "Empathy and Mirroring: Husserl and Gallese", in: Roland Breeur/Ullrich Melle (ed.), *Life, Subjectivity and Art: Essays in Honor of Rudolf Bernet*, International Centre for Phenomenological Research, Leuven 2011, 217–254, 220.
29 Martin, John, *Introduction to the Dance* [1939], New York 1965, 47.
30 Ibid.
31 Foster, *Choreographing Empathy*, 158.

neural mechanisms and “historical, cultural and other contextual factors”.[32] This is because although mirror mechanisms may be universal, their *activation* is context-specific. “Every instantiation of mirroring is always a process in which others’ behavior is metabolized by and filtered through the observer’s idiosyncratic past experiences, capacities, and mental attitudes.”[33] Similarly to Martin, then, Freedberg and Gallese argue that mirroring is a kind of translation of observed actions into the observer’s embodied response. In his recent book, *The Empathic Brain*, Christian Keysers states that “mirroring” is in fact a misnomer, because “what the mirror system really does is not so much mirror the neural state of whom we observe as *translate* and *reinterpret* what we see into the language of what we would have done or felt in that situation”.[34] Keysers uses “language” in the sense of a neurophysiological encoding of what we observe. This encoding is structured by our own neural systems and knowledge of the world rather than directly reflecting the neural state of others. It *can* reflect the state of others, but the extent to which it does so depends on how similar they are to ourselves. To encode or reinterpret a work of dance through our own neural systems and knowledge of the world is an act of translation, which literally means ‘to carry across’, where we embody the dance by ‘translating’ it into our own embodied experience.

Not only past experiences but also motivations particular to specific tasks and contexts inform our embodied responses to what we see. As physiologist and neuroscientist Alain Berthoz puts it, perception is organized by action, and how and even what we see is shaped by what we are looking for, what we are motivated to find. “In front of a landscape, the oculomotor path taken depends on what a person is seeking.”[35] We select what is useful for us to perceive, and we draw on different senses to acquire relevant information. The multisensory nature of perception was advanced by J.J. Gibson in the 1960s[36] and currently underpins much thinking in physiology and neuroscience. Gibson saw the

32 Freedberg, David/Gallese, Vittorio, “Motion, Emotion and Empathy in Esthetic Experience”, in: *Trends in Cognitive Sciences* 11.5 (2007): 197–203, 197.

33 Gallese, Vittorio, “Empathy, Embodied Simulation, and the Brain: Commentary on Aragno and Zepf/Hartmann”, in: *Journal of the American Psychoanalytic Association* 56.3 (2008): 769–781; cited in Foster, *Choreographing Empathy*, 168.

34 Keysers, *The Empathic Brain*, 2158.

35 Berthoz, Alain, *The Brain’s Sense of Movement*, London 2000, 196.

36 Cf. Gibson, James J., *The Senses Considered as Perceptual Systems*, Westport 1966. “We shall have to conceive the external senses in a new way, as active rather than passive, as systems rather than channels, and as interrelated rather than mutually exclusive. If they function to pick up information, not simply to arouse sensations, this function should be denoted by a different term. They will here be called *perceptual systems*.” (Gibson, cited in Berthoz, *The Brain’s Sense of Movement*, 58).

function of the senses as working together in an interactive system, in which kinesthesia plays an integrating role, in order to provide the organism with useful information on the environment and how to act in relation to it. "Affordance" is Gibson's terms for the degree to which the environment enables ("affords") actions, such as the degree of firmness of the ground which enables humans to walk. Perception can therefore be described as "multisensorial and action-related", and Berthoz points out that "multisensory convergences occur at all levels of the nervous system".[37] The more we are oriented towards an anticipatory, projective mode of perception, the more attentive we will be to the different modalities of the action we are observing – visual, auditory and somatosensory.

The so-called mirror system integrates perception and action and draws on different senses. Simply activating the intention to move through mirroring can itself bring about the neuromuscular changes associated with movement. Anticipation of movement produces "modulation of dynamic properties of the neuromuscular spindles according to the movement required by the task and the context".[38] As spectators of dance, we are plunged into the temporality of the dance's body, drawn into its movement and anticipating where and how it will move next. Not only in our minds but also in our bodies we are already moving ahead, projecting forward, and some choreographers will deliberately play with and disrupt our expectations in this respect, involving our imaginations more actively by challenging our kinesthetic habits. This aspect of projection figures strongly in the mirror neuron activity involving the premotor cortex, which is engaged in movement planning.

Moreover, the mirror system is not limited to vision, but involves different regions depending on context and motivation. Keysers writes:

> Although we have specialized brain areas for viewing the world, namely the visual cortex, when it comes to feeling what goes on in other people, we do not rely on a single, specific brain region. Instead, we seem to recruit those brain regions we would use to experience the same state, be it an action, an emotion or a sensation. The exact brain area activated changes from motor areas for actions, emotional areas for emotions and somatosensory areas for sensations, but the principle remains the same.[39]

For instance, as we watch another body touching or being touched, we can become involved in "sensing how our own body would move and interact with the

37 Berthoz, *The Brain's Sense of Movement*, 59.
38 Ibid., 29.
39 Keysers, *The Empathic Brain*, 2117.

object in the observed way".[40] Such sensing is itself a "simulation of action", involving both the "*motor output* that would be necessary for performing the observed action and simulating the *haptic somatosensory input* that would accompany the performance of such actions".[41]

Like emotional contagion, kinesthetic affects occur directly in the body, without conscious thought. Indeed, as neuroscientist Marc Jeannerod has argued, "we remain unaware of most of our actions, unless an unpredicted event interrupts their course and brings them to consciousness".[42] Also, in the case of affect or contagion, our attention usually focuses on the impact rather than the cause of the response. Writing on audience responses to music, philosopher Robert Davies has argued that "we tend to resonate with the emotional tenor of the music, much as we *catch* the emotional ambience emanating from other people"[43] but that "the listener subject to emotional contagion is usually oblivious to the way the affect is transmitted" (here, the musical stimulus).[44] However, intense affect can provoke our reflexive awareness, which leads us to connect our response with the stimulus that produced it. "The music affects her physiology, which makes her attend more closely to what it expresses and disposes her to account for her own reaction in similar terms."[45] Similarly, Thomas Clifton has argued that the properties we attribute to music in fact reflect our own embodied experience of listening to it. In other words, we attribute to music itself the affects and corporeal intentionality it has produced in us.

> The sound produced by an oboe is usually described as somewhat thin, nasal, rough, and slightly hollow. But this is not altogether accurate. Rather, these words are descriptive of our bodily behaviour: we have adopted an *attitude* of hollowness, thinness etc.[46]

Sometimes the words that describe corporeal attitudes adopted in response to sensory experiences also describe mental states, such as 'restful' (blue sky), 'edgy' (a movie), 'bouncy' (music). Returning to the discussion above concerning

---

40 Keysers, Christian/Kaas, Jon H./Gazzola, Valeria, "Somatosensation in Social Perception", in: *Nature Reviews Neuroscience* 11 (2010): 417–428, 424.
41 Ibid., my emphasis (D.R.).
42 Cited in Sheets-Johnstone, Maxine, "Movement and Mirror Neurons", 389.
43 Davies, Stephen, "Infectious Music: Music-Listener Emotional Contagion", in: Amy Coplan/Peter Goldie (ed.), *Empathy: Philosophical and Psychological Perspectives*, Oxford 2011, 134–148, 134; my emphasis (D.R.).
44 Ibid., 140.
45 Ibid., 143 f.
46 Clifton, Thomas J., *Music as Heard: A Study in Applied Phenomenology*, New Haven 1983, 68. I am grateful to Matthew Reason for drawing my attention to Clifton's book.

mirror neurons and translation, we could say that we 'translate' the music into our bodies in a context-specific way (depending on our own history, circumstances and motivations), in a process that occurs 'without a thought'.[47]

In response to Jeannerod's arguments that we remain unaware of most of our actions (movements), Sheets-Johnstone states that "any time we care to pay attention to our own movement – our own so-called 'action' – there it is".[48] She also states that "when we move, we feel the dynamics of our movement kinesthetically; we *feel* the dynamics of an unfolding form". In practice, however, most people don't pay attention to these dynamics, and our vocabulary for describing and analysing movement qualities is very restricted. Probably the most useful framework is that derived from movement theorist and practitioner Rudolf Laban, who analysed what he called the "effort" qualities of movement in terms of degrees of resistance to weight, space, time and flow.[49] This muscular activity is closely related to what Laban called "inner attitudes" on the part of the mover. Efforts could be both unconscious and involuntary, and their effects could be so complex as to defy description, such as in dance. "Different combinations of efforts create different moods in the dancer, which are analogous to the moods induced by combinations of colour in all their subtle varieties."[50]

Similarly to Davies' argument about listening to music, the dance spectator can 'resonate' with the effort qualities of the dance's body, by which I mean the choreographed body produced by the interactions between dancers, environment and spectator, which is brought into being by the dance.[51] In the case of film (or video), the dance's body is inseparable from the film's body, which Sobchack describes as "the material existence of the film as functionally embodied".[52] The engaged spectator may modify their own effort qualities in response to what they see

47 Sobchack, *Carnal Thoughts*, 65.
48 Sheets-Johnstone, "Movement and mirror neurons", 395.
49 Cf. Reynolds, Dee, *Rhythmic Subjects: Uses of Energy in the Dances of Mary Wigman, Martha Graham and Merce Cunningham*, Alton 2007, 7.
50 Laban, Rudolf, *Modern Educational Dance*, London 1976, 24.
51 Cf. Reynolds, Dee, "Kinesthetic Empathy and the Dance's Body, 121–136.
52 "I use the phrase 'the film's body' [...] to designate the material existence of the film as functionally embodied (and thus differentiated in existence from the filmmaker and spectator). The 'film's body' is not visible in the film except for its intentional agency and diacritical motion. It is not anthropomorphic, but it is also not reducible to the cinematic apparatus (in the same way that we are not reducible to our material physiognomy); it is discovered and located only reflexively as a quasi-subjective and embodied 'eye' that has a discrete – if ordinarily pre-personal and anonymous – existence." (Sobchack, *Carnal Thoughts*, 66). I am not differentiating here between film and video.

without even being aware that they are doing so. This can have physiological effects, as is the case with imagining an action. Carrying an imaginary weight on one's shoulders has been shown to lead in experimental conditions to "changes in cardiac rate and several other vegetative indices, as if the effort really were greater".[53] Moreover, while transmission of affect or emotional contagion can be subliminal, increased intensity and disruption to habitual patterns can trigger self-awareness, causing the viewer to reflect on the changes in their own embodied states and the causes of those changes. The extent to which any of these responses takes place of course depends on a number of factors that include the nature of the dance material and also the disposition of the spectator.

In order to examine aspects of these different kinds of responses I want to discuss audience feedback on the two films referred to above. The first is *5 SOLDIERS: The Body is the Frontline*, a 10-minute film choreographed by Rosie Kay and produced in collaboration with Aquila TV and Meshed Media. The live version premiered as a full-length dance production in the UK in 2010. The film version was launched online in March 2011 and is available in different formats,[54] including a director's cut and tracking shots from headcams worn by each of the five dancers. The second is a 3-minute film, *Loose in Flight* with Akram Khan, directed by Rachel Davies (2000), also adapted from a stage piece, and available online.[55] In discussing these films I am particularly interested in taking account of audience feedback. Rather than *implicitly* adopting a spectator's point of view while referring to a body of texts which provide a meta-language and a set of parameters through which to approach the analysis of a text (such as discourses on empathy or mirror neurons), I believe it is also important to enquire into the opinions of other spectators and to compare them with my own, which are informed by my research agenda. At the same time, it is important to emphasise that this is qualitative research, involving very small numbers, which provide case studies rather than quantitative evidence for any general claims. Also, the way I approach this feedback is informed by my particular interests.

Both films were viewed online. With *5 SOLDIERS*, feedback was by email, following a mailout request (five respondents) and with *Loose in Flight* it was given directly on the website of Watching Dance (thirty-five responses posted). It should be noted that respondents to *Loose in Flight* were not offered any incentive and gave their responses directly on the website without any personal interaction, while respondents to *5 SOLDIERS* were offered theatre vouchers and

53 Berthoz, *The Brain's Sense of Movement*, 32.
54 http://www.5soldiers.co.uk (19.12.2012).
55 http://www.racheldavies.com/tv/looseinflight.html (19.12.2012).

agreed to enter into personal correspondence by email. The *Loose in Flight* film and feedback can be viewed on Watching Dance.[56] Visitors to the website are invited to give feedback on the film, with the following instructions:

> Take a few minutes to watch the following video, Loose in Flight, performed and choreographed by Akram Khan and directed by Rachel Davies. Try to be conscious of how it makes you feel, and post your response below, using the questions as a guide. What was your experience of watching the film? How did it make you feel? Were you aware of any movement sensations while you were watching it? What aspects of the film did you connect to? What effect did the speed of the movement have? What was your response to the change in environment? What was the best moment for you?[57]

Email respondents to the callout for feedback on *5 SOLDIERS* were instructed as follows.

> Please go to the above website for the 5 Soldiers film.
> Watch the Director's cut (11.30 m) and at least one of the headcam films (10 m).
> You can also watch any other films on the site; this is up to you.
> Then please can you answer the questions below.
> Which headcam film(s) did you watch?
> Please give me up to 5 keywords that you associate with your experience of watching '5 Soldiers'.
> Now choose 2 of these keywords and explain a bit more what you mean by them.
> What for you were key moments in the film?
> Do you recall how you felt (emotionally, physically) as you were watching those moments?
> How aware were you of the soundtrack, camerawork and visual effects such as lighting, environment, costumes?
> Is there anything you would change in the piece?

The *5 SOLDIERS* film has a narrative structure and deals with a controversial and sensitive topic – war, with strong overtones of the wars prosecuted by the Labour government in the UK, especially as we hear Tony Blair's name mentioned in the voiceover at the beginning. It is set in an aircraft hangar and follows the story of five "soldiers" (four men, one woman) as they interact with each other while waiting for combat to commence. Some of the action is light-hearted (e.g. when they clown around to a soundtrack by the Black Eyed Peas), while sometimes the atmosphere becomes more tense, for instance when it seems that the male soldiers may sexually assault their female colleague. At the beginning, there is a lot of

56 http://www.watchingdance.org/discussion/discussion_2.php (19.12.2012).
57 These questions were formulated by Karen Wood in consultation with Matthew Reason.

marching and drilling, and close to the end they finally engage with the enemy (who remains invisible), with one of them being shot and wounded.

In the bleak and brightly lit environment of the hangar, there are no objects to 'cling to', nothing to provide stability. The shifting, blurred focus and speed of the camera's movement create a feeling of dizziness and confusion as floor, walls and ceiling whirl around, and in the absence of direct proprioceptive information the viewer doesn't know if the body of the 'I/eye' they inhabit through the camera lens is running, falling, crawling or lying. As stable vision is thereby disrupted we can become more attentive to contact with surfaces, hearing as well as feeling the dancers pull themselves across the floor. At one point the camera plunges our view down and in close to the body of the dancer in front, whose arm is right in front of our face. In this way, we are brought 'inside' the movement. Watching the hard shiny floor we can respond by sensing what it would be like to move across that surface and to impact on it with force, as the dancers do. At the same time, the constant shifting of the headcams foregrounds itself as an active framing which inflects and distorts our field of vision, thereby provoking reflexive awareness of our own perceptual processes.

I had asked respondents to watch at least one of the headcam films as these follow the movements of the dancers' bodies and are frequently out of focus. I was interested to know how people would respond to this and whether disruption of vision through the headcam footage would lead people to attend to other senses, particularly a kinesthetic sense of falling or dizziness, and the sense of touch. To borrow the words of Erin Brannigan, the dancing bodycams manipulate "our foundational, tonic stability", "drawing the audience into a field of movement that has no firm anchor".[58] Also, for me, the effect of being immersed in a field of movement without visual stability and control is that I feel that vision becomes more 'porous' to other senses, particularly touch through contact with the floor in this scene, and also hearing, perhaps because of needing to draw on all available information in order to orientate oneself.

One respondent passed over the question about "soundtrack, camerawork and visual effects such as lighting environment, costumes", and some comments were quite neutral, saying that "they were good and didn't detract from the performance" and "I was aware of different angles used with the camerawork and the lights". However, one said that the camera angles gave her motion sickness, continuing "but apart from that I really enjoyed the film". Another found both camerawork and lighting annoying:

---

58 Brannigan, Erin, *Dancefilm: Choreography and the Moving Image*, Oxford 2011, 111.

> The camerawork was omnipresent – it was a strong mediating presence – I craved a more conventional view by which I could see the dance overall. I realise the point of the headcam must be the dancing camera, maybe the filmed or CCTV experience of war, and hypersubjectivity – as in the film Lebanon? – but it did not let me the viewer see enough overall choreography. The floodlights were interesting but also annoying as they dazzled and further obstructed my view.

This spectator chose 'interference' as one of her key words and expanded:

> A sense of *interference* from the hazy camera shots and bits on the camera obscuring the view. And the interference of the camera movement and the obstructing body parts which sometimes were more in focus than the whole body or group. The handheld camera movements were potentially interesting, but I felt somewhat frustrated by them in the quantities in this piece as I could not get enough of the shots in which I could see or get a sense of involvement with the dance overall.

For this spectator, the relevant information was visual, as she was motivated by a desire to have an overview of the dance, which would have required greater stability and distance, rather than a sense of haptic visuality induced by the camerawork. Interestingly, she picked out as one of her key moments the disco music scene, for the reason that "there was a sort of built in narrative to that". Other viewers also focused strongly on the narrative and representational aspects of the film, judging it positively in terms of how accurate it was felt to be in representational terms and evaluating its depiction of the military situation. For instance, the lighting was judged in terms of its realism, being described as "fairly harsh throughout *which tallies with the subject*".[59] A member of the armed forces who had himself been wounded in combat commented that the film "depicted the emotions of, and relationships between soldiers well". One of his key moments was the Black Eyed Peas scene, which he said reminded him of "the first going out to night clubs with other soldiers and the relationships we had", continuing that it "made me feel happy". His other key moment was the final scene that "reminded me of my own experiences being wounded – made me feel uncomfortable". A female viewer picked out as her key moment "when the female dancer/soldier was pursued by the males" and commented that "the intimidation they created and the negotiation that followed were almost tangible". This made her feel "slightly anxious at first and then relieved".

A striking aspect of these accounts is the degree to which they are focused on empathically understanding the situation and intentions of others. Spectators use cognitive judgements to connect with the dancers/soldiers as 'characters'

59 My emphasis (D.R.).

and to understand their situation. For instance one spectator reflects on the role of discipline for members of the armed forces. "Discipline – key to armed forces. They must develop the response to obey in very difficult circumstances to protect themselves and others." Spectators also focus on connections with their own experience, either direct (as in the case of the soldier) or indirect, through background knowledge or through films (*Lebanon* and *Miss Saigon* were mentioned). Personal memories, experience and knowledge of the context are key factors in empathic response. There is a strong desire to see, and to understand, what is shown, rather than on sensory experience. In the case of the harsh light, for instance, which is evaluated in terms of 'tallying with the subject', its function could also be seen as emphasising the hardness and impenetrability of the floor by reflecting off it. Watching the dancers' bodies coming down on a hard floor we can respond by sensing what it would be like to move in that way, and also what it would be like to *feel* in that way. The spectators' responses articulate empathy much more strongly than affect or contagion. However, even though the headcam effects are sometimes experienced as negative, the disruption they produce leads spectators to focus on their own embodied reactions and how they relate to film techniques, noting their own frustration in not being able to see clearly enough or the motion sickness they feel and relating it to the camera work.

The website comments on the Khan/Davies film are very different from the email responses to the Kay film. The different contexts in which the research was conducted, as described above, make it impossible to directly compare the two sets of responses and any judgements about the reason for the differences can only be speculative. The films themselves are very diverse. Whereas the *5 SOLDIERS* film has five dancers and a narrative framework, and lasts for ten minutes, the online version of *Loose in Flight* lasts for three minutes and is a solo that does not have an obvious theme or narrative. Khan draws on traditional Kathak as well as contemporary movement vocabulary. The film is structured around contrasts between interior and exterior space, being partly shot inside an enclosed space in a disused building and partly outside in a post-industrial landscape, sometimes by a river. The editing is characterised by frequent jump cuts that play with the interaction between different spaces. It is interesting to note that the responses to this film are suggestive of affective encounters involving blurring of self–other boundaries rather than empathy as interpersonal understanding, as in the *5 SOLDIERS* film.

Again, the camerawork and editing have powerful kinesthetic effects, which a few viewers experienced negatively. For one, as in *5 SOLDIERS*, this was connected with the desire for an overall view of the piece.

> As I watched the film, I did not feel kinesthetically connected with the dancer himself. As a dancer, I expected to feel this empathetic sensation in my body, but instead I was consumed by a physical reaction of a different sort. I felt tension in my muscles, most likely initiated by the musical accompaniment, the video editing style, and the sharp intensity of the dancer.

Kinesthetic affect is generated here by the 'body' of the dance film rather than by the dancer as an individual. As noted above, with affect, the focus is on materiality of sensation rather than interpersonal relations between individual subjects. This spectator continues:

> I felt a bit overwhelmed by trying to visually process the quick camera cuts to different angles and various body parts. My mind kept trying to comprehend and visualize the choreography as a whole, so I felt myself getting slightly frustrated as my eyes were guided to various parts of the body instead. I felt this anxiety ease up as the environment transitioned to outdoors, when the viewer was allowed to see a wide angle shot of the dancer's entire body in motion.

As with the *5 SOLDIERS* respondent cited above, this viewer's desire to see clearly is connected with a strong motivation for intellectual comprehension, focused here on the choreography itself. However, many viewers found the camerawork pleasurable and appear to have been motivated more by kinesthetic affect than desire for visual clarity. "Kinesthetically, I was most moved by the editing of the work – though Khan's movements evoked bodily reactions in and of themselves, the quick cuts and sequencing of the various camera angles were extremely powerful." The camerawork also contributes strongly to a blurring of boundaries between the watching self and the dancer, as shown in this viewer's adoption of the movement intention she attributes to the dancer.

> When Akram was confined by four walls that looked to be almost closing in with the camera angles that were used, his movement looked as if he was trying to get out. He looked confined and his focus *gave me a sense of determination to get out* of where he was locked inside. Whenever the scene changed and he was dancing outside, *I felt much freer myself.*[60]

A similar sense of kinesthetic affect is conveyed in the following response:

> I immediately became extremely kinesthetically involved in this work. From the beginning of the film, I felt my eyes darting around and my centre of balance quickly felt off kilter. Through the extreme physicality and fast movements, aligned with the skilful videography, I constantly felt Akram Khan's energy directing mine. As the film progressed, I increasingly

---

60 My emphasis (D.R.).

> began to experience a kinesthetic empathy for Khan. My feet twinged, feeling the floor while watching him pirouette on concrete. I had a similar bodily connection while watching Khan fall. *I continually felt his movement in my own body.*[61]

Another spectator commented:

> I can best describe my experience of watching this film by saying that I was 'drawn forward'. I felt a pull toward the screen, almost to a physical point, where I was literally caught up in the momentum of the movement. I was captivated almost completely from the moment the film began. [...] I tensed and released with the tension in his body, responding as the flow of movement changed, almost wanting to move myself.

Kinesthetic affect is also distributed across the environment. "The environment of the film also enhanced the control over my physical response. I felt inhibited and confined when he was in the enclosed space, drawn forward towards Khan when he was outside, and dizzy with *the quick span of buildings*."[62] As in Davies' and Clifton's comments on how we listen to music, here the kinesthetic quality the viewer attributes to the editing reflects her own embodied experience of it.

Indeed, watching the dance film had an intense physiological impact. "Throughout this film I became more and more kinesthetically engaged and by the end, felt almost nauseous from the experience." The intensity of the experience led this viewer to reflect on it introspectively and analytically, seeking to connect her own responses to the stimuli.

> I am uncertain if my experience was directly attached to the dancer's performance of the movement, the film editing, the sound score, the environment, or the combination of them all, but the experience was dizzying and exhilarating. I often become incredibly physically engaged while watching dance, but this was an extreme level of physical involvement for me while watching a screen dance. I thoroughly enjoyed the work.

The sound of breathing in the soundtrack of the film contributed to the intensity of kinesthetic empathy which connects subjects while still preserving distinctions between self and other. "The sound of breath made me connect more to Khan's movements and feel them in my own body." However, it also contributed to what Sobchack calls "carnal subversion of fixed subject positions". A particularly striking aspect of embodied response, which we have also observed in other in-

---

61 My emphasis (D.R.).
62 My emphasis (D.R.).

stances of the Watching Dance research[63] is what could be called contagious breathing, where the rhythms of the dancers' breathing are transmitted to the spectator. Whereas breathing normally goes unnoticed, changes to one's pattern of breathing make one self-aware. Breathing is intimately connected with both movement and emotion, and such changes are normally motivated by one's own personal situation, such as being in a hurry, or feeling nervous. However, as in emotional contagion, where one is involuntarily affected by the emotion of others, spectators of dance can sometimes find themselves breathing with, not only the dancer's body, but also the dance's body (the body brought into being by the dance), which here includes the filmic medium. Also, I noted above the importance of the temporal dimension of kinesthetic empathy and how anticipation activates preparation for movement, and the effects of suspense in 'Loose in Flight' contributed both to altered breathing patterns and anticipation of moving.

In the following example, the spectator reports that her own breathing is impacted (constrained and then released) by the film's spatial location, camera angles and sound track.

> The breath at the beginning, the circling camera, the quick and repetitive film cuts, all served to magnify the tight, suspenseful atmosphere of the opening. The wide camera angle, the change to more melodic (less 'creepy') music, and the open air setting later in the film heightened the sense of breadth and ease when Khan dances outside. Interestingly, I found myself better able to breathe when watching him outside, than when the sound of breathing was present in the warehouse shots.

Other viewers also associated changes in breathing with the dancer's spatial restriction or liberation. "When the dancer moved outside the small room and his movements got bigger, I was relieved: I felt freer, my body loosened slightly and my breath moved easier"; "The fluidity of his movement contrasted the environment at times and at other times added to the sense of restriction I felt. I still feel I need to breathe out." As these examples suggest, the effect on breathing was sometimes experienced as uncomfortable, and indeed another spectator commented: "The breaths in the audio of the film immediately captured my attention and focus. In fact it made me feel uncomfortable. It made me connect with the entire film and dance in a personal, internal way." However, this discomfort is not seen as negative, on the contrary, she continues: "Very cool effect to achieve without being a live performance." Disruption to breathing causes viewers to re-

63 Cf. Reason, Matthew/Reynolds, Dee, "Kinesthesia, Empathy and Related Pleasures: An Inquiry into Audience Experiences of Watching Dance", in: *Dance Research Journal* 42.2 (2010): 49–75.

flect on the impact of their own affective encounters. "The video was suspenseful and intriguing. One of the feelings I had when watching it was the sensation of my breath being taken away. With the energetic movement enhanced by the actual sound of breath *I literally felt* my lungs pause for a moment."[64] Another viewer writes: "*I found* I was holding my breath, my body was tight and I was leaning toward the screen."[65]

What is striking about many responses to this film and also in some cases to the *5 SOLDIERS* film, is that while some can be described as empathetic in the sense of embodied responses which take the form of an 'emotional identification' with others, involving elements of perspective taking, many more take the form of 'affective encounters' which subvert fixed subject positions, where the corporeal boundaries between the spectator and the film's/dance's body become blurred. As one spectator cited above notes, "It made me connect with *the entire film and dance* in a personal, internal way".[66] Also, as spectators we can respond to the 'body' of the work, not by straightforward 'mirroring' but rather by 'translating' it into our own embodied experience, informed by our own motivation, experience and desires, to see, to touch, to be moved, and in so doing we can feel with, and even breathe with, the dance's 'body'. It may be that spectators motivated by a strong desire for visual distance and cognitive comprehension are less likely to be impacted by affective encounters than those who desire the porosity of vision and its openness to other senses, but this is a topic for further research.

References

Bennett, Jill, *Empathic Vision: Affect, Trauma, and Contemporary Art*, Stanford 2005.

Berthoz, Alain, *The Brain's Sense of Movement*, London 2000.

Blackman, Lisa/Couze, Venn, "Affect", in: *Body and Society* 16.1 (2010): 7–28.

Blair, Rhonda, "Cognitive Neuroscience and Acting: Imagination, Conceptual Blending, and Empathy", in: *The Drama Review* 53.4 (2009): 92–103.

Brannigan, Erin, *Dancefilm: Choreography and the Moving Image*, Oxford 2011.

Clifton, Thomas J., *Music as Heard: A Study in Applied Phenomenology*, New Haven 1983.

Coplan, Amy/Goldie, Peter, *Empathy: Philosophical and Psychological Perspectives*, Oxford 2011.

Davies, Stephen, "Infectious Music: Music-Listener Emotional Contagion", in: Amy Coplan /Peter Goldie (ed.), *Empathy: Philosophical and Psychological Perspectives*, Oxford 2011, 134–148.

64 My emphasis (D.R.).

65 My emphasis (D.R.).

66 My emphasis (D.R.).

Di Pellegrino, G./Fadiga, Luciano/Fogassi, Leonardo/Gallese, Vittorio/Rizzolatti, Giacomo, "Understanding Motor Events: A Neurophysiological Study", in: *Experimental Brain Research* 91 (1992): 176 – 180.

Foster, Susan, *Choreographing Empathy: Kinesthesia in Performance*, London 2011.

Foster, Susan, "Movement's Contagion: The Kinesthetic Impact of Performance", in: Tracy C. Davis (ed.), *The Cambridge Companion to Performance Studies*, Cambridge 2008, 46 – 59.

Freedberg, David/Gallese, Vittorio, "Motion, Emotion and Empathy in Esthetic Experience", in: *Trends in Cognitive Sciences* 11.5 (2007): 197 – 203.

Gallagher, Shaun, *How the Body Shapes the Mind*, Oxford 2005.

Gallagher, Shaun/Zahavi, Dan, *The Phenomenological Mind: An Introduction to Philosophy of Mind and Cognitive Science*, London, New York 2008.

Gallagher, Shaun/Sasma, Michal, "Phenomenology and Neurophenomenology: An Interview with Shaun Gallagher", in: *Aluze: Revue proliteraturu, filozofii a jinè* 2 (2003): 92 – 102 (http://pegasus.cc.ucf.edu/~gallaghr/GallINTERVIEW.htm; 3. 8. 2012).

Gallese, Vittorio, "Empathy, Embodied Simulation, and the Brain: Commentary on Aragno and Zepf/Hartmann", in: *Journal of the American Psychoanalytic Association* 56.3 (2008): 769 – 781.

Gallese, Vittorio/Fadiga, Luciano/Fogassi, Leonardo/Rizzolatti, Giacomo, "Action recognition in the Premotor Cortex", in: *Brain* 119 (1996): 593 – 609.

Gibson, James J., *The Senses Considered as Perceptual Systems*, Westport 1966.

Keysers, Christian, *The Empathic Brain*, 2011.

Keysers, Christian/Kaas, Jon H./Gazzola, Valeria, "Somatosensation in Social Perception", in: *Nature Reviews Neuroscience* 11 (2010): 417 – 428.

Laban, Rudolf, *Modern Educational Dance*, London 1976.

Leiberg, Silke/Anders, Susanne, "The Multiple Facets of Empathy: A Survey of Theory and Evidence", in: *Progress in Brain Research* 156 (2006): 419 – 440.

Lipps, Theodor, *Ästhetik: Psychologie des Schönen und der Kunst*, vol. 2: *Die ästhetische Betrachtung und die bildende Kunst* [1906], Leipzig 1920.

Lipps, Theodor, *Ästhetik: Psychologie des Schönen und der Kunst*, vol. 1: *Grundlegung der Ästhetik* [1903], Leipzig 1923.

Marks, Laura U., *The Skin of the Film: Intercultural Cinema, Embodiment, and the Senses*, Durham 2000.

Martin, John, *America Dancing: The Background and Personalities of the Modern Dance* [1936], New York 1968.

Martin, John, *Introduction to the Dance* [1939], New York 1965.

Massumi, Brian, *Parables for the Virtual*, London 2002.

McConachie, Bruce A., *Engaging Audiences: A Cognitive Approach to Spectating in the Theatre*, New York 2008.

Noland, Carrie, *Agency and Embodiment: Performing Gestures/Producing Culture*, Cambridge/Mass. 2009.

Ortony, Andrew/Turner, Terence J., "What's Basic about Basic Emotions?", in: *Psychological Review* 97.3 (1990): 315 – 311.

Reason, Matthew/Reynolds, Dee, "Kinesthesia, Empathy and Related Pleasures: An Inquiry into Audience Experiences of Watching Dance", in: *Dance Research Journal* 42.2 (2010): 49 – 75.

Reynolds, Dee, “Kinesthetic Empathy and the Dance’s Body: From Emotion to Affect”, in: Dee Reynolds/Matthew Reason (ed.), *Kinesthetic Empathy in Creative and Cultural Practices*, Bristol 2012, 121–136.
Reynolds, Dee, *Rhythmic Subjects: Uses of Energy in the Dances of Mary Wigman, Martha Graham and Merce Cunningham*, Alton 2007.
Sheets-Johnstone, Maxine, “Movement and Mirror Neurons: A Challenging and Choice Conversation”, in: *Phenomenology and the Cognitive Sciences* 11.3 (2012): 385–401.
Sheets-Johnstone, Maxine, *The Primacy of Movement*, Amsterdam 2011.
Sobchack, Vivian, *Carnal Thoughts: Embodiment and Moving Image Culture*, Berkeley 2004.
Stein, Edith, *On the Problem of Empathy* [Halle 1917], The Hague 1970.
Zahavi, Dan, “Empathy and Mirroring: Husserl and Gallese”, in: Roland Breeur/Ullrich Melle (ed.), *Life, Subjectivity and Art: Essays in Honor of Rudolf Bernet*, International Centre for Phenomenological Research, Leuven 2011, 217–254.

III Empathy

Marie-Luise Angerer

Affective Modulations in Politics, Theory and Art

"It seems to me", explained Brian Massumi in an interview 2003, "that alternative political action does not have to fight against the idea that power has become affective, but rather has to learn to function itself on that same level – meet affective modulation with affective modulation."[1] This implies that we should view politics from a more theatrical and aesthetic perspective. Politics has internalized a performative shift that forces it to fight its battles by different means. Key factors here include the ubiquity of media and their recent evolution: dwindling influence of the classical mass media of television and radio; growing strength of communications technologies like the Internet, smart phones, etc.; and the spread of technical agencies operating on an ahuman basis. This is paralleled by shifts in libidinal economy that are rewriting psychic and collective individuation in line with the corresponding economic and technological changes.

From cave painting to cuneiform script, from photography to film and television, and on to computers, the Internet and smart phones, human history must be understood, according to French media theorist Bernard Stiegler, as a history of the terms of the temporality that has shaped humankind, human memory and thus human history: past, present and future. Today, as Stiegler claims in *Philosopher par accident*, we are experiencing an "ecstasy of time",[2] due primarily to the industrialization of the 20th and 21st centuries. As epiphylogenetic beings, humans are defined by their mnemotechnics, and these are currently bringing about processes of temporalization and thus processes of individuation, described by Stiegler as the "industrialization of memory",[3] a "political and industrial economy based on the industrial exploitation of mental time".[4] In this view, consumer bodies as ego masses constitute the global marketplace of an industry that degrades and destroys the consciousness of these bodies, comparable to the wiping out of animal species and the destruction of natural resources. Never before, Stiegler claims, has such an industrial processing of libidinal economy taken

---

1 Zournazi, Mary, "Navigating Movements – With Brian Massumi," in *Hope: New Philosophies for Change*, Annadale 2002, 210 – 242, 234.

2 Stiegler, Bernard, *Philosopher par accident: Entretiens avec Élie During*, Paris 2004. Quotes here translated from the German edition: *Denken bis an die Grenzen der Maschine*, Berlin, Zurich 2009, 69.

3 Ibid., 70.

4 Ibid.

place as that currently underway via user profiling and other viral market strategies. Central to this restructuring of the libidinal economy is the altered position of sexuality, which for more than two centuries functioned as an instrument of biopolitical control, as analyzed by Michel Foucault in his books on sexuality and truth.[5] Today, sexual reproduction has mutated into a global medical business, while sexuality, in the guise of pornography and lust, has become a major market between advertising and media where all that is at stake is bodies and their maintenance. Cosmetic surgery, wellness farms, yoga and fitness, as well as online dating, chat rooms and Facebook all provide clear evidence of this.[6]

Since the 1950s, sexuality has been supplanted in culture and theory by affect. This shift is exemplified by the work of Silvan Tomkins, first presented to a wider audience in 1995 by Eve Kosofsky Sedgwick and Adam Frank under the title *Shame and its Sisters.*[7]

Tomkins developed his work in the 1950s and 1960s. He broke off his own psychoanalytical treatment and was familiar with the ideas of Jacques Lacan. Subsequently, Tomkins developed his affect theory as explicitly distinct from psychoanalysis because he considered a) that the system of drives it worked with was too small to act as an all-encompassing model and b) that it had ignored shame as the key affect. Tomkins defined his affects on the basis of systems theory, which was very popular at the time, beginning with a system of affect pairs: the positive affects are interest/excitement and joy/enjoyment, the neutral affects are surprise/startle, and the negative affects are distress/anguish, fear/terror, anger/rage, disgust/dismell and, above all, shame/humiliation. These affects stand in a dichotomous relationship to one another – depending on its intensity, neural stimulation causes the affective situation to swing into plus or minus. As we will see, this old model[8] of high and low speed as correlates

5 Cf. Foucault, Michel, *The History of Sexuality*, 3 vols., (Paris 1976 – 1984, English translations 1978, 1985, 1986).

6 Cf. Angerer, Marie-Luise/König, Christiane (ed.), *Gender goes Life: Die Herausforderung der Gender Studies durch die Lebenswissenschaften*, Bielefeld 2008.

7 Kosofsky Sedgwick, Eve/Frank, Adam (ed.), *Shame and its Sisters: A Silvan Tomkins Reader*, London 1995.

8 See, for example, Baruch Spinoza who also mentions a correspondence between a lessened ability to act and sadness, and between happiness and heightened activity. He writes: "By emotion [affect] I mean the modifications of the body, whereby the active power of the said body is increased or diminished, aided or constrained, and also the ideas such modifications." (Spinoza, Baruch, *Ethics*, Part 3: *On The Origin and Nature of the Emotions* [1677], New York 1951, 130).

of either good or bad mood is now to be found again in various definitions of affect – as the "missing half second" or as Benjamin Libet's "short delay".[9]

In 1993, Brian Massumi edited a book entitled *The Politics of Everyday Fear,*[10] focussing on an analysis of how fear is organized through mass media, especially in the United States, to legitimate political interventions. Since September 11th, 2001, if not before, the warlike consequences of this phenomenon have shown themselves all too clearly. And each of us knows from personal experience how fear works as a signal:[11] an empty suitcase at the airport or train station triggers uneasiness, we scrutinize passers-by, formulate suspicions, etc.

"Fear (The Spectrum Said)" is the title of another essay by Massumi on fear as a political instrument. In 2002, the Bush administration's Department of Homeland Security introduced a colour-coded terror alert system: green for "low", blue for "guarded", yellow for "elevated", orange for "high" and red for "acute" danger. "The nation has danced ever since between yellow and orange."[12] This example allows Massumi to play the register of affective modulation off against its politico-theoretical implications. Yellow-orange functions as a signal, allowing "central government functioning" to be "wirelessly jacked directly into each individual's nervous system".[13]

Foucault identified the crisis of government – of how people are guided and directed – as early as the late 1970s, anticipating the problematic shift from techniques of discipline to techniques of control long before September 11th, 2001. Whereas the disciplinary system presumes a prescriptive norm, "the starting point for the security technology is the empirically normal, serving as a second-order norm and allowing further differentiations".[14] Rather than distinguishing between the permitted and the forbidden, the "mechanisms of security" define an "optimal medium".[15] Applying Thomas Lemke's analysis of the government of risks – in which he traces the development from eugenics to genetic governmentality – to our subject of affective modulation, one can say that the use of fear as a

9 *Libet's Short Delay,* 2005, http://www.consciousentities.com/libet.htm (17.12.2012).
10 Massumi, Brian (ed.), *The Politics of Everyday Fear,* Minneapolis, London 1993.
11 Freud, Sigmund, "Die Angst" [1916/17], in: *Vorlesungen zur Einführung in die Psychoanalyse: Und Neue Folge,* Studienausgabe, vol. I, Frankfurt/M. 1969, 380–397, 382f.
12 Massumi, Brian, "Fear (The Spectrum Said)", in: *positions* 13.1 (spring 2005): 31–48, 31.
13 Ibid., 32.
14 Bröckling, Ulrich/Krasmann, Susanne/Lemke, Thomas, "From Foucault's Lectures at the Collège de France to Studies of Governmentality: An Introduction", in: Ulrich Bröckling/Susanne Krasmann/Thomas Lemke (ed.), *Governmentality: Current Issues and Future Challenges,* London, New York 2001, 1–33, 4.
15 Ibid.

signal installs an affective conjunctive that means the opening up of a chasm between promise and real diagnosis –affectization as a strategy of power.[16]

The History of the Half-Second

In the mid-1970s, students of media and communication studies in the German-speaking world heard from Hertha Sturm and her team that they had discovered the "missing half-second". Above all, Sturm wanted the results of her research to reach those responsible for making television, so that they could draw the necessary consequences. In her view, television needed to broadcast slower image sequences, audio and video needed to be more congruent; the text or spoken language should follow the images or vice versa, rather than supplying additional information. For as the researchers found, their test subjects (mainly children) were unable to process the excessive amount of information "properly" and their reactions were, quite simply, too slow for the abundance of images. As a result, children reacted "happily" to sad image sequences and "unhappily" to cheerful ones. The test subject's mood was gauged by measuring pulse, heartbeat and transpiration, giving a curve of physical arousal indicating mood – or rather allowing it to be deduced – with low frequency pointing to a depressive basic mood and high frequency pointing to high spirits. Surprisingly, these findings correspond quite clearly with the cybernetic theory of affect developed by Silvan Tomkins who also, as described above, equated lower-level activity with "sad" and higher-level activity with "happy".[17] The reason for the anomalous moods measured, according to Sturm and her team, was the "missing half-second" – an amount of time that passed between perception (signal, stimulus) and reaction without it being clear what occurred during this "lost time".

When her studies on the stressed television viewer[18] were (posthumously) published, however, they received little attention. Such an empirical approach to viewer research was scornfully dismissed (in the German-speaking world) in

---

16 Cf. Lemke, Thomas, "From eugenics to the government of genetic risks", in: Robin Bunton/Alan Petersen (ed.), *Genetic Governance: Health, Risk and Ethics in the Biotech Era*, London, New York 2005, 95–105.

17 Cf. footnote 8.

18 Cf. Sturm, Hertha et al. (ed.), *Wie Kinder mit dem Fernsehen umgehen*, Stuttgart 1979; Sturm, Hertha, *Fernsehdiktate: Die Veränderung von Gedanken und Gefühlen. Ergebnisse und Folgerungen für eine rezepientenorientierte Mediendramaturgie*, Gütersloh 1991; Sturm, Hertha, *Der gestreßte Zuschauer*, Stuttgart 2000.

favour of an ideology-critical, psychoanalytical *Theory of Visual Pleasure.*[19] With hindsight, one can say that Hertha Sturm tried at the wrong time (too early?) and by the wrong means to prove that media such as television have an emotional impact and that this is crucial to the way their verbal and visual content is perceived. What makes this emotional impact so strong, Sturm argues, is a half-second between stimulus and response that makes the (viewer's) response appear somehow "out of synch".

Twenty years later, however, this being "out of synch" makes a comeback in Brian Massumi's cultural theory of affect, contributing to a veritable *affective turn* within cultural studies and media theory. "The skin is faster than the word"[20] wrote Massumi in the mid-1990s, paraphrasing his definition of affect as an intensity belonging to a "different order": "Intensity is embodied in purely autonomic reactions most directly manifested in the skin – at the surface of the body, at its interface with things."[21]

Besides the definition of affect proposed by Gilles Deleuze, which is based essentially on Spinoza and his life force (*conatus*) and which in turn forms the basis for Massumi's work, something else was also at stake here – Massumi actually referred to Hertha Sturm's "missing half-second". For him, however, it became the terrain of affect. According to Massumi, affect is a virtuality which (as a dimension of the potential) facilitates actuality: "[P]astnesses opening onto a future, but with no present to speak of. For the present is lost with the missing half-second, passing too quickly to be perceived, too quickly, actually, to have happened."[22] Unlike Hertha Sturm, Massumi sees the missing half-second not as empty time, but as a space of time in which too much happens to be perceived.

In the mid-1980s, Deleuze's two books on cinema,[23] *The Movement-Image* and *The-Time Image*, initiated a major shift within film theory whose influence extends far beyond the discipline. In Deleuze's theory, perception is the amodal, asubjective part, while memory is a movement which (following Kant) affects itself, performing a kind of self-touching. Image and movement coincide and cannot really be separated. Besides Spinoza, what Deleuze was rediscovering for film and media theory here was above all Henri Bergson's theory of image and

---

19 Cf. Angerer, Marie-Luise, *Body Options: Körper. Bilder. Medien. Spuren*, Vienna 1999, 74–99.
20 Massumi, Brian, "The Autonomy of Affect", in: Paul Patton (ed.), *Deleuze: A Critical Reader*, Cambridge/Mass. 1996, 217–239, 219.
21 Ibid., 218 f.
22 Ibid., 224.
23 Deleuze, Gilles, *The Movement-Image, Cinema 1*, Minneapolis 1991; Deleuze, Gilles, *The Time-Image, Cinema 2*, Minneapolis 1991.

perception, a theory that has attained new importance in the context of recent developments in media technology.

With Bergson, we have arrived in the last years of the 19th century, whose second half was positively obsessed with missing time. In *A Tenth of a Second* (2009), Jimena Canales reconstructs the history of the search for and research into this missing space of time, documenting a huge interest within the disciplines of experimental psychology, astronomy, physics and metrology. Sigmund Freud was taken with it, as was Wilhelm Wundt at his institute of psychology in Leipzig. Others like Frances Galton saw the study of the missing split-second as a continuation of craniometry on a different level: those who react slowly have a sensitive personality, those who react quickly are aggressive, more intelligent. Gradually, this interest in measuring individual reaction times, "personal equating" or "personal error", also began to appear in art, with noteworthy early examples including Marey's chronophotography and Muybridge's proto-cinematography.

As Canales writes:

> The second half of the 19th century was marked by a burst of new research in these topics. [...] Many scientists in France and elsewhere publicised numbers for the speed of nerve transmissions not only in animals, but also in humans. [...] Various instruments came into use: Pouillet's chronoscope; Helmholtz's rotating drums; Arago's chronometers [...]; Donder's noematachometer [...], Marey's drums; [...] In the span of a few years, reaction time experiments shifted from being largely criticized by the scientific community to becoming foundational for a new discipline.[24]

All this began with Hermann von Helmholtz who wrote in 1850: "I have found that a measurable amount of time passes as the stimulus exerted by a momentary electrical current on the lumbar plexus of a frog is propagated to the place where the femoral nerve enters the calf muscle."[25]

Helmholtz was a student of Johannes Müller who, in 1826, formulated the law of specific sensory energy which states that each sensory organ always reacts to stimuli in its own way, whatever their nature. The eye, for example, reacts to mechanical pressure with a sensation of light. From this, Müller concluded that objective reality cannot be recognized, and that perception is something highly subjective, based as it is on and in the body. In his *Techniques of the Observer*,[26] Jonathan Crary accords a prominent place to Müller because he defined the eye

24 Canales, Jimena, *A Tenth of a Second: A History*, Chicago, London 2010, 28.
25 Quoted in: Schmidgen, Henning, *Die Helmholtz-Kurven: Auf der Spur der verlorenen Zeit*, Berlin 2009, 74.
26 Crary, Jonathan, *Techniques of the Observer: On Vision and Modernity in the 19th Century*, Cambridge/Mass. 1992.

and sight as being dependent on physical stimuli, thus, as Crary emphasizes, overturning the hegemony of a neutral visual apparatus.

But what Helmholtz had discovered with his measurements was not only the disappearance of time, but also and above all the delay of energy – the energy in a muscle is not exerted completely at the moment of the stimulus, "but to a large extent only after that stimulus has already ceased".[27] Between stimulation and contraction, then, time (and energy) passes – not much, but enough to be clearly identifiable. The immediacy on which previous assumptions had been based turned out to be "an interval, a period, a space of time both circumscribed and empty – an interim, *du temps perdu*".[28]

Now, the author of *À la recherche du temps perdu*,[29] Marcel Proust, had family ties with Henri Bergson, who was married to a cousin of Proust's. Lacking confidence in language, Bergson is said to have accepted only Proust as a writer, whose search for time went hand in hand with a search for its expression in words. At the height of his career, Bergson fought a never-resolved battle with Einstein on the question of time. The philosopher of *élan vital* and author of *Matter and Memory*[30] never abandoned his position that time is subjective, whereas Einstein famously defined time as independent of individual perceptions.

Henri Bergson understood the world as an image in which we move, ourselves a special kind of image. "There is", he writes, "no perception that is not prolonged into movement."[31] But precisely this moment of not-yet-movement – the interval placed by Bergson between one movement and another – is described by Gilles Deleuze as the moment of affect, and then interpreted by Massumi as the missing half-second:

> Affection is what occupies the interval, what occupies it without filling it in or filling it up. It appears suddenly in the centre of indetermination, that is to say in the subject. [...] There is therefore a relationship between affection and movement in general [...] but it is precisely in affection that the movement ceases.[32]

---

27 Schmidgen, *Die Helmholtz-Kurven*, 93.
28 Ibid.
29 Proust, Marcel, *In Search of Lost Time* (also translated as *Remembrance of Things Past*), 7 vols, originally published: Paris 1913–1927, published in English London 1922–1931.
30 Bergson, Henri, *Matter and Memory* [1896], New York 1991.
31 Ibid., 94.
32 Deleuze, *The Movement Image*, 65.

Affective Media Technologies

Up the present, technical and living processes have developed separately. Until far into the 20th century, life and technology trod separate paths and were also kept separate in the field of theory. But media analyses such as that delivered by Donna Haraway in the early 1980s, which have been developed since by N. Katherine Hayles, Alexander Galloway and Eugene Thacker and others, agree that media can no longer be defined as prostheses which amplify the senses, but that instead, they have attained a new immersive dimension, that they replace our senses, that they also make our senses more intense and more subjective, more intimate and more technical, that perception, memory and affect become a matter of technical modalities. With the cyborg, Haraway introduced a notion intended to render life's reliance on technology conceivable and theoretically graspable. Compared with the period of the "Cyborg Manifesto" in the mid-1980s, the ubiquity of technology has become many times greater: the net has, as Gallagher and Thacker write, become something elementary – an invisible, all-encompassing precondition for societal, social and mental processes.

Neo-cybernetic approaches today revolve around a question already addressed by George Canguilhem in his essay "Machine and Organism", where he advocates an understanding of technology as a universal biological phenomenon. In 1946/47, when Canguilhem was giving his lecture, he concluded by saying that for some years now, tests had been underway – at MIT under the name *bionics* – to research biological models and structures that could be used as models in technology. "Bionics is the extremely subtle art of information," writes Canguilhem, "that has taken a leaf from natural life."[33] Today, media are put on a level with insects, rays, instincts, stimuli and reflexes,[34] theories of imitation from the animal kingdom are transferred to the political and social crowd and swarm formations by humans. Not that comparisons between the animal and human worlds are anything particularly novel; what is new is the fact that today they are meant seriously, that the anthropological supremacy of the human is no longer capable of upholding itself in the current technical-organic overall structure.

When Canguilhem articulated his appeal immediately after World War II, warning against the reductionism of the rapidly expanding hegemony of cybernetics à la Norbert Wiener, it fell on deaf ears, not unlike Hertha Sturm's "miss-

---

33 Canguilhem, George, "Machine and Organism", in: Jonathan Crary/Sanford Kwinter (ed.), *Incorporations*, New York 1992, 45–69, 69.

34 Cf. Parikka, Jussi, *Insect Media: An Archaeology of Animals and Technology*, Minneapolis, London 2010.

ing half-second". Technology and biology, or technology as biology, was not a possible equation, for many reasons. Today, by contrast, one can observe a new liaison resulting from a linking of approaches in biology and information technology, a link established via time, life as time, and an original deferral. In this context, affect can be viewed as an interval that mediates between life and technology, or that facilitates life as technology.

Bodies in Motion[35]

It is no coincidence, then, that media and cultural studies have shown increased interest over recent years in fields and practices that have always dealt with a combination of body and movement, such as dance. As well as this striking interest in dance on the part of media and theory, dancers are looking to theory and technology for fresh stimuli.

One Flat Thing, reproduced by William Forsythe is a dance project which premiered in Frankfurt in 2000 and which provided the basis for a dance notation developed by Forsythe in cooperation with Maria Palazzi and Norah Zuniga Shaw. This notation is now accessible in an online version[36] and it offers interesting insights into the movements of the dancers' bodies, the paths and loops of their communication, their signal structures and gestures, the compression of their movements, and the volume of their bodies in motion. A basic element guiding dance movements consists of the so-called "cues" given by the dancers in order to be able to react to one another, pointers we as observers are usually unable to perceive because they are so minimal and exchanged so

---

35 This chapter is part of a broad-based debate that has also played a key role in discussions of feminism since the 1990s. Starting with Judith Butler's *Gender Trouble* to Elizabeth Grosz's *Volatile Bodies* through to Rosi Braidotti's *Nomadic Subjects*, an unbroken line can be identified that leads to the current dissolution, or rather refusal, of any fixed identity. Parallel to the rejection of signification and thus representation in the case of moving bodies in dance, the gender debate is seeing a rejection of subject and gendered identity. While Butler spoke of *doing gender*, leading to the emergence of a gendered subject, Grosz and Braidotti focus exclusively on the body and its ability always to become something different, capable of more than any culture can permit. Cf. Butler, Judith, *Gender Trouble. Feminism and the Subversion of Identity*, New York, London 1990; Braidotti, Rosi, *Nomadic Subjects: Embodiment and Sexual Difference in Contemporary Feminist Theory*, New York 1994; Grosz, Elizabeth, *Volatile Bodies: Toward A Corporeal Feminism*, Bloomington 1994. Cf. also Angerer, Marie-Luise, "Gender und Performance", in: Emmanuel Alloa/Thomas Bedorf/Christian Grüny/Tobias Nikolaus Klass (ed.), *Leiblichkeit: Geschichte und Aktualität eines Konzepts*, Tübingen 2012, 334–350.
36 http://synchronousobjects.osu.edu/content.html# (17.12.2012).

fast that the dancers' bodies appear to react intuitively – as if in a trance, they capture the movements of the other bodies that control their own.

At the opening of the 2010 International Symposium on Electronic Art[37] in Essen, Brian Massumi delivered a paper on dance, perception, affect and body movement. As part of the symposium, *One Flat Thing, reproduced* was presented in the form of an installation. The paper dealt, among other things, with subliminal perception, direct and non-sensuous feeling, and micro-perception.

These themes refer to the process philosophy of Alfred N. Whitehead, which has acquired a topical significance, especially for Massumi and other media theorists, as a way of theoretically tackling sensations and perceptions without consciousness and subject. Whitehead defines physical perception as always emotional, calling it a "blind emotion" that is "received as felt elsewhere in another occasion".[38] This involves not an accumulation of data but always a data relationship. The perceiving subject does not pre-exist the perceived world, but emerges through and in the process of perception: "feeling is subjectively rooted in the immediacy of the present occasion, it is what the present situation feels for itself, as derived from the past and as merging into the future".[39]

The degree to which the philosophy of Whitehead and Deleuze influences current discussions of body, movement and affect is reflected in Erin Manning's book *Relationscapes*. Manning, who works with Brian Massumi at the *SenseLab* in Montréal and publishes a series entitled *Technologies of Lived Abstractions*,[40] equates seeing with feeling, with feeling understood as movement-with:

> Affect passes directly through the body, coupling with the nervous system, making the interval felt. This feltness is often experienced as a becoming-with. This becoming-with is transformative. It is a force out of which a microperceptual body begins to emerge. This microperceptual body is the body of relation. While affect can never be separated from a body, it never takes hold on an *individual* body. Affect passes through, leaving intensive traces on a *collective* body-becoming. This body-becoming is not necessarily a human body. It is a conglomeration of forces that express a movement-with through which a relational individuation begins to make itself felt.[41]

This passage describes the entire process from perception via affect through to the moving and moved body, also making clear that it is not about individual

37 http://www.isea2010ruhr.org (17.12.2012).

38 Whitehead, Alfred North, *Process and Reality: An Essay in Cosmology*, Cambridge/Mass. 1979, 163.

39 Ibid.

40 Massumi, Brian, *Semblance and Event*, Cambridge/Mass. 2011.

41 Manning, Erin, *Relationscapes: Movement, Art, Philosophy*, Cambridge/Mass., London 2009, 95.

bodies, but bodies with other bodies, and that these must not necessarily be human bodies, or at least not exclusively.

If we now return to our point of departure, where Stiegler describes current technological developments, viewing the libidinal economy purely in terms of its industrial exploitation, then at the end of this paper we have arrived at a similar constellation, except that the developments have been given a positive spin. Whereas Stiegler views the body and the mind as being held in the vicelike grip of a pervasive technology, at the mercy of a negative modulation via affect culminating in a mass conformity of individuals, Manning, with reference to Deleuze and Whitehead, celebrates a body in movement and perpetual mutation whose reactions are controlled via intervals.[42] Here, too, the missing half-second makes an appearance. According to Whitehead, subjectivity takes place in this zone of lost time; life "lurks in the interstices of every living cell, and in the interstices of the brain".[43]

Bergson, too, described the brain as the place where the interval resides. In contrast to the scientific wisdom of his time, he declared the brain a *tabula rasa*, a "centre" or "zone" of "indetermination".[44] The brain is defined as a gap in time, as an "interval of varying length between stimulus and reaction".[45]

Conclusion

To conclude, I would like to quote a short film discussed by Erin Manning in her book as an example, or rather as proof that seeing equals feeling equals movement. In Amanda Baggs' *In my Language* (2007), a woman with her back to us flutters with her hands and sings three notes over and over. In the course of the film, various objects appear with which the woman interacts. The objects, the sound and the woman form a complex whole which is then interpreted in the second part of the film titled *My Translation*. The plane of feeling is transposed into a plane of articulation. In the artist's own words: "A lot of the way I naturally communicate is just through direct response to what is around me in a very physical sort of way. It's dealing with patterns and colors rather than

42 Ibid.
43 Whitehead, *Process and Reality*, 105f.
44 Cf. Schmidgen, Henning, "Leerstellen des Denkens: Die Entdeckung der physiologischen Zeit", in: Bernhard J. Dotzler/Henning Schmidgen (ed.), *Parasiten und Sirenen: Zwischenräume als Orte der materiellen Wissensproduktion*, Bielefeld 2008, 107–124, 108; Bergson, *Matter and Memory*, 23, 28.
45 Schmidgen, "Leerstellen", 109.

with symbolic words."[46] Now, what is important to know about this woman is that she is an autism rights activist. In January 2007 she published the video on YouTube describing her experiences as an autistic person.

However one chooses to interpret this work, it does make one thing very clear: language has become part of bodily expression and – like gestures, facial expressions, voice, etc. – it has no identifiable semantics, instead of which, to use an old-fashioned term, it is free-floating. Or as Manning puts it: it is no longer about signification, but responsivity. "Amanda Baggs feels the world. Watch her reading a book: she touches it, puts her face into it, listens to the pages rustling, smells it, looks at it. Becoming-bodies feel-with the world."[47]

This in turn means that the current interest in "the body in movement" is part of a broader shift outlined in my book *Vom Begehren nach dem Affekt* (*On Desire For/After Affect*),[48] where I argue that the *dispositif* of sex as defined by Foucault in his works changed at the end of the 20th century into a *dispositif* of affect in which the human body is understood as an affective organism, making humankind one species among others. In this process, language as the unique distinguishing feature of the human, as defined by anthropology, philosophy and psychoanalysis, has lost its persuasiveness. Instead, work is underway everywhere to overcome this anthropocentrism founded in language while the notion that matter, the body, and the senses have a will of their own is being pushed to the forefront.[49]

It is no surprise, then, that questions of *kinesthesia and empathy in dance*, as discussed by this conference, are the focus of ever-increasing attention, as such issues are part of a larger debate permitting the integration of the human subject into an ethological and ecological cycle. The focus on affect, as a first step, marks the end of the hegemonic status of sexuality, which, understood from a psychoanalytical perspective, is the decisive difference between humans and animals. Then, as a second step, affect equates animal and human beings – as an example one could quote Haraway's interest in her life with her dog Cayenne.[50] Empathy and isopraxis, bodies against bodies, bodies with bodies as communication between human and animal. Affects, then, in a third step, are understood as tuning the bodies with their "Umwelt", being signals (like fear) to warn the body – too hot, too cold, hungry and thirsty. This is affect as defined by Antonio Damasio.

---

46 Manning, *Relationscapes*, 216.
47 Ibid., 219.
48 Angerer, Marie-Luise, *Vom Begehren nach dem Affekt*, Zurich, Berlin 2007.
49 Cf. Alaimo, Stacey/Hekman, Susan (ed.), *Material Feminisms*, Bloomington 2008.
50 Haraway, Donna J., *When Species Meet*, Minneapolis, London 2008.

If we put these various levels together we get an impression of what makes affects so desirable – as bridging the gap between various species and the world. This is the expectation connected with the broad interest in affects, but, as I hope I have made clear, affects do not bridge anything – instead, they mark the unbridgeable.

Translated by Nicholas Grindell

References

Alaimo, Stacey/Hekman, Susan (ed.), *Material Feminisms*, Bloomington 2008.

Angerer, Marie-Luise, "Gender und Performance", in: Emmanuel Alloa/Thomas Bedorf/Christian Grüny/Tobias Nikolaus Klass (ed.), *Leiblichkeit: Geschichte und Aktualität eines Konzepts*, Tübingen 2012, 334–350.

Angerer, Marie-Luise/König, Christiane (ed.), *Gender goes Life: Die Herausforderung der Gender Studies durch die Lebenswissenschaften*, Bielefeld 2008.

Angerer, Marie-Luise, *Vom Begehren nach dem Affekt*, Zurich, Berlin 2007.

Angerer, Marie-Luise, *Body Options: Körper. Bilder. Medien. Spuren*, Vienna 1999.

Anon., *Libet's Short Delay*, 2005, http://www.consciousentities.com/libet.htm (17.12.2012).

Bergson, Henri, *Matter and Memory* [1896], New York 1991.

Braidotti, Rosi, *Nomadic Subjects: Embodiment and Sexual Difference in Contemporary Feminist Theory*, New York 1994.

Bröckling, Ulrich/Krasmann, Susanne/Lemke, Thomas, "From Foucault's Lectures at the Collège de France to Studies of Governmentality: An Introduction", in: Ulrich Bröckling/Susanne Krasmann/Thomas Lemke (ed.), *Governmentality: Current Issues and Future Challenges*, London 2001, 1–33.

Butler, Judith, *Gender Trouble: Feminism and the Subversion of Identity*, New York, London 1990.

Canales, Jimena, *A Tenth of a Second: A History*, Chicago, London 2010.

Canguilhem, George, "Machine and Organism" [trans. Mark Cohen/Randall Cherry], in: Jonathan Crary/Sanford Kwinter (ed.), *Incorporations*, New York 1992, 45–69.

Crary, Jonathan, *Techniques of the Observer: On Vision and Modernity in the 19th Century*, Cambridge/Mass. 1992.

Damasio, Antonio, *Looking for Spinoza: Joy, Sorrow, and the Feeling Brain*, Boston 2003.

Deleuze, Gilles, *The Movement-Image, Cinema 1*, Minneapolis 1991.

Deleuze, Gilles, *The Time-Image, Cinema 2*, Minneapolis 1991.

Foucault, Michel, *The History of Sexuality, vol. I: An Introduction*, New York 1978.

Foucault, Michel, *The Use of Pleasure: The History of Sexuality, vol. II*, New York 1985.

Foucault, Michel, *The Care of the Self: The History of Sexuality, vol. III*, New York 1986.

Freud, Sigmund, "Die Angst" [1916/17], in: *Vorlesungen zur Einführung in die Psychoanalyse: Und Neue Folge*, Studienausgabe, vol. I, Frankfurt/M. 1969, 380–397.

Grosz, Elizabeth, *Volatile Bodies: Toward A Corporeal Feminism*, Bloomington 1994.

Haraway, Donna J., *When Species Meet*, Minneapolis, London 2008.

Kosofsky Sedgwick, Eve/Frank, Adam (ed.), *Shame and its Sisters: A Silvan Tomkins Reader*, London 1995.
Lemke, Thomas, "From eugenics to the government of genetic risks", in: Robin Bunton/Alan Petersen (ed.), *Genetic Governance: Health, Risk and Ethics in the Biotech Era*, London, New York 2005, 95–105.
Manning, Erin, *Relationscapes: Movement, Art, Philosophy*, Cambridge/Mass., London 2009.
Massumi, Brian, *Semblance and Event*, Cambridge/Mass. 2011.
Massumi, Brian, "Fear (The Spectrum Said)", in: *positions* 13.1 (spring 2005): 31–48.
Massumi, Brian, "The Autonomy of Affect", in: Paul Patton (ed.), *Deleuze: A Critical Reader*, Cambridge/Mass. 1996, 217–239.
Massumi, Brian (ed.), *The Politics of Everyday Fear*, Minneapolis, London 1993.
Parikka, Jussi, *Insect Media: An Archaeology of Animals and Technology*, Minneapolis, London 2010.
Proust, Marcel, *In Search of Lost Time*, 7 vols, London 2002.
Schmidgen, Henning, *Die Helmholtz-Kurven: Auf der Spur der verlorenen Zeit*, Berlin 2009.
Schmidgen, Henning, "Leerstellen des Denkens: Die Entdeckung der physiologischen Zeit", in: Bernhard J. Dotzler/Henning Schmidgen (ed.), *Parasiten und Sirenen: Zwischenräume als Orte der materiellen Wissensproduktion*, Bielefeld 2008, 107–124.
Spinoza, Baruch, *Ethics*, Part 3: *On The Origin and Nature of the Emotions* [1677], New York 1951.
Stiegler, Bernard, *Philosopher par accident: Entretiens avec Élie During*, Paris 2004.
Stiegler, Bernard, *Denken bis an die Grenzen der Maschine*, Berlin, Zürich 2009.
Sturm, Hertha, *Der gestreßte Zuschauer*, Stuttgart 2000.
Sturm, Hertha, *Fernsehdiktate: Die Veränderung von Gedanken und Gefühlen. Ergebnisse und Folgerungen für eine rezepientenorientierte Mediendramaturgie*, Gütersloh 1991.
Sturm, Hertha et al. (ed.), *Wie Kinder mit dem Fernsehen umgehen*, Stuttgart 1979.
Whitehead, Alfred North, *Process and Reality: An Essay in Cosmology*, Cambridge/Mass. 1979.
Zournazi, Mary, "Navigating Movements – With Brian Massumi," in *Hope: New Philosophies for Change*, Annadale 2002, 210–242, 234.

Robin Curtis

Is the Movement of the Filmic Image a Sign of Vitality?

Absence would seem to be a definitive characteristic of the filmic medium. In contrast to theatre, film is defined by the *non-co-presence* of its audience with that which is represented. Film creates a temporal and spatial distance between that which was filmed and the film's eventual viewer with the viewer – on the surface of things – never able to bridge the gap between then and now, between here and there. Much of the film theory of the 1970s and 80s occupied itself, usually with a Lacanian foundation, with the paradox of film's power in the face of this 'lack' at the heart of the filmic image.

However, in the last two decades film scholars have begun to focus on the affective power of the moving image and the real traces these can leave on the viewer. Steven Shaviro, for instance, offered the following assessment in 1993 of that earlier response to film's uncanny power and in so doing marked out a new impulse for more contemporary investigations. He wondered, "But is it really *lack* that makes images so dangerous and disturbing? What these theorists fear[ed] is not the emptiness of the image, but its weird fullness; not its impotence so much as its power."[1] He continued: "The image is not a symptom of lack, but an uncanny, excessive residue of being that subsists when all *should* be lacking. It is not the index of something that is missing, but the insistence of something that refuses to disappear."[2]

In fact, the capacity of film to evoke the sensation of being touched, or the sensation of being moved have each been observed since the beginning of film's history. In 1937 Erwin Panofsky noted the following about these effects:

> In a movie theater [...] the spectator has a fixed seat, but only physically [...]. Aesthetically, he is in permanent motion, as his eye identifies itself with the lens of the camera which permanently shifts in distance and direction. And the space presented to the spectator is as movable as the spectator is himself. Not only do solid bodies move in space, but space itself moves, changing, turning, dissolving and recrystallizing [...].[3]

---

1 Shaviro, Steven, *The Cinematic Body*, Minneapolis 1993, 17.

2 Ibid., 17.

3 Erwin Panofsky, quoted in an abridged form in: Kracauer, Siegfried, *From Caligari to Hitler*, Princeton 1947, 6 (originally published in Panofsky, Erwin, "Style and Medium in the Moving Pictures", in: *transition* 1937, 124 f.).

In such a way as this, since the first films were shown in 1895, viewers have indeed felt that they *were* bridging the gap between then and now, between here and there. We do not merely recognize the representation of movement in the filmic image; we partake of that movement. We do not only react cognitively but also somatically. The filmic image does not simply represent objects in motion and does not merely attempt to evoke emotional response by exhibiting emotional situations. The movement that is a definitive component of the medium invites the viewer to partake of various forms of kinetic involvement.

That this sense of somatic and affective involvement in film is not dependent on involvement with an animate figure will be the topic of this text. Instead I will consider whether the real effects that are left on the body of the film viewer by the film viewing experience – despite the fact that the film does not actually touch that viewer – may come about as a result of an empathetic alignment that is evoked by pure motion within the image. This perspective on the relationship between affect and the effects of film viewing on the body of the viewer raises a number of questions: does this perspective on filmic empathy imply that the perceived movement of the film image is always taken as a sign of the vitality of the film image, and therefore is it this perceived vitality that invites a relationship of empathy? Or alternately can stasis, inertia or the lack of animation be viewed as an invitation to an alternate form of filmic empathy that is not based on a shared vitality? Does the recognition of a lack of animation within the filmic image bid the viewer to enter into a different kind of relationship with the image such as that which was suggested by Theodor Lipps' notion of negative empathy?

My prime objective in this brief discussion of the relationship between movement of the filmic image and its relationship to vitality and empathy is to point out the degree to which the moving image is often underestimated or ill understood in its affective and somatic impact, due to the fact that it is too quickly seen as having an effect analogous to that produced by and as thus comparable to other media. As such, modes of analysis differ too little. This, despite the fact that the reception of images and of texts differ substantially enough to require a corrective approach that may well consider, but must also move beyond an application of questions of narrative structure, character, event or even cultural context to include questions of line, movement, atmosphere or *Stimmung*. Such an approach would allow us to take up affect and empathy not as a direct result of a simple transposition of viewer into the shoes of a character or into a moment in historical time, but rather into an immersive space that may well be suggestive of a form of contemplation rather than distraction.

Much of the work in recent years that has considered the significance of empathy for film viewing has assumed, first and foremost, that processes of narrative and involvement with fictional characters are central features of filmic empathy. Situations, activities and the imaginative projection of possible future events occupy the majority of the viewer's attention in this account. The contributions of the film scholars Murray Smith and Ed S. Tan to this area of inquiry both fit into this category, although I would like to briefly mention the points at which they each offer a tiny window onto the kind of involvement that interests me here. In his book *Engaging Characters: Fiction, Emotion, and the Cinema* Murray Smith takes up Richard Wollheim's approach to imagination from the book, *The Thread of Life*, in which Wollheim distinguishes between "central" and "acentral" imagining.[4] Central imagining can be roughly aligned with the linguistic form, "I imagine", and as such suggests a visceral involvement with the acts or states in question such as "I imagine jumping from the roof" or "I imagine being revolted by the smell of rotten eggs". In each case the act of imagining demands a profound corporeal commitment, a view "from the inside",[5] as Smith terms it. He continues "I do not place myself 'in' the scenario, so much as entertain an idea, but not from the perspective (in any sense of the term) of any character within the scenario"[6] and as such central imagining represents a more generalized form of involvement. Its counterpart, acentral imagining, is aligned instead with the linguistic form "I imagine that" which allows for a more distant perspective on situations and events, one unconnected to a specific character. As such while Smith sees the experience of fiction as alternating between central and acentral moments of imaginative involvement, in his view, only central imagination can bring about empathy, combining "emotional stimulation, motor and affective mimicry and autonomic reactions *like the startle response*"[7] as he describes it here. Both Smith and Ed Tan, another recent commentator on filmic empathy, have returned to the work of Theodor Lipps in order to give con-

4 Wollheim writes the following about these two categories: "[...] the distinction is between iconic mental states that possess a point of view and those which don't: more specifically, between those which possess a point of view internal to that which they represent and those which don't. The point of view is the point of view of one of the dramatis personae: generally, but not necessarily, in the stricter sense of that expression. For it is generally, but not necessarily, the point of view of one of the dramatis personae who are persons. Iconic mental states with such a point of view I shall call centered and those without I shall call *acentered* [...]." And further: "[W]hen I centrally imagine an event, I imagine one person centrally – the person whom I imagine from the inside, the protagonist – and I imagine all the others peripherally: and that, when I acentrally imagine an event, I imagine everyone acentrally." (Wollheim, Richard, *The Thread of Life*, Cambridge 1984, 72, 74).

5 Smith, Murray, *Engaging Characters: Fiction, Emotion, and the Cinema*, Oxford 1996, 76.

6 Ibid., 77.

7 Ibid., 81.

tour to the popular notion of empathy and in particular to flesh out the role of both affective and motor mimicry in empathetic response. In both cases, however, they limit their examination of empathetic response to, as Smith puts it, “a narrative agent” and therefore “a person-like, human figure”.[8] But does the image of a human face or figure always lead to this form of engagement? If a human figure is visible in an image are we as viewers always obliged to only occupy ourselves with projections and hypotheses regarding questions of personality and narrative I would like to introduce some doubt about this clear-cut assessment of moving image involvement by means of an example, which explores the significance of movement for our experience of the animate and highlights alternative forms of engagement than those described thus far.

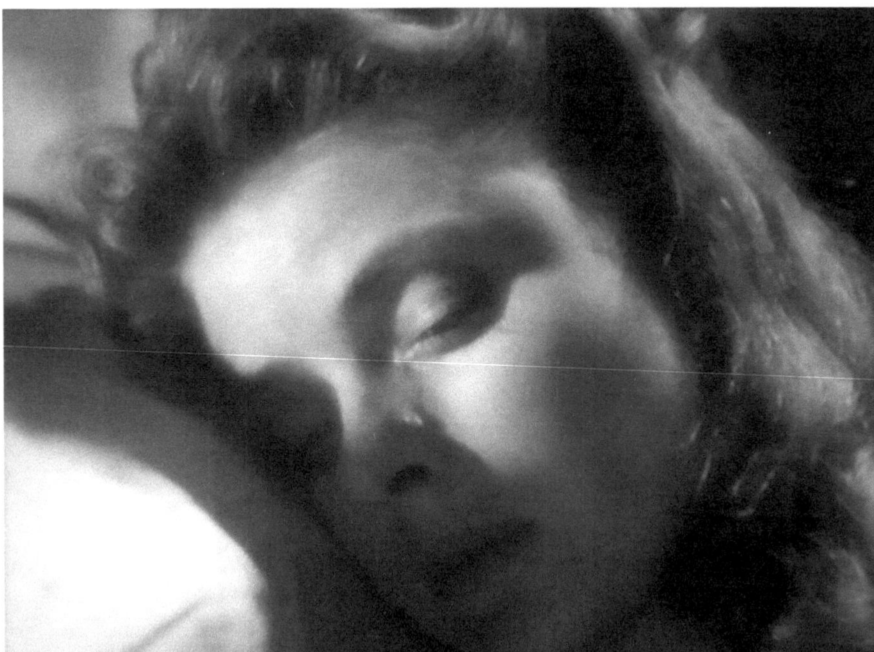

Necrologue by Christoph Girardet and Matthias Müller, Video, 1999.

8 Ibid., 111. It is significant to note that Smith emphasizes that, as a viewer of a narrative film, one is oftentimes occupied with identifying whether or not one is observing a human, non-human or inanimate object.

Necrologue (Christoph Girardet/Matthias Müller, 1999)

The above image hails from the film *Necrologue* and represents the entirety of its visual content, remaining on the screen for all of the film's three minutes and forty seconds. Although the image barely moves throughout this short film, it nonetheless becomes apparent to the viewer that it is a time-based image in a variety of ways that I will soon explore. *Necrologue* is the last film in the series of short films entitled *The Phoenix Tapes*, which were produced for an exhibition on the significance of the work of Alfred Hitchcock for contemporary art at the Museum of Modern Art Oxford. The work in its entirety, a collection of six found footage films, highlights a number of tropes in Hitchcock's oeuvre, among them the undercurrent of disgust and murderous rage throughout his work that is directed at women, which, when removed from the flow of its narrative context, is surprisingly intense and explicit. The image seen here is drawn from the little-known 1949 film *Under Capricorn*, set in 1831, in which Ingrid Bergman, a noblewoman, follows her husband to Australia after he is convicted of murdering her brother. In fact it was she who shot her brother when he threatened to kill her for marrying her working class husband. She remains with her husband, although their love does not survive the seven years he spends at hard labour, and she has become a reclusive alcoholic by the time an old friend arrives in Australia planning to save her by convincing her to leave her husband and thus rehabilitating her in society. She resists this plan. The viewer of the film *Necrologue*, however, might well be in possession of none of this information about *Under Capricorn*, and to my mind it adds only minimally to the *Necrologue*'s aesthetic power.

I suggest instead that there are two alternative forms of aesthetic and emotional involvement that are suggested by this moving image, and while the first may be described by means of a figure-based empathy, the second is more elusive and suggests the relevance of a more abstract form of *Einfühlung* (*pace* Worringer).[9] The other aforementioned film scholar who has concentrated on emo-

---

9 Despite the fact that it was widely criticized by his contemporaries in the field of art history, Wilhelm Worringer's doctoral thesis entitled *Abstraktion und Einfühlung*, published in 1908, has been continually in print since its first publication. It likely can be considered the best-known account of *Einfühlung* in the English-speaking world, since its translation in 1953. In what has come to be seen as one of the first accounts of abstraction in art, a feature which may best account for its lasting popularity, Worringer introduces an opposition between spatial involvement and *Einfühlung* that, despite being a largely unsupported hypothesis in Worringer's text, has had a lasting influence on understandings of *Einfühlung* (although it has been contested in

tional response and empathy in film, Ed S. Tan, differs in his approach in that he additionally considers the effects of what he terms the film's status as an "artefact". Or in other words, while he recognizes that it is one of the key features of what he terms a "traditional feature film" to not draw attention to itself as artefact,[10] and additionally that the focus of his study is that which he terms "the natural viewer of the fiction film",[11] he accepts (but only minimally develops) the idea that there may be a tension between the fictional world of a film and its own status as artefact that may be played out at the level of emotional response. He therefore distinguishes between "F emotions" (played out through the fictional world within the work) and "A emotions" (instilled by means of its particular form as an object made for aesthetic consumption).[12] The latter receive scant attention from him: Tan states, briefly but suggestively, that they "have to do with synthetic proprioceptive activity, such as mirroring a certain type of movement on the screen".[13] Nonetheless, the implications of this form of involvement are broad and belong to the lesser-known aspects of empathetic response as it was originally discussed in a series of debates around the notion of *Einfühlung* – that is, about the potential for proximity between the self and the other, the familiar and the strange – that took place within various fields of inquiry from the end of the Romantic period in Germany peaking in the early 20th century, only to fall off in the 1920s.

Both Herder[14] and Novalis[15] employed the verb "*sich einfühlen*" in the late 18th century to describe, in the former case, the attempt to vividly imagine

---

more recent scholarship, such as Jutta Müller-Tamm's *Abstraktion als Einfühlung: Zur Denkfigur der Projektion in Psychophysiologie, Kulturtheorie, Ästhetik und Literatur der frühen Moderne*, Freiburg/Br. 2005).

10 Tan, Ed S., *Emotion and the Structure of Narrative Film: Film as an Emotion Machine*, trans. Barbara Fasting, Mahwah 1996, 10.

11 Ibid., 11.

12 Ibid., see in particular the section "Fictional World versus Artefact", 64–66, and within the section entitled "An Overview of the Types of Emotion Experienced by the Film Viewer", 81–84.

13 Ibid., 82.

14 For Herder the act of "*sich einfühlen*" (or, literally, feeling into something or someone distinct from oneself) offered the possibility to "awaken" within one's own imagination the affective situation and perception of humans in foreign cultures or past epochs, that one would be able to take up such distant worlds in an appropriately complex manner. Cf. Herder, Johann Gottfried, "Auch eine Philosophie der Geschichte zur Bildung der Menschheit" [1774], in: *Schriften zu Philosophie, Literatur, Kunst und Altertum 1774–1787*, ed. by Jürgen Brummack/Martin Bollacher, Frankfurt/M. 1994, 32f.

15 According to Novalis, the act of "*sich einfühlen*" finally allowed one the opportunity to encounter Nature in the manner to which he and his contemporaries had long aspired: "So wird auch keiner die Natur begreifen, der kein Naturorgan, kein inneres naturerzeugendes und ab-

past eras and foreign culture, and in the latter, the desire to commune with nature in such an all-encompassing manner that a thorough intermingling occurs. However, while Herder's contribution is oftentimes overlooked, in an influential overview text from 1894 entitled "Zur Genesis eines Ästhetischen Begriffs", Theobald Ziegler, who was the first writer to employ the term "*Einfühlungsästhetik*", identified Novalis as the originator of the notion and his account of the term's origins has tended to be replicated since.[16] It was in this period in particular, in the latter third of the 19th century and the first years of the 20th, that *Einfühlung* rose to its greatest prominence, a product of the overlap between philosophical aesthetics and the nascent field of psychology, a dual, indeed even divided, allegiance that still reverberates to some extent within its successor, the term "empathy", today. Indeed it is particularly due to the tensions produced by this overlap that *Einfühlung* has gained something of a new lease on life (particularly in the context of research into mirror neurons)[17] and may offer new perspectives for interdisciplinary research into both affect and the perception of space.

Within the field of psychology the writings of the Munich philosopher and psychologist[18] Theodor Lipps are understood to provide the foundation for contemporary work on empathy (in fact, both Murray Smith and Ed Tan fallaciously

---

sonderndes Werkzeug hat, der nicht wie von selbst, überall die Natur an allem erkennt und unterscheidet und mit angeborener Zeugungslust, in inniger mannigfaltiger Verwandtschaft mit allen Körpern, durch das Medium der Empfindung, sich mit allen Naturwesen vermischt, sich gleichsam in sie hineinfühlt." (Novalis, *Die Lehrlinge zu Sais*, in: *Schriften*, ed. by Ludwig Tieck/Friedrich von Schlegel/Karl Eduard von Bülow, Berlin 1837, 99). *Die Lehrlinge zu Sais*, written between 1798 and 1799, remained unfinished upon Novalis' death and was published posthumously in 1802.

16 Ziegler, Theobald, "Zur Genesis eines ästhetischen Begriffs", in: *Zeitschrift für vergleichende Literaturgeschichte* 7 (1894): 113–120, 116.

17 The initial first publication announcing the discovery of "mirror neurons" can be found in: Rizzolati, Giacomo et al., "Action recognition in the premotor cortex", in: *Brain* (1996): 593–609. The connection between the mirror neuron and empathy (with an explicit link to Theodor Lipps' work) has been promoted by one of the researchers who belonged to the team of researchers in Parma, Italy, that discovered the existence of mirror neurons in macaque monkeys, Vittorio Gallese. See, for instance: Gallese, Vittorio, "The Roots of Empathy: The Shared Manifold Hypothesis and the Neural Basis of Intersubjectivity", in: *Psychopathology* 36 (2003): 171–180.

18 Lipps is referred to as both a philosopher and a psychologist, due to the position he held at the university in Munich, which underwent a transition from inclusion within the field of philosophy to that of psychology. Lipps was known officially as the *Inhaber des Lehrstuhls für systematische Philosophie* (systematic philosophy) and later as *Inhaber des Lehrstuhls für Psychologie und Philosophie* (or for psychology and philosophy) at the Ludwig-Maximilians-Universität München.

cite him as the "originator" of the concept). Despite this honour, the breadth of both his claims and those of the scholars who preceded him in their investigation of empathetic experience is often ignored. Lipps sought to describe an inclination he believed was common to all humans to quite literally *feel into* (i.e. *sich einfühlen*) or empathize with other creatures and object, whether they be animate, inanimate or indeed phenomena such as atmospheres, colours or sounds.

Lipps's famous formula that aesthetic enjoyment should be understood to be objectified self-enjoyment,[19] made this broad application possible. In other words: in the form of things we perceive an analogy to the expressive quality of the vitality of the human body. According to Lipps the basic inclination to this kind of broad engagement with the world (i.e. this kind of *Einfühlung*) is based on an involuntary, instinctive mimicry of others. He describes its effect on the viewers of dancers who themselves begin to sway or rock, or else in what he terms the sympathetic tension, or internal mimicry felt by the observer of a tightrope walker. Although there are many social prohibitions in place that prevent such uninhibited mimicry from being displayed openly, it nonetheless often emerges from the body unbidden and quite unconsciously and as such is experienced, according to Lipps, as a projection into another. He writes:

> Even this, my own striving, I perceive as part of the movement that I am viewing. I experience it as something that belongs to that movement implicitly. I feel myself to be striving with this movement, even striving with the kinaesthetic movement image that corresponds to the visually observable movement and thus feel myself to be striving in effect together with that movement. To put it in more general terms, I feel myself to be striving to complete a movement.[20]

It is this automatic form of sympathetic experience that is at the heart of *Einfühlung*. The *experience* of *vitality* is key here in that Lipps understands every object to be a potential focus for the experience of *Einfühlung:* "thus every object is, for

---

19 Cf. Lipps, Theodor, "Empathy and Aesthetic Pleasure", in: Karl Aschenbrenner/Arnold Isenberg, *Aesthetic Theories: Studies in the Philosophy of Art*, New Jersey 1985, 403–412, 403. First published as: Lipps, Theodor, "Einfühlung und ästhetischer Genuß", in: *Die Zukunft* 54 (January 1906): 100–114. This remains the sole text by Lipps to have been translated into English.

20 Lipps, Theodor, *Grundlegung der Ästhetik*, Hamburg 1903, vol. 1, 120: "Eben dies mein Streben fühle ich in der optisch wahrgenommenen Bewegung. Ich erlebe es als etwas unmittelbar dazu Gehöriges. Ich fühle also mich in dieser Bewegung strebend, nämlich strebend nach dem kinästhetischen Bewegungsbilde, das der optisch wahrgenommenen Bewegung entspricht, und damit zugleich nach dieser letzteren. Ich fühle, allgemeiner gesagt, mich in einem Wahrgenommenen strebend nach Ausführung einer Bewegung." If not otherwise noted, all translations by the autor.

me, an individual, perhaps not as a consequence of a logical perspective but rather as a psychological fact. A multifaceted kind of activity is exuded by the object and as such a unified self is felt to be within that object."[21] This perspective has consequences for the manner in which one perceives the forms of objects. He continues,

> I give expression to this kind of *Einfühlung* in everyday life when I say that the line stretches or bends, surges up and away again, confines itself; and when I say that a rhythm strives or refrains, is full of tension or resolution etc. This is all my own activity, my own vital, internal movement, but one that has been objectified.[22]

Thus it would not be correct to say, with Lipps, that one is concerned with the objective, given world or with the characteristics of the things in it, but rather one is occupied with *the characteristics of one's own embodied engagement with that world* and its things, in short: with the sensations, activities and atmospheric moods that come about through this engagement.

In Lipps's book *Raumästhetik und geometrisch-optische Täuschungen* spatial experience is examined in detail and architecture is introduced as an object of *Einfühlung*. There, Lipps links the Doric column to a striving energy, claiming that,

> the forceful act of composure and erectness demonstrated by the Doric column is pleasing for me, just as it is pleasing to me to compose and straighten myself, an activity that I recollect. [...] I sympathize with the bearing exhibited by the Doric column or with its internal vitality, because I recognize a mode of bearing that is both natural and pleasurable to me. Thus all enjoyment of special forms, and we might add all aesthetic enjoyment in general, is a product of an uplifting, sympathetic sensation.[23]

---

21 Lipps, Theodor, *Leitfaden der Psychologie*, Leipzig 1906, 196: "So ist jedes Ding für mich ein Individuum, nicht in der logischen Betrachtung, aber als psychologische Tatsache. Es ist in ihm das in ein mannigfaltiges Tun auseinandergehende und darin sich zur Einheit zusammenfassende Ich enthalten."

22 Lipps, *Leitfaden der Psychologie*, 196: "Solcher Einfühlung gebe ich schon im gemeinen Leben Ausdruck, indem ich von der Linie selbst sage, sie strecke sich, beige sich, woge auf und ab, begrenze sich; und vom Rhythmus: es sei in ihm ein Fortstreben und Zurückhalten, Spannung und Lösung usw. Dies alles ist meine Tätigkeit, meine lebendige innere Bewegung; aber eben objektiviert."

23 Lipps, Theodor, *Raumästhetik und geometrisch-optische Täuschungen* [1897], Amsterdam 1966, 7: "Das kraftvolle sich Zusammenfassen und Aufrichten der dorischen Säule ist für mich erfreulich, wie das eigene kraftvolle Zusammenfassen und Aufrichten, dessen ich mich erinnere, und wie das kraftvolle Zusammenfassen und Aufrichten, das ich an einem Anderen wahrnehme, mir erfreulich ist. Ich sympathisiere mit dieser Weise der dorischen Säule sich zu verhalten oder eine innere Lebendigkeit zu betätigen, weil ich darin eine naturgemäße und mich beglückende

The Doric column is thus a key example for the positive striving that leads to what Lipps terms "positive" *Einfühlung*, one that is always perceived as a life affirming aesthetic impulse. Life and activity are synonymous for Lipps: "for what I empathize with [or *feel into*] is life in general. And life is energy and internal activity, striving and achieving"[24] The opposite of this is what Lipps termed "negative Einfühlung," a force that is, to his mind barely conceivable as an aesthetic practice and is thus only touched on fleetingly. I however would like to take up this possibility and thus challenge the assumption that *Einfühlung* can only offer a positive, vital and life-affirming association.

Necrologue by Christoph Girardet and Matthias Müller, Video, 1999.

Returning to our filmic example, *Necrologue*, we view a close-up of Ingrid Bergman's face that has been slowed significantly for roughly four minutes; after two minutes the motion of the image is nearly imperceptibly reversed

eigene Verhaltensweise wiedererkenne. So ist alle Freude über räumliche Formen, und wir können hinzufügen alle ästhetische Freude überhaupt, beglückendes Sympathiegefühl."

24 Lipps, "Einfühlung und ästhetischer Genuß", 100: "Denn was ich einfühle, ist ganz allgemein Leben. Und Leben ist Kraft, inneres Arbeiten, Streben und Vollbringen."

and the tear that rolled down her face begins to roll upwards. The excruciatingly slow pace of movement in this film causes one to begin to doubt initially whether the image will move at all; the grain of the emulsion on which the image itself is imprinted, is a more prominent source of movement at the beginning than the figure depicted there is. Indeed it seems at first as if the tear were animated independent of the face on which it rolls. While it is certainly clear that the image of this face is not abstract – obviously one can identify it as belonging to the actress Ingrid Bergman – it is equally clear that the mere identification of the face's owner does not begin to describe the aesthetic operations in place here. The planes of the face and the grain of the emulsion that compose its light and dark sections compete for attention to the extent that the face as a marker of subjectivity recedes.

If one were to take up the model offered by Murray Smith through Richard Wollheim, could one speak here of central or acentral imagining? I would argue that either way, the implicit focus on narrative agent suggested by that model is misplaced here. We are not concerned with events as they can or will unfold or with the protagonists in them. Our involvement is more properly here with an abstract image, with lines and planes and with matter that only minimally suggests subjectivity. It seems to be only obfuscating to insist on the 'event' of the rolling tear being central here and that the face offers any insight into the motivation of the 'event' in question; all of this material, which otherwise would be highlighted in a discussion of filmic empathy and emotional response is – I would argue – peripheral to this film. There is a tension here between movement and stasis, activity and inertness, a beauty that is a product of precisely that tension.

The title of the film, *Necrologue* is a rather seldom used word for an obituary notice that additionally denotes the one who writes them, the documenter of the dead. The figure in this film is not dead, it moves, but the slowing and reversal of movement *does* suggest a change in the status of what we see. Thus the theme of death is in the air here, even if there is no 'event' that explicitly denotes that transition. In an extended discussion of the depiction of death in documentary film, which I admit raises different ethical questions from those presented by fictional representations, but nonetheless can be instructive in this case, Vivian Sobchack has offered some insight into the disquiet inherent in representing the transitional moment, or indeed the intangibility of that moment if it is in fact represented; as an example she takes up the home movie footage shot by chance by Abraham Zapruder on super 8 of Kennedy's assassination. She writes that,

> [t]he transformation of a being into nonbeing, its location at what T.S. Eliot in 'Burnt Norton' describes as the 'still point of the turning world' where being is 'neither flesh nor fleshless' and eludes us, is only perceptible by way of contrast with what is representable. That

> is, death can only be represented in a visible and vigorous contrast between two states of the physical body: the body as *lived body*, intentional and animated – and the body as *corpse*, a thing of flesh unintended, inanimate, static.[25]

However, what we have here in the example offered by *Necrologue* could not really be viewed at any moment as an intentional, animated body, despite the fact that we observe movement; instead our attention is drawn to the fact that the human figure represented on film can take on the quality of mere matter, aesthetic material. Sobchack continues:

> [...] conventionally the corpse signifies, first, the deterioration of materially embodied being into absolute 'thing-ness' and then into absolute 'no-thing-ness' – both experientially unknown (and unknowable) states of 'being.' This is not to say that we do not respond physically, emotionally, and cognitively to the sight of what is believed to be a real corpse on the screen but rather that we respond to it always as other than we are and an object. Indeed the horror of the corpse is precisely that it is not perceived as a subject – even as its objectivity confronts us with and reminds us of the limits and end of human subjectivity.[26]

Would this preclude *Einfühlung?* I challenge Sobchack's either/or approach implicit here that, like Murray Smith's before it, insists on the involvement of a narrative agent for empathy to occur and is therefore confounded (and thus empathetically hindered) by the transition from materially embodied being into "thing-ness".

While this strict division between active agents that invite empathy and matter that cannot, may well be upheld in most contemporary accounts of empathy that seek, first and foremost, to describe the operations of narrative, I ask to what extent did Lipps and his many forerunners in the conceptualization of *Einfühlung*, such as both Herder and Novalis, but also Rudolf Hermann Lotze, Friedrich Theodor and Robert Vischer and August Schmarsow, have a different objective in mind – as do I? The compassion that often a significant component of aesthetic experience may well generally have its roots in the striving and vitality felt within the "positive" form of *Einfühlung*, that was described by Lipps. However what this short film makes accessible to us should more properly be aligned with "negative" *Einfühlung* in that it allows us, despite its depiction as a moving image, to contemplate precisely this shift from embodied being to "thingness" – which is only as unknown and unknowable as nature itself. It is nonetheless a constant source of aesthetic stimulation, one that can too easily be buried and rendered invisible

25 Sobchack, Vivian, "Inscribing Ethical Space: Ten Propositions on Death, Representation and Documentary", in: *Carnal Thoughts: Embodiment and Moving Image Culture*, Berkeley 2004, 236.
26 Ibid.

Necrologue by Christoph Girardet and Matthias Müller, Video, 1999.

by an analysis that focuses on representation and narrative rather than abstraction and the more elusive forms of involvement it engenders.

References

Gallese, Vittorio, "The Roots of Empathy: The Shared Manifold Hypothesis and the Neural Basis of Intersubjectivity", in: *Psychopathology* 36 (2003): 171–180.

Herder, Johann Gottfried, "Auch eine Philosophie der Geschichte zur Bildung der Menschheit" [1774], in: *Schriften zu Philosophie, Literatur, Kunst und Altertum 1774–1787*, ed. by Jürgen Brummack/Martin Bollacher, Frankfurt/M. 1994.

Kracauer, Siegfried, *From Caligari to Hitler*, Princeton 1947.

Lipps, Theodor, *Leitfaden der Psychologie*, Leipzig 1906.

Lipps, Theodor, "Einfühlung und ästhetischer Genuß", in: *Die Zukunft* 54 (January 1906): 100–114.

Lipps, Theodor, "Empathy and Aesthetic Pleasure" [1906], in: Karl Aschenbrenner/Arnold Isenberg (ed.), *Aesthetic Theories: Studies in the Philosophy of Art*, New Jersey 1985, 403–412.

Lipps, Theodor, *Grundlegung der Ästhetik*, Hamburg 1903.

Lipps, Theodor, *Raumästhetik und geometrisch-optische Täuschungen* [1897], Amsterdam 1966.
Müller-Tamm, Jutta, *Abstraktion als Einfühlung: Zur Denkfigur der Projektion in Psychophysiologie Kulturtheorie, Ästhetik und Literatur der frühen Moderne*, Freiburg/Br. 2005.
Novalis, *Die Lehrlinge zu Sais*, in: *Schriften*, ed. by Ludwig Tieck/Friedrich von Schlegel/Karl Eduard von Bülow, Berlin 1837.
Rizzolati, Giacomo et al., "Action recognition in the premotor cortex", in: *Brain* (1996): 593–609.
Shaviro, Steven, *The Cinematic Body*, Minneapolis 1993.
Smith, Murray, *Engaging Characters: Fiction, Emotion and the Cinema*, Oxford 1995.
Sobchack, Vivian, "Inscribing Ethical Space: Ten Propositions on Death, Representation and Documentary", in: *Quarterly Review of Film Studies* 9.4 (Fall 1984): 283–300.
Sobchack, Vivian, *The Address of the Eye: A Phenomenology of Film Experience*, Princeton 1992.
Tan, Ed S., *Emotion and the Structure of Narrative Film: Film as an Emotion Machine*, Mahwah 1996.
Wollheim, Richard, *The Thread of Life*, Cambridge 1984.
Worringer, Wilhelm, *Abstraktion und Einfühlung: Ein Beitrag zur Stilpsychologie* [1908], Munich 2007 (translated as: *Abstraction and Empathy: A Contribution to the Psychology of Style*, trans. Michael Bullock, Chicago 1997).
Ziegler, Theobald, "Zur Genesis eines ästhetischen Begriffs", in: *Zeitschrift für vergleichende Literaturgeschichte* 7 (1894): 113–120.

Ann Cooper Albright

Feeling In and Out

Contact Improvisation and the Politics of Empathy

In an extraordinarily profound and poetic short essay on Rembrandt's paintings, art critic John Berger traces the differences between the artist's drawings and his paintings – particularly the late portraits. Whereas in his drawings Rembrandt is a master of proportion, in his paintings this realistic perspective is radically altered. Berger asks: "Why in his paintings did he forget – or ignore – what he could do with such mastery in his drawings?" Alluding to the historical context of Rembrandt's time, Berger suggests: "He grew old in a climate of economic fanaticism and indifference – not dissimilar to the climate of the period we are living through. The human could no longer simply be copied...the human was no longer self-evident; it had to be found in the darkness."[1] Berger searches for language to address what is not directly visible in Rembrandt's painting, and postulates that "Something else – something antithetical to "real" space must have interested him more."[2] Vital yet elusive, palpable yet not immediately visible, this "something else" present in Rembrandt's work is defined by Berger as a "corporeal space." By distorting a part or parts of the bodies he was painting, Rembrandt was able to give them what Berger calls a "special power of narration." Tellingly, this corporeal space is incompatible with architectural, measured space. It is connected to energy, not geometric lines. Berger writes: "corporeal space is continually changing its measures and focal centres, according to circumstances. It measures by waves, not metres. Hence its necessary dislocations of 'real' space."[3]

In order to give his readers a sense of the different orientations of this corporeal space, Berger charges us to "leave the museum" and go the emergency room of a hospital. It is there, Berger insists, that we will find

> [t]he space of each sentient body's awareness of itself. It is not boundless like subjective space: it is always finally bound by the laws of the body, but its landmarks, its emphasis, its inner proportions are continually changing. Pain sharpens our awareness of such space.

1 Berger, John, "Rembrandt and the Body," in: *The Shape of a Pocket*, New York 2001, 103–112, 105.

2 Ibid., 106f.

3 Ibid., 109.

> It is the space of our first vulnerability and solitude. Also of disease. But is also, potentially, the space of pleasure, well-being and the sensation of being loved.[4]

For Berger, this corporeal space can be felt by touch more clearly than it can be seen by sight, which is why it is the space that nurses occupy more often than doctors. "[O]n each mattress, within each patient, it takes a different form."[5]

I am intrigued by Berger's notion of a corporeal space, one that requires another "way of seeing" to register its potency. In the writing that follows, I want to explore how this space prioritizes touch and "feeling" rather than seeing, shifting the traditional subject/object dynamic of these exchanges. Of course, I am writing not only about the social and political relationship between painter and model, or even that of an art critic and the work of art, but also of the relationship between one's self and an "other." I will argue that by attending to the *practice* of feeling rather than its *affects*, Contact Improvisation can help us revise Western notions of empathy that are based on a psychological conception of the individual *subject* and an *object* of sympathy.

In English, *feeling* is both a noun and a verb form. Its many definitions span the gamut from the strictly material – such as to finger, palpitate, or touch something – to the highly cerebral. It can be used to describe a physical sensation (I feel something sticky), an intellectual perception (I have a feeling that...), or an emotive state (feeling blue). *Feeling* can refer to both the surface of the body and the interior self. *Feelings*, of course, are closely linked to empathy, and nowhere is this more obvious than in the German term *Einfühlung*, which can be translated as *feeling in* or *feeling into*. As Susan Foster outlines in her recent genealogy of empathy, this term was originally coined in 1873 by German aesthetician Robert Vischer and subsequently translated into English as "empathy."[6] In its late 19th-century German context, *feeling into* (or empathy) was primarily used to describe the experience of contemplating, moving into and merging with a work of art, something that John Berger does very well in his perceptive writing. In an early 21st-century context, however, empathy usually refers to the experience of relating to someone else's circumstances, and constitutes the stuff of daytime talk shows à la Oprah Winfrey. As *feeling* moves from a verb to a noun, from the physical sensing of touch to a projected image of another's experience, it can take on the colonial baggage of sympathy and the psychic mantle of emotion.

---

4 Ibid., 107.

5 Ibid.

6 Foster, Susan, *Choreographing Empathy*, London, New York 2011, 127.

But what if we were to refuse this stabilizing of a verb into a noun – of an active experience into a passive object? What if we kept *feeling* at the surface of the body, rather than letting it sink into what Foster describes as the late 19th century's "newly constructed interiority whose proclivities for repression, identification, transference, and sublimation were just beginning to be explored and whose defining consciousness could be fathomed only through intensive introspection"?[7] What if we approached *Einfühlung*, or feeling into, as a kinesthetic practice rather than a psychological state? By holding our attention to the physical, I am not trying to suggest that this realm is any more authentic, natural, "real," or less culturally grounded than the psychological. On the contrary, I am quite interested in foregrounding the socio-political moorings of corporeal training. But it is crucial for us to recognize just how quickly and easily we tend to elide feeling with emotions, setting up a subject position based on possession (I have emotions) rather than one based in sensation (I am feeling).

Picture by John Seyfried, taken at a Jam at Oberlin College.

Contact Improvisation has been around for almost four decades and I have been involved with the form for three of those decades. I have participated in and taught workshops in many different communities all over the world. Although

7 Foster, *Choreographing Empathy*, 154.

the form has changed and moved as it adapts to different historical circumstances and geographic locations, there are some fundamental elements that comprise the core of the physical training, no matter whether it is conducted in German or English, Mandarin or Tamil. One of these is a focused attention to sensation at the level of the skin.

As we all know, skin is one of the largest and most sensitive of our organs. It covers our entire bodies and it is impossible to exist in the world without one's skin. Ironically, however, many people go through their everyday lives with little awareness of their skin as a perceptual faculty. This is because our current post-industrial culture reifies the visual almost to the exclusion of our other senses, including those of sound and smell. Most of us use sight to navigate the world – off-line as well as on-line. Generally speaking in the West, seeing is believing, and feeling is suspect. We tend to become aware of our skin only in extreme situations such as fear (the skin crawling up the back of my neck), awe (it gave me goose bumps), or pleasure (the tingling sensation of a lover's caress). Much of the foundational training in Contact Improvisation attempts to reverse this cultural hierarchy by reducing our dependency on the visual and bringing awareness to the nuances of the tactile. In Contact, one's skin becomes a primary site of communication.

Picture by John Seyfried, taken at a Jam at Oberlin College.

The first step in this process of retraining our corporeal habitus is to release the tension that is a direct result of what I call a territorial approach to the body's integrity. We can conceive of our skin as either a boundary or a conduit and this shift in perception leads to a radically different understanding of the relationship between myself and the world. If my skin is seen as a barrier to disease, infection, or any kind of "otherness," I might well approach life with a certain cold war mentality, shoring up any breeches in my defense system and using my skin as a wall or a container meant to keep me safe from the outside world. If, on the other hand, I experience my skin as the porous interface between myself and the world, then I will be more apt to engage my skin as a permeable, sensitive layer that facilitates that exchange. As Corey Spiro, one of my students in a recent Contact class, suggests:

> I feel as though we live in a world where the boundary between self and "other" is constantly being defined, labeled, and monitored. This is especially apparent in our perceptions of the ownership of space. MY PROPERTY, MY ROOM, etc. Nowhere is this line more clearly drawn than at our skin [...]. It's all too easy to convince oneself that the skin represents the ultimate energetic boundary between self and other. Of course, this barrier works both ways, just as it stops the world from coming into us, it similarly prevents our conception of self from expanding beyond the limits of our physical bodies.
>
> I would expect then, that opening the pores of my skin wide enough to let the world in would be a frightening experience. Rather than an upsetting intrusion, however, I was surprised to find out that it was actually extremely refreshing. My energy in class was perhaps lagging a little bit today, but I felt that by opening myself up I was able to simultaneously expand outward into the energy of Wild Main Space and also feel more acutely the electromagnetic fields of everyone else standing around me. In short, opening my pores did more than just "let the world in," it also let me out. The feeling was one of freedom and relief, as I was no longer alone within the prison-like confines of my injured and fatigued frame.[8]

In this dialogue between the self and the world, one becomes aware of the intriguing possibilities of interdependence, including a deeper sense of responsibility. I think of responsibility not as an oppressive duty towards others, but rather as an ability to respond, an ability to be present with the world and as a way of being present with oneself. This is the fruit of kinesthetic attention, a physical mindfulness that prepares one for improvisation. It is also a kind of somatic engagement which leads to a profound psychic reorganization as well. If the world is already inside one's body, then the separation between internal and external – self and other – is much less distinct. The skin is no longer the boundary between the

8 Spiro, Corey, "Journal Entry," in Ann Cooper Albright (ed.), *Encounters with Contact: Dancing Contact Improvisation in College*, Oberlin/Ohio, Oberlin Theater and Dance Program (distributed by Contact Quarterly), 2010, 40.

Picture by John Seyfried, taken at a Jam at Oberlin College.

world and myself, but rather the sensing organ, which brings the world into my awareness. Given the anxiety swirling around boundaries and bodies in contemporary society, however, this latter sensibility requires a bit of practice.

One of the earliest exercises that I give in my improvisation classes is referred to as "the small dance" or "the stand." First developed by Steve Paxton in the early seventies as he explored the physical skills that would lead towards defining the form of Contact Improvisation, the stand allows one to focus on the internal movements created by the shifts of bones, muscles and breath required to stand "still." After they have been warming up, moving through the space for awhile with big, vigorous movements, I ask the students to chose a spot and stand in a relaxed, but active manner. Engaging one's peripheral vision is crucial to this process, and I tell the dancers to try and release the fronts of their eyes, allowing images and colors to come into their head instead of straining their eyes in order to go out and grab the visual image. Often, I will call their attention to the sensation of the moisture on their skin, asking them to feel the difference between air and clothing. Next, I ask them to concentrate on opening the pores of their skin so that it becomes like a screen window, allowing air, smells and sounds to come in from the outside. I ask them to try to breathe through the pores of their skin. Only once they sense the responsiveness of their own skin, are my students ready to work with a partner and feel their weight shifting back and forth between two people. I emphasize the homonymic connections between pore (of the skin) and pour (as in pouring water from a pitcher), asking the students to reflect in writing on what it feels like to open the pores of your skin wide enough to let the world pour in. Here is how Isabel Roth, another of my recent Contact students, responds to this physical practice:

> I think the idea of opening pores as being similar to the idea of opening your mind. It's not as if you can actively think to open pores and actually feel the individual pores opening. But it is a palpable feeling of release, of spreading and opening your skin to the physical space and people around you [...]. Just like opening the pores of the skin allows you to be ready to receive, it also makes you ready to give. Skin is such a pliable and ever-flexible organ, constantly shifting and regenerating, depending on movement and contact. By opening the pores you prepare the skin for contact and for the willingness to open up to another's touch. Now ready to accept that touch, it is easier to reciprocate pouring weight from open pores to a partner.[9]

As you may have noticed, each of my students' responses uses feeling as both noun and verb – an active state of sensing and also a reflection of that experience. These two meanings of the word resonate with one another, vibrating in an ambiguous space between a subject (who feels) and an object (of feeling). Reading the students' descriptions of their experience, I am reminded of Berger's sense that corporeal space is measured in "waves, not metres," and is predicated on touch, not sight. The somatic state of responsiveness that these students articulate is crucial in preparing the body to enter safely into a Contact duet. But before I move into an analysis of the physical dimensions of touch and sharing weight, I want to look at two different ways of thinking about empathy by making a distinction between introspection and interoception.

---

9 Spiro, "Journal Entry," 38.

Etymologically, introspection means to look into one's self, which is usually specified as one's own mind or feelings. This interior space is the site of empathy, envisioned as contained within one's self until it is drawn out by the object of one's gaze, sympathy, or even pity. As Foster demonstrates in her study cited earlier, introspection is implicated in the scopic economy of the 19th-century self. Interoception, on the other hand, replaces the visual emphasis (spect) with the more tactile sensibility of cept. Used mostly in neuropsychology, the term "interoception" references one's ability to feel sensations arising from within the body, specifically one's visceral organs, giving us the term "gut feelings." Advances in brain imaging over the last decade have helped scientists locate interoception in the right frontal insula, a part of the brain also identified with emotional intelligence. It could be easy to collapse these two terms into an overall feeling of empathy. But as any Zen master will tell you, feeling does not necessarily have to evolve into emotion. In fact, I want to suggest that the physical mind of interoception can produce an entirely different kind of empathetic exchange, one that stays with *feeling* without getting stuck in the emotional baggage of *feelings*.

Once my students are comfortable with opening the pores of their skin, we begin the infinitely interesting process of learning to pour our weight, like water, into one another's bodies. Starting with two hands, one partner will firmly, yet openly, touch another person on the back or shoulder, kinesthetically "asking" their partner to pour their weight into the receptacle of their hands. The asking partner can regulate how much weight is given by resisting and pouring back even as they accept the responsibility for the other person's weight. This mutual pouring creates an energetic dialogue that continuously loops between the partners. Eventually, the partners begin to pour their weight back and forth, using different body parts as their physical contact revolves around the space and across their bodies. As the dancers gain fluidity in the giving and receiving of weight, the dancing tends to speed up. This is the moment when the responsiveness of one's body is critical. There is no time for the lengthy processing of emotions here; one has to focus entirely on keeping up with the point of contact.

This point of connection is sometimes referred to in Contact parlance as the "third mind." Allowing their dancing to be led by this "third mind," the two partners endeavor to follow its spatial and rhythmic journey throughout the studio space. At first it may seem clear which partner is leading and which one is following, but eventually those roles evolve into such a fluid and subtle exchange that the categories of leader and follower lose their oppositional moorings. This does not mean, however, that all difference is collapsed. For me, this 'third mind" marks an intersubjective space in which one is aware of sensations both internal and external without necessarily categorizing those feelings into socially recognizable roles. The notion of a "third mind" directs attention away from the oppositional poles of self and other, stretching a single line into a more open field of play. Contact trains for a physical interconnectedness that is akin to what Deirdre Sklar calls "empathic kinesthetic perception."

> Emphatic kinesthetic perception suggests a combination of mimesis and empathy. [...] Whereas visual perception implies an "object" to be perceived from a distance with the eyes alone, empathic kinesthetic perception implies a bridging between subjectivities. This kind of "connected knowing" produces a very intimate kind of knowledge, a taste of those ineffable movement experiences that can't be easily put into words. Paradoxically, as feminist psychologist Judith Jordan points out, the kind of temporary joining that occurs in empathy produces not a blurry merger but an *articulated perception of differences.*[10]

It is this "articulated perception of differences" that I want to focus on in these last few pages. When I am teaching Contact and I use terms such as "interconnected," "feeling one's partner's experience," or "moving together," I emphasize that this "going with the flow" does not mean one becomes a neutral container, nor does it suggest a "blurry merger" of energies such that the dancing homogenizes into one long fluid chain of rolls and lifts. Quite the contrary. The sensitivity to another's experience also creates an awareness of subtle differences, differences that can be celebrated within the improvisation. While I do not have time to fully engage with Merleau-Ponty's ideas about intersubjectivity and touch in this context, I do think it is important to point out that in French the verbs for "touch" and "feel" are both transitive and reflexive verb forms. That is to say that one feels an "other" at the same time that one feels oneself feeling. Similarly one can touch something and feel oneself being touched at the same time (such as Merleau-Ponty's famous example of one hand holding the other). This looping across to another and then back to oneself intrigues me, for it loosens up the psychological patterns of always already relating to an "other" as an object (of empathy, scrutiny, or desire...). This play of difference can be accentuated in another dance score, which I give to my students. Here are my instructions:

> This is a duet, not an exercise. A dance, not an activity. To begin, one person lies down, completely passive, allowing their weight to sink fully into the floor. Their partner begins to move their body with attention to giving the passive person an experience of the weight of their bones and the mobility of their joints. As any one who has ever done any kind of body work or physical therapy knows, a passive body allows one to feel sensations unavailable to a body that is self engaged, even the most released one. Focusing on their breaths, the partners establish a vibration of energetic exchange. Bit by bit, percentage point by per-

10 Sklar, Deidre, "Five Premises for a Culturally Sensitive Approach to Dance," in: Ann Dils /Ann Cooper Albright (ed.), *Moving History/Dancing Cultures. A Dance History Reader*, Middletown 2001, 30–32, 31f. (emphasis added).

> centage point, the passive partner becomes increasingly active, engaging first the core of the body's structure and working outwards to mobilize the limbs – arms, legs, head and tailbone. Both partners dance together in a fully active state. Eventually, the originally active partner becomes progressively passive until they are lying on the floor, enjoying the sensations of their own body through the manipulations of their partner's.

The implications of this score are pretty obvious. Over the course of this duet one experiences the entire continuum of possibilities of being active or passive. Normally in our culture, these various positions of active and passive are pathologized into power dynamics, where the passive figure is seen as not having control, as being either infantile or lazy, rendering them an object of pity. But my experience and that of many of my students is that the experience of being totally passive, rather than feeling powerless, actually opens up a great deal of feeling that can create its own pleasures and sense of agency. Experiencing both extreme ends of these positions can be truly revelatory. For instance, Heather Sedlacek writes:

> I also found novelty and enjoyment in being able to dance at a different level than my partner. [...] It was clearly stated that we were at different levels, that this was okay, and that the high intensity partner would take care and responsibility for the low intensity partner. Thus, for the first time I didn't have to resist when my partner resisted or attempt to match her intensity. I didn't have to be fire when she was fire, or wind when she was wind. I could simply revel in the percentage that our teacher called out every few minutes. [...] Reaching 100 % intensity and then helping my partner down to 0 % provided another new and powerful experience. [...] I felt a sense of responsibility that I have not felt before in Contact. Instead of moving with my partner and following the point of contact, as my partner decreased in intensity, I began to control her movements and direction. I had a unique sense of agency in the dance that for me is usually left up to the Third Mind, not to an individual partner.[11]

Throughout this paper I have tried to articulate how Contact Improvisation creates a corporeal space in which *feeling* allows for an interconnectedness with another person without solidifying that relationship into the subject/object dyad implicit in classic conceptions of empathy. I have highlighted how attention to skin as porous and open to the world can facilitate a dancing based on an interchange and multiplicity of subject positions. Moving with the point of contact requires a willingness to stay engaged with *feeling* (verb) in the present moment, refusing to allow any one kinesthetic exchange to get stuck in a particular *feeling* (noun). This is not to suggest that relationships in Contact Improvisation are so fluid as to be meaningless. Quite the contrary. But we need to enter something

11 Sedlacek, Heather, cf. Spiro, "Journal Entry," 17.

like Berger's corporeal space with the dancers in order to read the meaning of their connection differently. Watching two people explore the continuum of energies available in Contact, we become aware of the basic generosity at the core of the form. To dance with you, I need to first feel you, recognizing that this feeling can change. The improvisational possibilities of this dancing can teach us that *Einfühlung* does not have to be only an introspective process, but rather can open us up to *feeling* both in and out.

References

Berger, John, "Rembrandt and the Body," in: *The Shape of a Pocket*, New York 2001, 103–112.

Foster, Susan, *Choreographing Empathy*, London, New York 2011.

Sklar, Deidre, "Five Premises for a Culturally Sensitive Approach to Dance," in: Ann Dils/Ann Cooper Albright (ed.), *Moving History/Dancing Cultures. A Dance History Reader*, Middletown 2001, 30–32.

Spiro, Corey, "Journal Entry," in Ann Cooper Albright (ed.), *Encounters with Contact: Dancing Contact Improvisation in College*, Oberlin/Ohio, Oberlin Theater and Dance Program (distributed by Contact Quarterly), 2010.

Sabine Zubarik

"Touch Me If You Can"

The Practice of Close Embrace as a Facilitator of Kinesthetic Empathy in Argentine Tango

Premise: Tango as a Practice of Empathy

Argentine tango stands out among partner dances: in its social (as opposed to performance or competitive) version, it is an improvised dance and is predicated more than any other partner dance on the principle of leading/following. Indeed, not only are combinations of steps improvised in their sequence, e.g., in swing and salsa, but each single element may be danced or interrupted at any moment, in varied quality and cadence. The leader decides on the sequence of elements, the direction in relation to the line of dance, and the quality of the movements according to the music, while the follower reads the lead from the partner's body and translates it into corresponding movements. To do this, the partners must constantly focus on each other in order to locate each other in the couple's shared space and read the forthcoming move, while concurrently focusing on themselves in order to be located and read correctly by the other. This requires certain abilities of the practitioners in terms of body skills, such as good balance, awareness of one's shifting weight on the floor and the direction of movements (e.g., linear or spiral), control over the beginning and the ending of a movement or the exact amount of energy put into a movement, and so on. Psychological skills are equally required, such as the willingness to follow somebody else's suggestions and the openness to listen to the partner's proprioception as well as to the experience and abilities she or he brings to the dance in terms of balance, grounding, or movement dynamics.

How can we think of kinesthetic experience in the case of a social partner dance such as Argentine tango? How can one single out awareness of one's own body position and body motion while engaging in movement with someone else as one unit? Is it possible to distinguish the proprioception of one constituent of the movement relationship from the other, when the very premise of the dance is for partners to merge as much as possible? In a shared dance, can there be kinesthetic experience without empathetic experience? Because in tango the two bodies are so deeply listening to each other *while* locked in an embrace *while* moving together to the music, any sense of self is always already an act of reaction to sensing the other.

For example, the tension of my partner's hand on my shoulder blade makes me realize my own degree of relaxation or tenseness; the tonicity of an embrace will make me aware either of a matching or of a differing degree of muscle tone in my own body; my body's grounding into the floor or upward extension can be felt because of my partner's grounding or lifting. Before I even listen to all of this in myself, I have already listened to it in my partner's body.

Theodor Lipps coined the term "Einfühlung," for which there is no other English word than "empathy" – its literal translation would be "feeling into." The German word "Empathie" is in fact the back-translation of the English word "empathy."[1] In Lipps' understanding, *Einfühlung* or *feeling into* is based on the principle of involuntary, instinctive, unmediated mimicry of an external Other (inanimate object or living being). However, what is mimicked is not so much the physical, outward form of the movement itself, but rather the emotional, inner experience of the movement. My body imitates what it perceives as the intent or impulse behind the movement. The mimicry is that of a kinesthetic image, in the sense of an aggregate of sensations that evolves within myself.[2]

To what extent can Lipps' research on *Einfühlung* prove useful in our investigation of the phenomena of translation and transmission in a couple dancing tango? Is what Lipps describes as mimicry also at work in the interaction and responsiveness of tango dancers?

To begin with, however, we must account for two distinctions: first, Lipps' *Einfühlung* operates by means of visual perception, whereas in tango the main sense is the tactile sense, and the medium is touch. Second, Lipps describes the situation of an observer and an observed, whereas we are interested in situations involving two actors, both of them at the same time perceiving and being perceived.

One point, however, is consistent with Lipps' notion of *Einfühlung:* the dancers do not imitate the outward form of their partner's movement, but their own movement emerges from what they feel to be the movement their partner is striving for. The degree of torsion of the upper body preparing to turn, the lightness of quick rhythmic steps, the length of languid steps, the closeness or openness of an embrace, and so on – all of them are rather a *feeling into* the impulse of a movement than a mere imitation of the partner's movement itself. Could this help us establish a phenomenological theory of responsivity in dance, does it

---

1 Curtis, Robin, "Einführung in die Einfühlung," in: Robin Curtis/Gertrud Koch (ed.), *Einfühlung: Zu Geschichte und Gegenwart eines ästhetischen Konzepts*, München 2009, 11–29, 11.
2 Lipps, Theodor, *Ästhetik: Psychologie des Schönen und der Kunst*, vol. 1, Hamburg, Leipzig 1903, 115. What Lipps was not able to explain at the time was later discovered to be the function of mirror neurons. See also Curtis, "Einführung," 12–17.

help us describe the relationship between the perceived movement of an Other and one's own performed movement?

The Significance of the Embrace in Argentine Tango

In tango, the medium of transmission is touch, taking the form of an embrace.[3] The type of embrace and its implications for one's own position as regards axis and mobility are thus decisive for the functioning of nonverbal and reciprocal communication within the couple – a fact that is due to the above-mentioned improvisational character of Argentine tango. At each and every moment, using targeted and mostly minute movements, the leader conveys to the follower motion impulses, directional indications, and suggestions for speed and modulation, which the follower reads by means of the contact points resulting from the embrace, and implements in her own movements. Each shift in body tension or in the position of individual body parts triggers a modification of the embrace, which in turn elicits a change in the transmission of motion impulses.

Over the course of tango's slightly more than one hundred years of history, different styles and thus different types of embrace have evolved. One example is the very close embrace in the so-called *milonguero style*, in which the partners' upper bodies lean against each other and, in order to maintain the forward-tilting static equilibrium, weight is transferred onto the partner (by the follower more so than by the leader). The forward tilt of the torso allows for greater spacing in the hip area, in which contact almost never occurs. Another version of close embrace can be attained by the two partners remaining within their own axis, i.e., not transferring any weight or exerting any pressure; in this case, the docking of the upper bodies occurs by means of a softening in the knee joints, triggering a slight shift forward in weight towards the ball of the foot and making the entire upper body move slightly forward. Unlike *milonguero style*, contact with the partner is established not only in the upper chest, but also in the upper abdomen.

The so-called *salon style* features a half-open embrace, in which a sort of triangle is created: on the closed side of the embrace – i.e., the man's right side

---

3 See Sarah G. Cant's essay, in which the author deals specifically with the issue of the embrace in Argentine tango: Cant, Sarah G., "In Close Embrace: The Space between Two Dancers," in: Mark Paterson/Martin Dodge (ed.), *Touching Space, Placing Touch*, London 2012, 211–230; in particular 211: "Argentine tango is a dance shaped through and by touch."

and the woman's left side – the upper bodies join in close contact, touching directly. On the more open side of the embrace – the man's left, the woman's right – the upper bodies open up slightly and do not touch.

In *open embrace*, both sides of the upper body are equally free of touch, only the arms and the hands on the partner's torso find direct contact. In *tango nuevo*, a modern and more athletic version of classic tango that has been popular for some years especially in Europe, the embrace is wide and elastic: depending on the figures, which can be quite expansive and sweeping, the arms disengage and lock as required. As if connected by hinges, in a full turn the woman's left hand can thus slide down to the man's forearm, or his right hand can let go of her torso and only touch her upper arm or, in a 90-degree position, slide diagonally across her back to the hip. In some figures, the embrace even breaks up entirely (so-called *soltadas*), when one partner dances around the other or executes a turn on his or her own.[4]

In a recent fusion of classic and modern tango, which some call the "new salon style" (*nuevo tango de salón* or *tango nuevo de salón*),[5] the close embrace is increasingly taken up again, yet handled in a more elastic manner, allowing dancers to establish distance as needed when the integration of modern figures requires it. Mostly when walking, couples come back into close or very close embrace. While the boundaries are blurry, it is possible – as can also be seen in workshop and course announcements – to distinguish roughly between an open (= wide) and a close (= tight) embrace.[6]

4 See Petridou, Elia, "Experiencing Tango as it Goes Global: Passion, Ritual and Play," in: Gabriele Klein (ed.), *Tango in Translation: Tanz zwischen Medien, Kulturen, Kunst und Politik*, Bielefeld 2009, 57–74. In this essay, Petridou very aptly connects the three main styles of embrace to tango's historical evolution in Buenos Aires and Europe, linking each style with a different concern and thus a different discourse, such as the quest for authenticity (*estilo milonguero*), the display of passion and grandeur (*open* or *salon position*), or the wish for play and sport (*tango nuevo*).

5 Both terms are in use – depending on whether the instructor views the salon style as undergoing an innovation (*nuevo tango de salón*) or links back *tango nuevo* to the salon style (*tango nuevo de salón*).

6 *Open* / *wide* resp. *close* / *tight* are used as synonyms.

Tango dancers in open embrace.

Tango dancers in close embrace.

What distinguishes the close embrace from other types of embrace? What are its implications in terms of body techniques, psychology, and social interaction? What are its preconditions and its consequences? This investigation aims to provide answers to the question of how a change in the style of embrace, from open to close, affects the dancers' perception, transmission, and processing of feelings.

Experiment with an Introductory Workshop on the Close Embrace

Set-up of the Experiment

In December 2009, this set of questions was explored with a group of 18 tango dancers in Erfurt (Germany); all of them had some years of dancing experience, but had trained exclusively in open embrace and had, until then, not practiced the close embrace.[7] In a four-hour workshop, this group was introduced to the close embrace by a tango instructor couple specialized precisely in this field.[8] They were shown where the points of contact lie, where weight is transferred

7 Experienced dancers were chosen because: a) for beginners, at first even the open embrace appears close and represents a break from the everyday distance kept between individuals; b) beginners are too overwhelmed by learning the steps, spatial coordination, and the new demands they face in terms of body control to perceive much else.
8 Andrea Streubel and André Artmann from Hamburg (Germany).

and where one's own axis is maintained, how partners can connect, and which body posture is required. Dancing in close embrace was practiced using easy figures, which participants already mastered without difficulty in open embrace. The instructors provided explanations and corrections, and participants were occasionally asked to change partners.

In a subsequent one-hour group interview, participants were asked what differences they had perceived and how they would describe them, especially with regard to emotional aspects such as body contact and sense of well-being, connection to the partner, and unity as a couple.[9]

Findings

a) Intensity and quality of feeling: One often-mentioned difference to the open embrace was the subtlety of transmission. The smallest of signals is enough for the follower to implement even large movements, while conversely even the smallest deviation of the leading or the following upper body holds potential for major mistakes. A high degree of body awareness, the ability to fine-tune one's movements, and an even greater isolation of upper and lower body are thus preconditions. As a consequence, concentration and the readiness to receive minimal signals from the partner are tremendously heightened, which also fosters the perception and transmission of feelings (such as excitement, joy, well-being, anxiety, and tenseness).

Participants agreed that both the intensity and the quality of feelings were enhanced in close embrace, due to the unmediated contact with the partner's torso and neck-head area. On the one hand, it is no longer possible to conceal certain uncontrollable emotional and physical responses in oneself, which express themselves directly through heartbeat, breathing, skin temperature, perspiration, and muscle contraction or release in the shoulder and upper arm area. On the other hand, the perception of these factors in the partner is equally inescapable. The sense of harmony provided by a well-functioning dance can thus arise on the mere basis of a synchrony of breathing or the pleasantly perceived warmth of the skin, regardless of the execution of steps and figures. For this very reason, some participants also stated that this type of embrace felt too close to them, as the dance partner is able to perceive things that one may not even be

9 This material is available as video recordings of the workshop and subsequent group discussion. There are also transcripts of the discussion and an attendant comprehensive analysis report.

aware of, or that lie beyond direct control, and one is furthermore submerged by intensities of feeling that then need to be handled. The worry of being an imposition or of triggering an unpleasant experience for the other was significantly higher than in open embrace.

b) Kinesthetic and emotional sensations: The kinesthetic experience was compared to that of a friendly embrace; most respondents acknowledged that dancing could very well be understood as "cuddling," since after all one finds oneself arm in arm with somebody for several minutes at a time without letting go. The boundaries of personal space and avoidance of touch, so distinct in everyday life, are transgressed and one perceives highly intimate characteristics, such as the smell of the body and the breath, the texture of the skin, body hair etc. Depending also on the success of the dance, these factors are assessed differently, in extreme cases as an imposition and a confrontation (also with the limits of one's own tolerance), but equally as a privilege of being allowed so close to someone.

A striking visual difference between open and close embrace that was mentioned was the restricted field of vision both for the leader (on his right side) and for the follower (also on her right side or, depending on the difference in height, on both sides). This was said to reinforce a sense of insecurity and caution, resulting in an attempt to minimize step length, scope of movement, and figures, as well as to lead and follow in an even more highly controlled manner. The leap of faith towards the partner was described as incomparably higher than in open embrace, where followers in particular are able to play a greater part in protection.

At an emotional level, a major difference to open embrace was viewed in the extent of the sense of responsibility on the part of the leader and the sense of trust on the part of the follower. Whereas, in the event of a misunderstanding, in open embrace it is still possible to recover by going over into a step and re-stabilizing before focusing once more on the partner, in close embrace such a misunderstanding triggers an immediate break in physical connection and may lead to stumbling or loss of balance, with immediate effect on the mutual granting of trust and security, hence also implying a (brief) emotional break in connection.

c) Blurring and mutual influence of feelings and sensations: All respondents agreed that physical sensations and emotional processes are mostly directly linked in perception and difficult or barely possible to separate in description. For example, the formidable closeness was named as a challenge to "let the other one in." Bad posture, such as rounded, forward-arching shoulders in the leader, or a recoiling upper body in the follower, triggered the feeling that the partner was not fully allowing contact, and were appraised not only at a physical but also at an emotional level, as "not being accepted" or "not being able to give oneself over" – more so than in open embrace.

It is thus noteworthy that purely physical errors immediately trigger emotional responses, such as loss of confidence in the other and in oneself, distancing, helplessness, and the experience of an unpleasant external dependency. For these very reasons, some of the more experienced dancers view the close embrace as a "higher art" of tango and find the harmony that may potentially be achieved, the subtlety of signals, and the precision of movement even more fulfilling. They experience the close physical contact as extremely pleasant, because when it is mastered, it is less prone to disruptions.

The more intense perception of physical characteristics that are not as such relevant to the dance, e.g., pleasant or unpleasant smell, damp skin, or breathing noises, was repeatedly named as a strong influence on the liking felt for the partner and the associated flow of communication within the couple. Opinions greatly diverged as to the importance of avoiding these "disruptive factors" (such as sweat or direct skin contact). It was furthermore observed that, once a functioning dance contact has been established, tactile or olfactory perceptions that would otherwise be viewed as unpleasant are met with a great degree of tolerance and considered irrelevant compared to the more important pleasure of dancing (as remarked by one female participant: "If someone dances well, nothing else matters").

d) Nonverbal communication, negotiation, and adjustment: While dancing, the partner must be given to understand in a nonverbal way whether one is feeling comfortable or even enjoying oneself, and suggestions as to what could be modified must be made by bodily means. One noticeable aspect is the balancing out of pleasant contact and excessive pressure. Overly long-lasting and firm brow or cheek contact, initially perceived as "cuddling," if stepped up in intensity due for example to emotional involvement (one wants to display even greater dedication, be even more fervently absorbed, achieve even greater intimacy), may quickly turn into disruptive or even painful pressure, leading to tension in the neck and disturbing emotional harmony within the couple or interrupting it until a more relaxed head posture has been "negotiated." According to participants, the shift from "nice" to "too much" is so gradual that dancers themselves often notice it only "too late," when the disruptive factor has already caused a deterioration in communication. It is therefore necessary to constantly shift the focus between physical well-being and emotional involvement. When getting too carried away emotionally, one loses control and does not display the right degree of intensity of contact, self-momentum, and initiative.

A good example is the balancing out of mutual transfer of weight at the beginning of the dance, so-called "docking" – a motion sequence to which much time and care was devoted in the workshop experiment. It becomes apparent how the sudden tilting forward of one partner (too fast approach) can lead to a displacement in axis in the other partner, especially if the latter is lighter. If, for example, the man tilts forward into "docking position" faster than the woman can react, this causes her to tilt backwards or to hollow out her back to act as counterpoise. The timing of mutual transfer of weight is thus a question of the relative weight of the bodies, whose joint static equilibrium must be negotiated bit by bit.

Statements Relating to the Close Embrace in a Questionnaire Study and in Interviews

Setting

From December 2009 to January 2010, a questionnaire study was conducted with a total of 30 tango practitioners of varying level (beginner to professional), age (29–66 years), and gender (14 men, 16 women). The questionnaire study dealt with the sensations and feelings that respondents observed within themselves while dancing or watching others dance, and with the dependence of these feelings on factors such as type of embrace, partner, type of touch, dance experience, surroundings etc.[10] The following only addresses the findings related to the type of embrace and associated instances of touch.

Over the same period of time, a total of 19 experienced tango instructors were interviewed and questioned about their feelings while dancing tango. Conversations mainly revolved around questions regarding outstanding experiences of feelings of flow or fusion in all their aspects. Since respondents kept addressing the differences between close and open embrace, some of the interview findings are used for the present investigation.[11]

---

10 The study is available in form of 30 questionnaires and an attendant analysis.

11 The study's precise data is as follows: 17 tango instructors were interviewed on 29 and 30 January 2010 at a Germany-wide tango instructor meeting in Erfurt. They were divided into four, mixed-gender (with one exception) groups of varying sizes (6/3/3/5). The digitalized interview material has a duration of around 160 minutes and its transcript numbers 19 DIN A4 pages. A further interview was conducted with two tango instructors from Hamburg on 5 December 2009, yielding 30 minutes of film material and 11 pages of text. These 5 group interviews with a total of 19 respondents, as well as the resulting findings report, provide the basis for the following analysis.

Findings of the Questionnaire Study

a) Preferences in terms of type of embrace: Answers to the question "Do you prefer dancing in open or in close embrace?" bring to light distinct differences between men and women: while 7 out of 14 male respondents prefer dancing in close embrace, 3 like both equally, and 4 prefer dancing in open embrace, 12 of the 15 female respondents expressed no preference and said the preferred type of embrace depends strongly on the partner, 2 prefer dancing close, and only one prefers dancing open.

Those women dancing both types of embrace and making the style dependent on their partner gave as a reason that, for close embrace, external circumstances (height, appearance) and inner circumstances (liking and trust) must be right. If it "all fits," they greatly enjoy dancing close. In such circumstances, it "does not matter" whether they dance close or open, or rather they seek constant shifts from one to the other in order to experience the different qualities of the two types of embrace.

Female proponents of the close embrace named as its advantages that it allowed them to better "come into their own," experience better "nonverbal exchange," "feel more at ease," and "get more of a kick." Female respondents named as advantages of the open embrace that it allowed them to "better feel their own weight, their axis, and their scope for movement" and that in general it was "more free and independent."

Asked why they favored the close embrace, male respondents named the following reasons (in different variations): "more fusion," "better contact," "allows me to feel my partner more strongly," "dance is more intense." Those favoring the open embrace said they did so mostly because they were beginners and lacked practice with close embrace. Among the men who dance both styles with equal pleasure, one perceives the open embrace as more athletic, and another does not wish to establish "closeness" with every partner.

b) Pleasant and unpleasant instances of touch: Asked which instances of touch were perceived as pleasant and which as unpleasant, a newcomer to the close embrace stated that chest and stomach contact still felt too close to him, but that this was slowly changing. An experienced dancer said exactly the opposite and described a very agreeable feeling in the stomach/chest area.

Amongst women, there is such a thing as "too close" in close embrace: instances of touch that "force them into certain positions" (pincer-like grips etc.) are perceived as distinctly unpleasant. The frequent use of the semantic field of "force," such as "coercive grip" or "constricted," is conspicuous. The women explain their unease with the loss of "mobility" as well as "too great proximity." This is motivated by a sense of loss of control, the inability to disengage from an embrace as they wish. Coercive grips are described by adjectives such as "firm," "too close," "hard," and "rough." Movements that are too "fast," "jerky," or "sudden" are also considered unpleasant, triggering "anxiety" and "fearfulness." Movements in which the man attempts to "press" himself against the woman are also viewed as inappropriate and perceived as "smothering," to the extent of feeling "nailed" to the partner's chest.

However, men also frequently mention the aspect of "mobility" and substantiate this by talking of "clinging to one's neck and shoulder" and "heavy pressure," which cause them physical pain. Both genders also often point to the predisposition felt towards the partner: whether or not she or he is viewed as likeable from the outset plays a part in the way touch is then perceived.

Asked whether an instance of touch is considered as pleasant or not, both genders seem less concerned with the "what," i.e., the specific body part being touched, than with the "how," i.e., the quality of touch. Men named as pleasant: touch "that occurs by mutual agreement," that is "vivid and fusing," that comes from the "heart," and "signals affectionate care." Women predominantly mentioned as pleasant: touch that is "caressing," "soft," "gentle," "tender," and "warm." It also matters to them that touch or embraces be "precise," and they wish men to provide "confident yet careful" physical contact. Their touch should be "calm." Clarity, calm, and caution lead to a sense of trust and safety in female dancers. They feel "respected" and are granted "the possibility to actively provide input," "freedom," and "latitude," notwithstanding and indeed precisely in close embrace.

Interview Findings

a) Significance of the close embrace for the transmission of motion impulses and the sense of harmony within a couple: Asked whether the often described feeling of fusion of the two partners occurs rather in open or in close embrace, most respondents concede that it may happen in both styles of embrace, yet also say that they personally feel it much more or even exclusively in close embrace. It is not an impossibility in open embrace, where "it can also be a great experience, but it is different, this physical contact is missing," there is "less intent." A man says: "This physical contact is very important, otherwise there is no kick. A lot is about touch [...]. For me, this does not function at a distance. Distance can be danced in a very athletic manner, it can also be very beautiful, but the real kick is touch."

The feeling of "flowing into each other" occurs more through contact in the solar plexus area, less via the arms and shoulders, which is why it is more likely to happen in close embrace. One woman names the transmission of breathing frequency, which can occur in close embrace, as an important criterion.

For one man, however, the type of embrace is not at all relevant:

> For me, it doesn't depend at all on how you embrace, whether close or open, it is simply about how you make contact with each other. How open and willing one is, so that you simply meet on a certain wavelength. At the end of the day, this is something that also happens without touching, and nevertheless you immediately feel something, even without moving through space together, and nevertheless there is a connection. In tango, it is possible to act it out and express it together.

One woman confirms this experience: "I don't notice such a great difference in the quality of contact. [...] When somebody remains at a distance, it doesn't matter whether we dance in close or in open embrace, it is simply not nice."

However, these two interviewees constitute rather an exception. Another man even goes so far as to refuse to define the wide embrace as an embrace. This statement is highly interesting, as it leads *ad absurdum* the generally accepted yet in effect not very considered notion of "close vs. open embrace." For him, either there is an embrace or there is not, in which case one should speak of an "open style" but not of an "embrace":

> In tango, we always speak of the embrace, and I think an embrace only occurs when you dance close, and even that is not necessarily enough, because an embrace means [...] that the arm is involved. When I am not dancing close, my hand is somewhere on her shoulder blade, which is why there really is a big difference between dancing close or not close, simply on grounds of the existence or non existence of an embrace. Naturally you feel the other person anyway, but I think the embrace is a very important part of tango [...] there are women whom I can dance close with, but no embrace occurs. I notice this physically, when I cannot rest my arm properly. [...] either it is really a matter of bad posture. But I think it is more a question of attitude. In that case, the embrace just doesn't happen and as for the touch – in my opinion, the touch is the embrace and feeling emotionally touched stems from this embrace.

Without real embrace there is no situation of touch and, according to this dancer, no *being touched.*

Asked to describe the precise emotional quality of close embrace, the nature of this "being touched," many interviewees provide answers that revolve entirely around the notion of "feeling." One female respondent's statement is typical:

> Close dancing is much more intimate, or rather: can be more intimate. It is also possible to dance close without much *feeling touched.* When I sense there is still a defense, a lack of engagement. This can happen. But when it is gone, it can be very intimate. And for me that is not the case in open dancing. For me this is a rather "more athletic dance." More

> fun with expansive movements, also very nice, a bit of a different dance in my view. For me, dancing close is about more feeling, more being touched, feeling more closeness, sensing the other one directly: do I feel excitement or release, or the pleasure taken in movement, or is there sometimes a change in embrace? Everything is much more direct and I also find it nicer in terms of the embrace. Close embrace makes me feel intimacy.

Feeling oneself and each other physically, explains a male dancer, also means "*allowing* the other to feel oneself physically." Intimacy, which can induce emotion, hence strongly depends on tactile, kinesthetic sensing and seems not to be triggered visually.

b) Unpleasant experiences in close embrace: Clear aversion was expressed against the feeling of "being used" or being treated as a lifeless "device," as indicated by the following statement by a female respondent, representative of many others:

> When somebody who actually is a good dancer forces me like that, leads me so forcefully, I notice, now he is using me, like a device, I notice, I am not at all here, he doesn't even perceive that this is unpleasant for me – then I become so aggressive that I would like to just leave him standing there, I will also push myself away and try to shift into open style, I need space then.

Asked how a partner should react, one man says: "When the woman wants to have that distance, one has to react. When it becomes brittle." Asked what exactly he means by "brittle," he explains: "The opposite of this soft, rounded sense one gets when it feels really good – tension actually builds up in the muscles." Another woman compares this "brittle" physical tension with an unwanted embrace in everyday life: "It is just like in a normal embrace. You notice whether someone is about to push you away. All the aspects are already there, without music, without dancing, without tango."

However, she goes on to say, the aggravating factor in tango is that one is subjected to this helplessness for an indefinite length of time:

> You are stuck, there are men who pull you up from behind or seem to have a cast-iron arm. As a woman, you can work and strain against this, but it is just really unpleasant. I have the feeling that men partly realize it, but obviously that is not always the case. Which leaves you waiting for it to finally be over.

As already gathered from the replies given in the questionnaire study, the interviews confirm that close dancing is only desired by women when the leader knows what he is doing and is able to do so "full-heartedly" and make his body available:

> I am allergic to half-close dancing. There are men who want to dance close, but would prefer not to touch. So they pull you close, but they offer no surface. They find it unpleasant, embarrassing, or scary and then they go about it half-heartedly. I hate that. In that case, I prefer dancing open.

This statement explains the great discrepancy between male and female responses in the questionnaire study to the question as to whether they favor close or open embrace. Indeed, owing to the distribution of forces and the constellation of the embrace, a leader will not have to face situations in which a follower pulls him towards her while pulling back her own body. Followers, however, frequently have to deal with being pulled close by the leading arm without being met in return by a present upper body. Often, leaders even only withdraw from contact in the center of their body, while hunching forward so strongly in the shoulder and head area that they pull a smaller woman underneath their center of gravity, bend over her, and bring her into a painful lasting swayback position, making it impossible to transmit impulses or move the legs conveniently. Faced with this experience often enough, women's initial reaction to the suggestion of close embrace tends to be one of caution, and they will favor an open embrace which may offer less intensity of feeling, but will also involve less pain.

c) The close embrace as fulfilling a need for intimacy: Dancers seem to be aware and to accept that many of them use the close embrace as compensation for a lack of intimacy with the other gender or with other people in general. They also view it as a life-enhancing opportunity allowing them to enter into unfettered physical contact with other partners even when involved in a committed relationship. A middle-aged male interviewee describes this as follows:

> I think the 35–50 age group, […] is mainly an age in which people find themselves in committed relationships. Physicality with other women and men is not really on the cards. In everyday life, there are few opportunities to be intimate with members of the other sex. In tango, it is allowed. I find that very beautiful. That we give ourselves license to do so. I call this soft infidelity. This is exactly what happens for me in tango. I enjoy other women, physical contact with them, especially when they also enjoy it.

One woman says that after a major breakup she found solace in tango, where she could be touched by men without incurring the obligation of getting sexually involved. She says that "during those three minutes, I was lacking for nothing" and that the closeness helped her to gradually engage with men again. As she describes it, "as a woman, one can feel protected in tango." Because in tango the close embrace in particular does not function without physical contact and the experience of intimacy, it automatically represents a legitimization to seek this very contact and turn it into an end in itself. However, this intimacy is significant only for the duration of the dance. Anything beyond is not covered by this legitimization.

One woman mentions that the practice of the embrace in tango has also altered her everyday experience of embraces, in that she carries them out more consciously and more deliberately: "I notice this around other people, friends, be it only when I embrace people to greet them. There is another quality to it. Either I embrace or I don't. But when I do, then it is an embrace, not a banal gesture."

Summary

The findings of the three studies may be summarized as follows regarding kinesthesia and empathy.

The close embrace requires more subtle signals from both partners, and minimal impulses can trigger major changes in movement: as a result, on the one hand it is more prone to disruption, and on the other hand it entails a particularly sophisticated form of communication, which is why it is regarded as a high art of tango. Dancing in close embrace requires training, practice, experience, and an inner disposition for intimacy.

The constant focus on oneself and on one's partner enhances the perception and the transmission of feelings and sensations, especially since uncontrollable emotional and physical responses in oneself, or the perception thereof in the partner, cannot be suppressed. For practitioners, the perception of physical and emotional sensations and feelings is often directly linked and can barely be told apart in description. Purely physical facts are also given emotional value.

External dependency on an Other is experienced to a great degree. Compared to other dance styles, the close embrace requires a greater sense of responsibility in leaders and a greater degree of trust in followers. In general, men tend to favor the close embrace, while women only favor it when the partner is practiced, otherwise preferring the open embrace. Their choice of type of embrace strongly depends on the partner. Reasons for preferring the close embrace include the greater degree of contact and transmission, the heightened feeling of self, and the possibility of "fusion" with the partner. Drawbacks, especially for women, are the constriction and dependence on the partner, the relinquishing of mobility and discretion, and the helplessness in the event of an unpleasant lead. Dancing involves a constant balancing out of pleasant contact and unpleasant pressure.

Pleasantness of contact is barely influenced by the question of "what" (the very same contact points may lead to unease in newcomers and great well-being in practiced dancers), but much more so by the question of "how." Followers seek precise yet soft touch, in order to feel safety and trust. Leaders seek touch that is lively, affectionate, soft, and round.

An inner sense of being moved occurs kinesthetically through touch, not visually; in any case, the close embrace restricts the field of vision of both partners. For most practitioners, a sense of intimacy while dancing requires touch via contact of the upper bodies, especially in the solar plexus area.

The boundaries of personal space and avoidance of touch, so distinct in everyday life, are transgressed. The quest for intimacy with unknown partners of the opposite gender is legitimized by close embrace in tango and is perceived as life-enhancing and free of lasting obligation both by singles and by people living in committed relationships. For some, tango is of therapeutic value in helping them to better engage with other people.

In relation to Lipps' notion of *Einfühlung* and the different ways of understanding the transmission of feeling in tango, the initial question was that of the separability of kinesthesia and empathy. The argument that, in a shared dance, the feeling of self is *per se* a reactive act (of the self in relation/reaction to the other), is borne out by the outlined research data. Transmission of feeling occurs via kinesthetic sensing of the partner and the subsequent response of adjustment and alignment, combined with the willingness to offer oneself up for sensing by the partner. Empathy does not follow from passive reception, being emotionally stirred is no motionless state, but rather it is an active interplay of "move and let move." Physical sensing as a question-and-answer process between two dancers is the engine through which and the space within which a transmission of feelings can occur.

Einfühlung, says Lipps, is an activity, an inner one; it is an active effort of free inner attention, the act of appropriating or acquiring something. Since the perception of the partner's state of being-feeling is already an active process, it seems only consistent that the perceiver should also react with an outward action, responding physically to what she or he feels while experiencing herself or himself in the partner's movement. Is it really possible to tell whether the reacting person is enjoying her or his own movement as such, or as the physical fulfillment of what she or he sensed while *feeling into* the partner's movement? As Lipps claims, aesthetic enjoyment is to be understood as objectified enjoyment of the self: "I enjoy myself in a sensory object that is distinct from me."[12] One does not enjoy

---

12 Lipps, Theodor, "Einfühlung und ästhetischer Genuß," in: Emil Utitz (ed.), *Aesthetik*, Berlin 1923, 152–167, 152: "Ich genieße mich selbst in einem von mir unterschiedenen sinnlichen Gegenstand. Dieser Art ist der ästhetische Genuß. Er ist objektivierter Selbstgenuß."

the movement of the partner as such (since one cannot perceive it neutrally as such, but only through the subjective looking glass of personal processes of perception),[13] but one enjoys the sense of being oneself, of being alive in the movement of an Other. Since in tango both partners are constantly performing this action, they are at any given moment sensing themselves (in the other) and sensing the other (in their own manner of reacting to her/him). As the interviewed dancers repeatedly state in the surveys, they are very much with the other and very much with themselves. Could it be this mutual and iterative process of convergence that lies at the core of what we might call the desire in and for tango: the desire to become yourself while becoming someone else who becomes herself or himself in becoming you, so that both at the same time become none and the two together?

Translated by Maud Capelle

References

Cant, Sarah G., "In Close Embrace: The Space between Two Dancers," in: Mark Paterson/Martin Dodge (ed.), *Touching Space, Placing Touch*, London 2012, 211–230.

Curtis, Robin, "Einführung in die Einfühlung," in: Robin Curtis/Gertrud Koch (ed.), *Einfühlung: Zu Geschichte und Gegenwart eines ästhetischen Konzepts*, München 2009, 11–29.

Lipps, Theodor, *Ästhetik: Psychologie des Schönen und der Kunst*, vol. 1, Hamburg, Leipzig 1903.

Lipps, Theodor, "Einfühlung und ästhetischer Genuß," in: Emil Utitz (ed.), *Aesthetik*, Berlin 1923, 152–167.

Petridou, Elia, "Experiencing Tango as it Goes Global: Passion, Ritual and Play," in: Gabriele Klein (ed.), *Tango in Translation: Tanz zwischen Medien, Kulturen, Kunst und Politik*, Bielefeld 2009, 57–74.

13 Lipps, "Einfühlung," 158.

Kornilia Chatzimasoura

Lost and Found in Interpretation

Senses and Emotions in Contexts of Argentine Tango[1]

Senses and emotions are often perceived as absolute, truthful, and meaningful by the individuals who experience them. A number of questions arise on that matter when conducting fieldwork among the amateur dancers of Argentine tango. Argentine tango is considered a sensuous and sensual dance that creates numerous sensations and emotions not only in the dancers but also in the spectators.

Emotions, sensations, and cognition are closely interlinked,[2] and even the memory of one sense is preserved in another.[3] The ideal would be to be able to conceive and study the experience, the person, and life as a whole. This venture is difficult if not unattainable. Nevertheless, when in our research we subdivide in order to study one aspect of human life, we should always keep in mind that it is a more or less arbitrary division and that the studied aspect is just a part of a bigger unity. The objective is not to break up the experience or the person but to put together the pieces of the puzzle.

In this article we will see that sensations and emotions cannot be considered outside of their various social and personal contexts. In order to understand and perceive sensations and emotions, we need to consider the status and involvement of the person in the situation when they are sensed. We will examine how the emotions and sensations felt while dancing tango are perceived and interpreted both by dancers and spectators and how the absence of a sense can become an emotional and sensual experience, even the model of an experience, for a spectator. In fact, we are going to see that the "sensori-affectivo-motricity"[4] of an individual can become an experience appropriated by others.

The way that dancers experience sensuality in tango reveals the fundamental nature of one's perspective, and provides an insight into the different sensory worlds through which they construct their own reality of the dance.

---

1 I am thankful to Elia Petridou for her support and feedback.

2 Laban Hinton, Alexander, *Biocultural Approaches to the Emotions*, Cambridge 1999, 164.

3 Seremetakis, Nadia, *The Senses Still: Perception and Memory as material Culture in Modernity*, Chicago 1996.

4 Julien, M.-P./Rosselin, C./Warnier, J.-P., "Le Corps: matière à décrire," in: *Corps/revue interdisciplinaire* 1: *Écrire le corps* (2006): 45–52.

The specific example[5] of the women dancing with their eyes shut will allow us to examine the sensations and emotions which result from this act not only for the actors but also for the male and female spectators. We distinguish between female and male spectators because their respective roles in dancing tango take influence on their interpretation of what they see, and consequently on their sensations and emotions.

The Distinct Roles of Men and Women in Dancing Tango

In tango, the roles of the partner who is leading and the one who is following are distinctive, with different steps and duties. Even though this article is written in English, I prefer to use the French terms *cavalier* and *cavalière* to describe the leader and the follower respectively. I find that using the terms "follower" and "leader" does not reflect the embodied feminization and masculinization of each role as strongly as the terms *cavalier* and *cavalière* do. In France, these terms are sometimes used alterably with the terms "man" and "woman" by tango dancers. For example, when a woman decides to take up the leading role, she might say "I will be the man" instead of "I will be the *cavalier.*" Leading seems to be a synonym of manhood.[6] Indeed, according to Butler, gender is not what one is, but what one does. It is through the actions that the idea of gender exists. We become women, men, feminine, masculine by our behavior, by our ways of acting.[7]

The partner who is leading, usually[8] the man, guides his dance partner with the movements of his chest. Even though the response to every step proposed is

---

5 The case studies and the reflections that are going to be presented in this article are drawn from the ethnographic fieldwork that I have conducted with amateur dancers of Argentine tango in the south of France. The fieldwork lasted four years and it was conducted in the context of my PhD research. Thirty interviews were transcribed and innumerable informal discussions were analyzed. Participant observation was conducted at tango courses, festivals, *milongas* (events for dancing exclusively Argentine tango) and other activities related to dancing tango.

6 For further reading on masculinities in tango see Tobin, Jeffrey, "Models of Machismo: The Troublesome Masculinity of Argentine Male Tango-Dancers," in: Gabriele Klein (ed.), *Tango in Translation*, Bielefeld 2009, 139–169; Tobin, Jeffrey, "Tango and the Scandal of Homosocial Desire," in: William Washabaugh (ed.), *The Passion of Music and Dance: Body, Gender and Sexuality*, Oxford 1998, 79–102.

7 Butler, Judith, *Gender Trouble: Feminism and the Subversion of Identity*, New York 1990.

8 In some cases women lead, too. This mainly happens when in a *milonga* there are more women than men, so some women tend to invite each other in order to have the opportunity to dance. In some rare cases, the women choose to lead by their own initiative. Of great interest

codified, the order in which some of these figures are presented is up to the will and the inspiration of the *cavalier.* Consequently, tango is an improvised dance, which requires a significant degree of concentration to the slightest movement of the dance partner. Not only does the *cavalière* have to concentrate on her partner, but the inverse is also true. The *cavalier* has to take into consideration the responses of his partner in order to read just his next movement or his whole dancing style if needed.

The *cavalier* and the *cavalière* envision their roles differently. The *cavalier* is guiding, he ought to maintain an awareness of the space around him, not only of the room itself but also of the continually transforming space created by the movements of the other dancing couples. The *cavalière* usually does not need to keep such a visual contact with the surroundings, so she can choose to look where she wants or even to shut her eyes.[9]

The Normativity of Gestures and Postures: Contextualization of Movements, Senses, and Feelings

In tango, two people get intimate while dancing. This is not necessarily experienced as a sensual or erotic relationship. On the other hand, the imaginary and popular representations automatically send messages of a romantic relationship when it comes to two people being in each other's arms and moving together to the rhythm of the music, especially if the partners look into each other's eyes.

It should be noted, however, that it is rather unusual to see two people looking each other in the eyes while dancing tango. The beginners, in their first classes, when standing in front of their partner, wonder where to look. Looking their dance partner straight in the eyes is pretty intimidating and embarrassing for novice dancers. The instructions of the tango teacher as to the direction of the gaze contextualize the gaze and the situation in which it can be exchanged or not. Managing the gaze becomes part of the dancer's technique. Therefore, it

---

regarding the stakes and negotiations of gender, identity, and power are the queer *milongas* (see Savigliano, Marta, "Notes on Tango (as) Queer (Commodity)," in: *Anthropological Notebooks (Slovene Anthropological Society)*, 16.3 (2010): 135–146.

9 The tango dancer and instructor Adriana Pegorer is actually working on an interesting and innovative project concerning visually impaired people dancing tango (see http://www.adagioconbrio.co.uk/, 15.7.2013).

should be considered as a tango posture. Marcel Mauss was the first to make the point that the physical postures and techniques are also cultural ones.[10]

Seduction is a part of all social relations,[11] and it is certainly part of tango. The dancers want to seduce their potential or actual dance partners in order to be invited or to have their invitation to dance accepted.

The game of seduction in tango sometimes perturbs the dancers. Irène (eighteen months of dancing experience) – who, by the way, seems happy with her life partner – explains the confusion that she sometimes feels due to the ambiguity of the dance:

> It's true that it is ambiguous, and the problem when dancing in couples is that, I think, there is a thing that is created artificially all the same, it's automatic, I mean, you don't dance all alone, so, automatically, you become intimate with another person and at the same time the dance brings your body into play, and even, in a rhythm, so [...]. Yes, that's the ambiguity of the dance in itself and it's true that this is what makes me feel uncomfortable. I like dancing. I like even what is artificially created during the dance, but that's all. I do not go dancing in order to meet a man.

Later in the interview, Irène suggests that the agitation that she may sometimes feel also has to do with the fact that she is not yet an experienced dancer. She isn't familiar with all the steps and she doesn't go to *milongas* frequently. Her lack of experience makes her live the ambiguity of the dance even more intensively. It is quite common among novices, who do not know the repertoire of all the different steps and figures of tango, to ask themselves whether the steps suggested by a more experienced leader are really tango steps, or improvised steps in order to profit of the situation and feel their body closer and in a more intimate way than the one that is normally accepted in tango dancing. In conclusion, it is not really the act itself that is disturbing. It is rather the context and the intentionality. The female dancers with whom I have discussed are ready to accept or even feel some pleasure with the physical contact, the gestures, and the sensations that they may share with their dance partner, on the condition that these movements are recognized and accepted as dance movements, as being part of the dance. The same movements and postures can generate very different feelings according to the context within which their interpretation takes place.

We will now see how the experience of sensuality in dancing tango also depends on the perspective and status of each person in the hierarchy of tango.

---

10 Mauss, Marcel, *Sociologie et anthropologie*, Paris 1985.

11 Baudrillard, Jean, *De la séduction*, Paris 1979.

The Sensuality under Conditions: Plurality of Sensory Worlds

Argentine tango is supposed to be a sensual dance. To understand how sensuality can be experienced while dancing tango, we need to be aware of the contexts in which the dancers may experience it. These contexts are quite different for beginners and experienced dancers.

Novice male dancers are usually terrified at the idea of having the responsibility not only to invite a woman to dance but also to ensure that the dance is going to be satisfying for both of them.

On the dance floor, and under the gaze of others, they have to dance without disturbing the other dancing couples, combine the steps that they have learned in class, think how to guide their dance partner, and keep a continuous flow without stopping to check what was wrong in every false movement. Sensuality is not quite the word to describe their experience of dancing. Adrien (eighteen months of dancing experience) is conscious that the feeling of a dance is related to the experience and seniority of the dancer:

> I am not a dancer with 2,500 years of experience so I can't yet feel the sensuality of the dance, or if a partner wants to communicate to me a desire, etc. What is more, the people with whom I dance, because of their inexperience, cannot make me feel all that by their movement either.

The inexperienced dancer does not necessarily feel this famous sensuality. Furthermore, he does not have yet adequate means in order to receive and understand the subtle messages of sensuality that his *cavalière* may try to transmit to him.

The beginners are too attached to the technique to be able to fully live the experience of "sensuality" in tango. If they feel something sensual, this is usually related to the fact that they have someone in their arms or that they practice some "provocative" figures. So, Jean (one year of experience in tango) recalls that he was very impressed by the fact that he danced with around twenty different women in his first tango lesson.[12]

> I passed from the arms of twenty women, eh, I have never done that before..., I got out of the class a little bit dizzy, it seemed so strange to me. Ah, after some time I realized that this feeling faded away. Even the close embrace does not bother me anymore. There is the time of the dancing, where there is an important physical proximity but, I do not know how to

12 During the classes, tango instructors invite the dancers to regularly change their dance partners. Thus, in only one lesson, all the *cavaliers* dance with all the *cavalières* of the class.

> explain that, eh, when the song is over, the two dancers split, and I do not think that the shared intimacy is that strong. On the exterior it seems very intimate but finally now I experience it in a very inoffensive way. [...] I had heard that it was super sensual, and I realize now that to a spectator, who does not practice this dance, it can seem sensual, but a dancer knows that it is very codified. [...] There is the step where the girl can make embellishments and slides her leg back up to the leg of the guy, etc. Yes, on the outside, depending also on the girl, this can seem hot. As a dancer you know that it is a codified moment and that that's the only thing that can happen at that very moment and so you do not put any intention of sensuality.

For Jean, at his stage of apprenticeship and dance experience, dancing tango made him revise his ideas about the sensuality of this dance. He realized that all the movements that make the tango seem sensual to the spectators are just codified steps.

Indeed, often people who are outsiders to this dance, who are just spectators, are impressed by the sensuality that springs from the couples dancing tango.[13] The spectator judges by what s/he sees: a couple who dance closely, a game between two partners whose legs interweave in suggestive figures. It is different for the initiated spectator. He, or she, by having his or her own experience in dancing tango, evaluates differently the sensuality or the seduction that is emanating from the dancing couple. As Hess puts it, "ordinarily for the dancer, when trouble may be experienced, it comes from the touch and not from the view. The perturbed gaze is the one of the spectator."[14]

Of course Jean's experience, described earlier, is not necessarily representative of the experience of all beginners, even though his feeling is quite common among novices. It should be noted that Jean is a beginner who has demystified the sensuality of tango but has not yet mastered the technical components of the dance in order to be able to savor all the joys and the subtle sensations that can be felt while dancing tango with ease. His insight of sensuality is associated to his seniority in dance and depends on his degree of technical competence.

Once the dancers acquire the technical skills and the experience to dance in a *milonga*, novel emotions and sensations emerge that define what is "sensual" about tango. Free from the worries of having to think of the next step and how

---

13 Tango shows use in an ostensible way the image of tango-"passion" (for an analysis of "passion" in tango see Savigliano, Marta, *Angora Matta. Fatal Acts of North-South Translation*, Middletown 2003). According to Elia Petridou, these shows have played a major role in the creation of the aesthetics of social tango dancers (Petridou, Elia, "Experiencing Tango as it Goes Global: Passion, Ritual and Play," in: Gabriele Klein (ed.), *Tango in Translation*, Bielefeld 2009, 57–74).

14 "D'ordinaire, pour le danseur, lorsque trouble il peut y avoir, cela vient plutôt du toucher que de la vue. Le regard troublé est celui du spectateur" (Hess, Rémi, *Le moment tango*, Paris 1997, 138).

to indicate it to his partner, the *cavalier* can now concentrate on his sensations and make the most of them. It's the same for the *cavalière*. Having confidence in her dance, and without the anxiety caused by lack of experience, she can let herself go and appreciate the dance, and the unique moment that is created in every dance, which rests on her partner but also on the music and on the general ambiance of the *milonga*. While beginners often interrupt their dance because of misunderstandings between the two partners, experienced dancers can enjoy the accumulation of their sensations and the fusion with their partner during the dance. These dancers may achieve moments of grace[15] during which they experience "shivers" and "mini-ecstasies," and feel cut off from reality. Some people describe these moments as "tango-orgasms." These moments can be reached when the dancers are disconnected from what goes on outside their dance and they "let go," while staying concentrated on their partner and on their own feelings at the same time. Experienced dancers consider tango to be a "sensual" dance but not necessarily an erotic one. Tango is lived as a "sensual" dance because of the intense emotions and sensations that are felt and shared between the dance partners.

The feeling of the dance is different depending on the seniority of the dancers. According to E.T. Hall, people from different cultures inhabit different sensory worlds.[16] The same assertion is transposable to the world of tango. Beginners live in a different sensory culture from the experienced dancers when it comes to the dancing moment. Of course, everyone lives his/her experiences in a unique way, which also depends on the circumstances. Nevertheless, having practiced and experienced tango for a longer period of time allows the more advanced dancers to gain access to a broader and different range of sensations. They become more open and they savor intensely all the sensations and pleasures that emanate from the dance, with their partner, in the rhythm of the music, and in the midst of the other dancing couples. However, the more the dancers progress in their expertise in tango, the more demanding they become, and the number of partners with whom they can feel and share extraordinary emotions dwindles down.

15 See also the communication of Sabine Zubarik's paper: Zubarik, Sabine, *Sublime Feelings: The Experience of "Flow" in Dancing Tango*, delivered at symposium *Tango: Creation, Identification, Circulations*, Paris, 26–29 October 2011 (http://globalmus.net/?Communications-audio-mp3, 15.7.2013).

16 Hall, Edward T., *La dimension cachée*, Paris 1971, 15.

Being in and out of Tango

As we saw in the case of Jean, the image of tango that the dancers-to-be have before starting to practice is very different from the one that they acquire after their immersion in the dance. Every tango dancer I have talked to agrees that the impression that she/he had of the dance when they were "outsiders" has nothing to do with what they experienced once they were immersed in tango. "Outside" the tango, they just had a visual experience of the dance as spectators. When they started dancing it, tango became an actively lived experience. The different ways in which tango is represented and appreciated by those "in"[17] tango and those "outside" of tango are situated mainly at the level of senses and sensations. In general, spectators who are not familiar with tango often find this dance "sensual," "slow,"[18] with pre-established figures so complicated that they are surprised by the synchronization of the dancing partners. On the other hand, the initiated dancers know that it is a codified[19] dance, intense, rich in sensations and emotions, improvised, with figures which all act according to the logic of being together and being-one with the dance partner. The spectator is usually impressed by the spectacular figures and does not pay attention to the quality of the connection of the dancing couple and the perfect and harmonious execution of the simplest steps. Inevitably, the "outsider" (uninitiated spectator) focuses on the spectacular, whereas the "insider" (dancer) is interested in the feeling and the small details.

The "inside" and the "outside" are also valid on other levels of involvement. There is an "inside" referring to the moment of dancing and to the connection that is created between the dancing couple and an "outside," felt by the dancer who is not dancing at that very moment but is still immersed into the "culture of tango."

It is worth noting that the concept of the "spectator" is ambiguous and thus interesting to think about. Roland Huesca talks about the "desire of the other's gaze" that the spectators can experience.[20] They know that they are being looked at (by other spectators). They are not just observers, they are also being observed. They do not reduce their actions only to seeing but they also dispose themselves to be seen. What are the borderlines of being a spectator? Is the status of the spectator a passive one? Absolutely not, but this is to be discussed on another occasion.

---

17 "In" involves dancing tango and participating in the sociality of the *milongas*.

18 It also depends on where they have watched tango being danced. Whether it was in the context of social dance or a show which included acrobatic figures of tango.

19 With pre-established figures.

20 Huesca, Roland, "Le théâtre des Champs-Élysées à l'heure des ballets russes," in: *Vingtième Siècle, Revue d'histoire* 63 (1999): 3–15, 7.

The people who attend *milongas* are those who are familiar with the dance: they are "insiders." So, usually, the notion of the "spectator" at a *milonga* is not the opposite of the notion of the "actor." At a *milonga*, most of the people who are present are potential dancers, who have a personal experience of the dancing moment. In this way, we could suppose that the spectators' gaze on the dancing couples is filtered through their own personal experience of dancing. Yet we will see that when a dancer changes his or her status and becomes a "spectator," his or her perception and judgment of the dance change as well.

Tango, and dance in general, is linked to seduction. Not necessarily to a seduction that would have sexuality as an objective. It's just that for multiple reasons, the dancers want to be likeable and to seduce. The image of two people dancing tango, especially when they dance well, can promote the sensuality. So the spectators "can't really say if the couple that is dancing in front of them know each other for three minutes or for thirty years, if they are going to separate after the tango dance or if they are going to stay together till the end of their days."[21] Even the experienced dancers can't say with certitude if the couple that they are observing is also a life couple, unless of course they already know them.

Women Dancing with Their Eyes Shut

The *cavalières* often say that the act of closing their eyes allows them to dance better. They can concentrate better on steps proposed by their partner and they can get rid of the exterior distractions. Only the movement of their partner's body guides them: "Sometimes I happen to close my eyes. [...] I dance much better this way... I can really concentrate on what is happening within the dance and I dance much better. I do not have exterior distractions, you see?" (Irène). With closed eyes, followers can feel and better follow the rhythm of the dance and the movements of their partner, in short, they can abandon any exterior attachment in order to better concentrate on the movements of the *cavalier*.

---

21 "Ceux qui regardent ne pourraient pas dire si le couple qui danse devant eux se connaît depuis trois minutes ou trente ans, s'ils vont se séparer après le tango ou rester ensemble jusqu'à la fin de leurs jours" (Niiniluoto, Maarit, "La mélancolie dans la musique populaire finlandaise," in: Michel Demeuldre (ed.), *Sentiments doux-amers dans les musiques du monde – Délectations moroses dans les blues, fado, tango, flamenco, rebetiko, p'ansori, ghazal*, Paris 2004, 243–250, 243).

Even though the *cavalières* close their eyes in order to cut off the "outside world," keeping the eyes shut can also create bridges of communication, often involuntary ones, with the people who are outside the dance, the spectators.

Looking at the Women Dancing with Their Eyes Shut

Tango can be danced either in open position (with some distance between the two bodies) or in close position (when the torsos are touching). In both cases, there are various ways of positioning oneself against the partner. If a cavalière chooses to close her eyes, it will be in close embrace. In this case, the posture of the two partners is such that it is difficult to see the other's face. So usually, the *cavaliers* are not aware if their partner dances with her eyes open or shut.

The feedback that I got from the spectators when talking about the women dancing with their eyes shut was different depending on whether the spectator was a man or a woman, a *cavalier* or a *cavalière*. Furthermore, the particular situation and status of the person also had a role to play.

The Female Novice Spectators' Perspective

The main place to watch tango danced with eyes shut is at a *milonga*. The women who close their eyes are dancers who already have some experience of the dance. The beginners are mainly concerned with the proper execution of the steps and with correctly following their partner. The possibility to dance with closed eyes does not really cross their mind before seeing it or before someone, like their tango teacher, brings it to their attention.

For the female beginners, keeping their eyes shut while dancing seems to be a procedure of normalization and of imitation of the established dancers: an imitation with the objective either to dance better by doing what others do (if the good dancers do so, this means that shutting the eyes really helps to dance better), or to appear to the others as an experienced dancer.

For the novice dancers, the closed eyes are the symbolic and postural asset of the experienced dancers. Therefore, female dancers may regard this practice as a further step of their apprenticeship. Shutting the eyes represents, deliberately or not, a proof directed to others or to themselves of their capital of tango experience. The act of closing the eyes becomes a sign of prestige, a "semiological code."[22]

The Male Spectators' Gaze

Whereas for most female spectators the closed eyes of the female dancer is a sign of an experienced *cavalière* or a way to dance better and enjoy the dance, for the male spectators, it is usually regarded as a sign of the dancing ability of her male partner. Women are translating what they see from their point of view as *cavalières* and are focusing on the sensations caused by this act, while men are interpreting the situation from the viewpoint of the *cavalier*. Everyone is interpreting the situation from the perspective of his/her own role in dance. This difference of approach implies a difference not only in the felt emotions but also in their interpretation.

Take, for example, Michael, an experienced male tango dancer. Michael feels a hint of jealousy when his companion shuts her eyes while dancing with another man: "[...] if I see her with the eyes shut, surrendering [...]."[23]

"With the eyes shut, surrendering": a woman who keeps her eyes shut while dancing would, thus, for Michael be a woman who lets herself be carried away by her male partner. Dancing Argentine tango with the eyes shut can be perceived as a situation of great intimacy. Michael describes the paradox of his own experience: when he dances with different *cavalières*, he feels that there is nothing ambiguous about it and that they share the same pleasure of the dance, the communication, the harmony, and the rhythm. He doesn't feel anything equivocal in the dance. However, he becomes suspicious of the innocence of dancing with the eyes closed when he sees his own companion closing her eyes while dancing with other male dancers. He interprets the same act in a different way from what he knows from experience. His jealousy and imagination take over. Out of the sphere of the dancing moment, Michael becomes a possessive jealous man who watches his beloved dancing tango with her eyes shut, in the arms of another man. He is not a dancer any more claiming the "innocence" of tango. As a spectator, his perception changes because his status changes as well.

Le Breton explains that "we are not moved in general or by the sudden release of a biological process. Rather, we are moved according to our particular involve-

22 Guiraud, Pierre, *Le langage du corps*, Paris 1980, 4.
23 "*[...] si je la vois les yeux fermés en train de s'abandonner [...].*"

ment in a given situation."[24] So, Michael's feeling of jealousy in the previous example is not provoked by a particular and objective stimulus, i.e., closed eyes. Rather his jealousy is provoked by his personal involvement in a specific situation.

Looking at something or someone triggers memories, past experiences, emotions, ideas, and representations. That's because the senses, the emotions, the thoughts, and the memories are inseparable. Gombrich and Joyal assert that "whether we are producing a representation or we are contemplating it, we represent the object through 'our memorized acquisitions' and through what we expect to perceive, which means that we hear or see certain things rather than others according to our bias."[25] Both the "memorized acquisitions" and "what we expect to perceive" are extremely complex. The dancer who claims that he doesn't think of tango as a sensual dance and who had experienced and categorized his dancing moments with different dancing partners as "innocent," can feel jealous if he sees his life partner dancing with another man. The memories and images that are activated do not depend strictly on experience or information obtained on the same subject. The sources of our memories and representations are various, and the ways in which they are formed, complicated. The type of representation that will be activated in any given situation depends not only on the situation but also on the way the person involved relates to the specific situation.

In the case of the jealous dancer, it wasn't his experience of the dance that took over when he saw his companion dancing with another man with her eyes closed. Probably, he reactivated the stereotype of the "sensual tango". His jealousy was due to the way he evaluated the situation.

> This evaluation depends on situational factors but also on other factors that are linked to the culture and personality of the subjects: a person interacts with his (or her) social and psychological environment according to the values, beliefs, engagements, and objectives which create a predisposition to the emotion by making him (or her) susceptible to some

---

24 "On n'est pas ému en général ou par le déclenchement inopiné d'un processus biologique, mais face à une implication particulière dans une situation donnée" (Le Breton, David, *Les passions ordinaires – Anthropologie des émotions*, Paris 1998, 102).

25 "Que nous soyons en train de produire une représentation ou en train de la contempler, nous (nous) représentons l'objet, à partir 'd'un acquis mémorisé' et de tout ce que nous nous attendons à percevoir, ce qui fait que nous sommes disponibles pour entendre ou voir telles ou telles choses de préférence à d'autres" (Gombrich, E.H., *L'art et l'illusion: psychologie de la représentation picturale*, Paris 1971, 239) (Joyal, France, "Parler tango, quelques considérations méthodologiques autour d'un phénomène métaphorique," in: France Joyal (ed.), *Tango sans frontières*, Québec 2010, 129–144, 133).

of the elements of the situation. The emotion depends therefore on the way that the person evaluates and analyzes the situation.[26]

Sensory Communication

A *cavalière*'s choice to keep her eyes closed is an act subjected to multiple interpretations. The spectators' imagination plays a significant role in these interpretations. Besides, according to Baudrillard,

> the real doesn't seem to interest anyone because the real is the place of disillusion [...]. We are only absorbed by the signs that are empty, absurd, elliptic, with no reference, which make no sense. [...] the spirit is irresistibly bewitched by the space which is left empty by the senses.[27]

The closed eyes of the female dancer arouse the imagination of the spectator and create sensations. The spectators are bewitched by the woman who shuts her eyes because she becomes mysterious. The absence of her gaze, as every absence, leaves a free space that the spectators fill with their own meaning. It becomes a *lieu* of experience, both for the female dancer and for the spectator. The sensorial is not only experienced by oneself, on a personal and individual level, but it can also be shared or appropriated by others.[28]

The reasons provided by women for keeping their eyes shut and the interpretations given by spectators can sometimes be very different. The disparity between the feeling of the female dancers and that of the male spectators offers an example of appropriation of the sensory experience of others. The sensorial invades the body of the person with shut eyes – but also the body of the person who is watching her. There is a sensory communication between the two beings.

---

26 "Cette évaluation dépend de facteurs situationnels mais aussi d'autres facteurs liés à la culture et à la personnalité des sujets: une personne interagit avec son environnement psychologique et social en fonction de valeurs, de croyances, d'engagements et d'objectifs qui la prédisposent à l'émotion en la rendant susceptible de réagir à certains éléments de la situation. L'émotion dépend donc de la manière dont la personne évalue et analyse la situation" (Despret, Vinciane, *Ces émotions qui nous fabriquent – ethnopsychologie des émotions*, Paris 2001, 218).
27 "D'ailleurs le réel n'a jamais intéressé personne. Il est le lieu du désenchantement. [...]. Seuls nous absorbent les signes vides, insensés, absurdes, elliptiques, sans références. [...] l'esprit est irrésistiblement envoûté par la place laissée vide par le sens." (Baudrillard, *De la séduction*, 69).
28 Seremetakis, *The senses still*.

The Tactility of the Intangible

In this article, I do not intend to present the whole range of viewpoints concerning the sensuality in tango, the personal experiences of the women who dance with their eyes shut, or all the perceptions and interpretations of the spectators.

The personal experience is lived by every individual in a different and particular way. The human experience should not be reduced to a summation of facts and already constructed representations. The perception that everyone has of his or her own experience is a complicated construction.

The representations and the emotions aroused depend on the evaluation of a situation and are linked to the way a person relates to it. Consequently, the experience and the judgment of a specific situation can be different when being judged from the inside or the outside, even when it concerns the one and same person. Complex interlinked mechanisms intervene: various emotions, sensations, and representations, which can often conflict with each other and depend on the situation and the involvement of the person.

Discussing the sensuality felt while dancing tango and the ambiguity of some dance figures, we saw that emotions depend on the interpretation of a specific situation. At the same time, the emotion and the sensation may be transformed by the interpretation and the meaning given to the felt emotion-sensation.

The sensations experienced on a personal level may be appropriated by others and even become a model of experience. The closed eyes of the *cavalière* link her feeling of intimacy, her feeling of the dance, with the exterior world, with the people who are watching her. The *cavalière*'s personal sensory experience becomes shared with the spectators who are investing in it their own interpretations and emotions. This connection, even if it is often involuntary, always interpretative and pretty subjective, still remains a source of creation of meaning and of senses, and becomes a *lieu* of experiences and sensations. We realize once more that the body is not just a sign to decipher but also, and mainly, a *lieu* of experiences and a source of actions. The dance is not just a spectacle. The dancing bodies are not simply texts to be deciphered by the spectators.[29] The persons, the bodies who are dancing, are also and primarily actors: actors who feel and fabricate sensations. They are producers of experiences not only for themselves but also for the others.

Even though the closing of eyes often carries a different meaning for cavalières and for spectators, there is still some sort of communion between them, a shared experience between the dancer and the spectator. The senses of the

---

29 Cowan, Jane, *Dance and the Body Politic in Northern Greece*, Princeton 1990.

woman become an experience for the person who is watching her: it interferes with his own senses, emotions, and imagination.

Dancing, as well as doing fieldwork in dance, is an intense personal and social experience, which offers rich insights for the comprehension of synesthesia, empathy, senses, and emotions, among others. The use of concrete ethnographic examples offers the possibility to "touch" these intangible subjects. Nevertheless, I feel that I should stay loyal to my informants and conclude with the point that they all insisted on: dancing is, above all, pleasure.

Glossary

cavalier: the leading partner
cavalière: the following partner
milonga: the word *milonga*, as used in this text, is a term denoting the place/event where Argentine tango is danced exclusively
tanguero/a: tango dancer

References

Baudrillard, Jean, *De la séduction*, Paris 1979.
Butler, Judith, *Gender Trouble: Feminism and the Subversion of Identity*, New York 1990.
Chatzimasoura, Kornilia, "Danser les yeux fermés," in: *CORPS, éditions Dilecta* 7 (2009): 51–56.
Chatzimasoura, Kornilia, *Devenir* tanguero *dans les années 2000 en France: une quête sensible de l'authenticité*, PhD thesis, Université de Nice-Sophia Antipolis, Nice 2011.
Cowan, Jane, *Dance and the Body Politic in Northern Greece*, Princeton 1990.
Despret, Vinciane, *Ces émotions qui nous fabriquent – ethnopsychologie des émotions*, Paris 2001.
Gombrich, E.H., *L'art et l'illusion: psychologie de la représentation picturale*, Paris 1971.
Hall, Edward T., *La dimension cachée*, Paris 1971.
Hess, Rémi, *Le moment tango*, Paris 1997.
Huesca, Roland, "Le théâtre des Champs-Élysées à l'heure des ballets russes," in: *Vingtième Siècle, Revue d'histoire* 63 (1999): 3–15.
Guiraud, Pierre, *Le langage du corps*, Paris 1980.
Joyal, France, "Parler tango, quelques considérations méthodologiques autour d'un phénomène métaphorique," in: France Joyal (ed.), *Tango sans frontières*, Québec 2010, 129–144.
Julien, M.-P./Rosselin, C./Warnier, J.-P., "Le Corps: matière à decrier," in: *Corps/ revue interdisciplinaire* 1: *Écrire le corps* (2006): 45–52.
Laban Hinton, Alexander, *Biocultural Approaches to the Emotions*, Cambridge 1999.
Le Breton, David, *Les passions ordinaires – Anthropologie des émotions*, Paris 1998.

Mauss, Marcel, *Sociologie et anthropologie*, Paris 1985.
Niiniluoto, Maarit, "La mélancolie dans la musique populaire finlandaise," in: Michel Demeuldre (ed.), *Sentiments doux-amers dans les musiques du monde – Délectations moroses dans les blues, fado, tango, flamenco, rebetiko, p'ansori, ghazal*, Paris 2004, 243–250.
Petridou, Elia, "Experiencing Tango as it Goes Global: Passion, Ritual and Play," in: Gabriele Klein (ed.), *Tango in Translation*, Bielefeld 2009, 57–74.
Savigliano, Marta, *Angora Matta. Fatal Acts of North-South Translation*, Middletown 2003.
Savigliano, Marta, "Notes on Tango (as) Queer (Commodity)," in: *Anthropological Notebooks (Slovene Anthropological Society)*, 16.3 (2010): 135–146.
Seremetakis, Nadia, *The Senses Still: Perception and Memory as Material Culture in Modernity*, Chicago 1996.
Tobin, Jeffrey, "Models of Machismo: The Troublesome Masculinity of Argentine Male Tango-Dancers," in: Gabriele Klein (ed.), *Tango in Translation*, Bielefeld 2009, 139–169.
Tobin, Jeffrey, "Tango and the Scandal of Homosocial Desire," in: William Washabaugh (ed.), *The Passion of Music and Dance: Body, Gender and Sexuality*, Oxford 1998, 79–102.
Zubarik, Sabine, *Sublime Feelings: The Experience of "Flow" in Dancing Tango*, paper delivered at symposium *Tango: creation, identification, circulations*, Paris, 26–29 October 2011, http://globalmus.net/?Communications-audio-mp3 (15.7.2013).

Parts of this article have already been published in Chatzimasoura, Kornilia, "Danser les yeux fermés," in: *CORPS, éditions Dilecta* 7 (2009): 51–56; or in Chatzimasoura, Kornilia, *Devenir* tanguero *dans les années 2000 en France: une quête sensible de l'authenticité*, PhD thesis, Université de Nice-Sophia Antipolis, Nice 2011.

Sonia Abadi

Dancing Tango

The Weave of an Emotional, Corporal and Social Network

Tango is not just another partner dance. It developed as a popular expression of the diversity of immigrants in a foreign land, and it turned into a social and cultural web. In tango, the erotic scene of the dancing couple is closely related to the social scenario.

Following a double movement, from deep-rooted to rootless, from the individual identity to a sense of belonging, tango has come to enjoy great popularity around the world, as a net weaver that strings together space and time, linking personal feelings with the dancers' immediate environment and the world, the past with the present and the future.

In the beginning, tango was just a way of dancing to the music that came to Buenos Aires from diverse cultures and countries, using improvised figures as well as sensual and expressive movements. It was a critical and transforming assimilation of foreign influences, a cultural blend.

But what is the real meaning of tango?

Unlike other popular dances that follow a rhythm, a tango dancer moves freely between the rhythm and the melody, depending on his personality, style and mood. Tango also introduces temporary stillness into movement, called *corte* ('stop') and *quebrada* ('break'). These figures, in which the man holds the woman tight against his body, sways and bends her in a provocative way, create an element of surprise, fun and freedom.

The dancer performs to the music the way he wants to. He uses the space freely, stopping when he wants and starting again when he feels that the music is telling him to do so.

Tango musicians created a special music to be danced in that way. Initially, this was based on intuition alone, but later accompanied and described by musical theory.

Tango music was influenced by Argentine folklore, black rhythms, Spanish music, the *canzonetta napoletana*, English music (brought to Cuba by pirates and influential in the south as a spice of the *habanera*), the baroque music taught to the native population by missionaries and, of course, the European dances of waltz and polka.

Later on, tango lyrics would appear, sassy and spicy at first, then more descriptive and dramatic.

A metaphor of a great erotic, social and cultural get-together, tango is the result of an encounter of bodies, souls and roots of those men and women who were both the sons and daughters of the rootless and the parents of a new cultural identity that developed from the late 1800s to the early 20th century.

European raw material that the alchemy of crossed loves and their heirs has transformed into tango, a bastard in a changing world, tango was born in Buenos Aires, but more specifically in its *bordellos*.

This is why, for the 'good' society of the time this accursed dance, born in the underground, was a forbidden thing. However, little by little, tango ceased to be perceived as the 'waste' of the city and began to develop its own identity and beauty, becoming a complex and moving art.

This is how it left the underworld, reached the city centre and projected itself onto the world. Yet up to this day it has kept its 'out-of-bounds', transgressive imprint.

Since then, tango has been danced in places called *milongas*, where a complex net of codes is woven, defining the interactions between the dancers and conditioning their attitudes, their contact and their relationships. This is the human experience that I want to take a closer look at, the *milonga* as a popular celebration, as a social performance, and not just as a show.

We call *milonga* one of the three rhythms of tango, beside the tango waltz and tango itself. Yet *milonga* also refers to the dance hall where tango is danced and to the net that shapes the tango dance universe.

We say 'to dance a *milonga*' when we talk about rhythm, 'to go to a *milonga*' when we go to a dance hall and 'to belong to the *milonga*' when we refer to ourselves as members of the worldwide net of people who share the passion of dancing tango.

I Empathy and Body Language: Matches and Mismatches, Contact and Communion

Two bodies that are just one. An animal with two heads and four feet. A mythological creature, half man and half woman. A monster that embraces itself. A mingling of legs that avoid and rub against each other. A mosaic of pale and dark skin, covered and naked legs, strong and delicate arms.

From head to toe, dancing tango engages the anatomy of the entire body. Not just individual bodies but perfect fits, ensembles of two bodies of complementary signs.

Their arms shape a frame that holds them in the embrace. Yet she may barely rest the back of her hand on his shoulder, in a gesture of pretended indifference. Or he may keep her in a rigid hold that feels more like domination than protection.

Dancing tango is not just about coordinating steps to the rhythm of the music. There is a feeling, a personality that is expressed in the body, in the attitudes, moves and stillness of the real tango dancer.

Every dancer engages with his or her own identity and experiences, and is, at the same time, ready to 'read' the partner's experience empathically. The tango couple puts on stage the living experience of being oneself with each other.

And in the *milongas*, tango is danced with different partners, creating a sequence of encounters, communions and separations.

At first, there is the invitation. The tie is created by staring. For both men and women, the gazes avoid each other or meet. They might be shy, brazen or insolent.

The man is the one who approaches the woman. She stands up, and he goes to meet her. Sometimes there is a 'hello', even a kiss on the cheek if they already know each other, yet if not, they just embrace each other without saying a word.

They meet as if they had never seen each other before, and they will part as if they were to never see each other again. Like two strangers, even if they have danced together many times before, they will start the ritual once again.

And in the experience of a new partner... the soothing embrace of a well-known body or the disquieting embrace of a stranger. Breath, warmth and shapes that are assembled. Distance and nearness, matches and mismatches. At times a magical communion, at other times an uncomfortable wrestle.

To lead, to be led, empathy appears in different ways. A complex and subtle form of perception and knowledge, the empathic skill requires a particular mindset, and it achieves its richest nuances in the union between a man and a woman.

In any partner dance, and especially in this particular dance made of corporal and emotional questions and answers, this ability can be detected, trained and developed.

And this appears in each physical dialogue, showing the different styles of communication between man and woman.

In tango, as in life, the only control of time a woman has over a man is to slow him down, never to rush him. That is her art. While the man moves forward, the woman resists, although without much conviction.

Yet sometimes, the empathy fails. Some women are hasty and answer before their partners finish asking, or they leave their questions unanswered. Some men express themselves with difficulty or indecision. She then has to 'translate', and the meaning reveals itself to her delayed by a tenth of a second. As she is too alert, she gets tired and enjoys herself less.

There are some laconic dancers whose movements are bare and austere. Some dazzle you with their art. Others are so flowery that they appear cloying. Not to mention the amateurs who dance a monologue that they have learned by heart, do not know how to lead and who, when their partner cannot follow, say patronising things like: 'You didn't know that step, did you?'

Women's poetry deserves a chapter of its own. They are supposed to let themselves be led. Yet some resist, and who knows if that is because they are shy or because they act in an ill-timed attack of feminism. Others are merely dragged along, displaying a passivity that looks more like resignation than commitment.

However, some women manage to keep up the dialogue while putting their own energy into the dancing, introducing an embellishment from time to time, playing subtly with distances and gestures, and giving the floor, their accomplice, caresses they do not dare give the man.

In fact, the experienced *milonguero* does not even need to lead. He takes her firmly in his arms and cradles her on his chest. He lifts her and carries her off, 'asleep', guiding her with the rhythmic bellow of his own breathing.

II The Embrace, the Transmission of the Emotional Experience and the Moves: This Is How You Tango

An encounter that begins by looking, that continues by embracing and displays itself by dancing. A counterpoint between experience and creativity, balance and sensibility, friendly communication and elusive seduction.

From the moment of the embrace, the quality of the surrender is silently accorded. How close, how tight, the way the heads come together, the pressure of his arm around her waist, the weight of her arm around his neck. Enveloping, caressing, or merely a hand on the partner's shoulder, barely brushing against it.

The contact of the heads is the first indicator of intimacy. In general the woman defines where and how she places her face in relation to the man's. If she turns it in the same direction as his, she approaches his mouth. With her head to the left, on his shoulder, she is closer to his ear, and he to hers. Although in that same position she could also shift slightly backwards and move nearer his mouth.

The man proposes the type of contact between the torsos, whether he faces her with his whole chest or whether they form an open angle, whether he offers her a rigid 'wall-like' plane or shapes a hollow for her that she can settle in comfortably.

Straight away, when taking the first move, he defines the length of the steps and the strength of his own energy. She takes the bet and answers back with her own energy.

They dance together, sharing full and empty spaces. They both listen to the other's body, guess where the other's feet are, register their emotions, sometimes their anxiety and occasionally their surprise. They convey their experiences in a secret dialogue of questions and answers. Begging, bargaining or demanding, with reserve, modesty or distrust.

They neither look at each other nor talk. If words are needed, it is because the body language is failing.

She never takes the initiative, she can just slip into a whim that does not disrupt the continuity of their movement. Feeling his intention, she holds back a bit, creating suspense and a light tension, just to prove that she is there and that he is not dancing alone. This brief delay assures the emotion that does not exist in stage tango where the synchronicity is perfect because the choreography has been carefully rehearsed and memorised. At the *milonga*, an unforeseeable time and space are at stake, that form the secret of tango, where the good dancer improvises his moves and his way of passing across the music.

They look like one being, body and soul. But, as it is often said, 'it takes two to tango'. And yet, two is not enough. In this celebration, the couple is not dancing alone.

They dance with the music, slow or wild. With the orchestra and its unique style, they follow the rhythm or the melody, the bandoneon or the violin.

They dance with the singer who whispers snippets of dreams or nightmares in their ears.

They dance with other couples, in a shifting circle that multiplies its own energy. They also dance with the external gaze of a real or imaginary audience that holds and supports them.

It is a subtle balance of relations none of which should prevail. The selfish man who dances alone deprives his partner of the union she longs for. The couple closed in on itself gets isolated from the others and the sacred fire without adding their own flame to the tribal dance. Those who only show off betray their privacy.

But when all the parties have been called together, the communion is perfect. Mystery of the bodies in harmony, magic of the tango, the emotion is intense and complete, embracing body and soul.

In a strange contradiction, they all wish that the tango will go on forever and, at the same time, that it will soon come to an end, fearing that a stumble might break the spell.

The last note dies out; they extend the embrace for a few more seconds. When the experience is extraordinary, words are out of place; they hesitate to look at each other, moved and frightened by such surrender.

And this miracle of creativity reappears in every couple, in every piece, in a unique and original expression.

III Men and Women, the Differences Game, Eroticism, Seduction and Show-off: the Meaning of the *Milonga* Experience

The man, crouched, handcuffed and 'tamed' for many hours behind the steering wheel, at the desk or the counter, arrives at the dance hall in order to open up, 'spread out' and express himself. It is his chance to be unique and break with the rules of the herd.

The woman, chasing the kids, the man, running around the supermarket and going after her long proclaimed liberation, finds in the dance the time to dream and abandon herself in the arms of someone else, released from the duty of making her own decisions for a while. Cradled, protected and guided along, it is absolutely legitimate for her to evade the obligation to be independent.

At the same time she acquires new rights: she can sit alone and stare openly at the man she wants to dance with, and she gets to embrace a stranger, and another and another...

Last bastion of intimacy and sensuality, dancing tango means recovering the experience of depending physically and emotionally as well as fulfilling the dream of feeling and knowing in one's own flesh the emotions of the partner.

And eroticism always forms an implicit background, otherwise there is no real tango. To start with, there is the secret autoeroticism in the intimate contact with one's own body. Eroticism can, again, be found in the couple, allowing the recovery of what they have missed or the discovery of what they cannot find in everyday life. The man feels even manlier, and the woman feels even more feminine.

But because of this open display of sensuality, the *milonga* needs strict codes.

As in every activity performed outside social conventions, codes replace laws. In tango, a consented ritual of social transgressions, precise codes to regulate the relationship between the individual members of the group.

And in the *milonga*, the physical contact between individuals who have only just met is the shared transgression that is only protected by the space of the dance floor and the time a tango lasts.

Good manners require that the hand can slip only briefly before it reaches its proper position – unless there is a joint, consented intention to caress and to get intimate with each other.

Hardly ever will these unwritten codes be broken, but there is still a risk of non-consented touching: seeking the contact of the pelvises, sliding the torsos and rubbing them against each other, slipping one's arm like in a caress, bringing one's mouth closer to the partner's mouth, putting the arms around the woman's back in order to brush her right breast. All of these transgressions are forms of harassment where the explicitness disrupts the harmony of the invisible bond.

But if all goes well, the man may, for a brief moment, fulfil his secret fantasy of owning a woman's body. And she can also have the experience of being a part of the man's body.

Passion or friendship may come later, but dancing is, in itself, an act of unconditional love, not only of technique, not even of sensuality. She tries to please him in every way. He tries to make her feel good. They are simply there, aware of themselves and of their partner, and they enjoy it, whether they are being watched or nobody sees them.

However, the true passion is the tango itself. For her, dancing with many men is like constructing from them all the figure of the ideal man, a generous and lavish lover. That is why it is hard for any single man to keep exclusive possession of a woman.

For the man, dancing with many women is like hugging in his arms all the women on earth.

A game for adults, but also the continuation of childhood games, when any boy could be a noble knight or a daring pirate and every girl could become a shy princess or a cruel witch.

Incorporating male chauvinism and the eternal feminine, tango is a generous and perfect partner that allows for the creation and display of every fantasy. It is an ally, an accomplice, a lover and a source of endless pleasure and new sensations.

It is a great stage of two bodies for which you can write your own script, by breaking old prejudices and long-standing codes. The dream of being a man and a woman, tender and cruel, docile and dominant. Tango stands between the allowed and the forbidden, sex and tenderness, body and soul.

IV The Dance Codes, the Ballroom, the Ties

In the seventies, during the dictatorship in Argentina, tango was not banned, but it 'went missing'. Even if tango music was played, the repression, as always, had its effect on the body. As the basis of eroticism, aggression, instinct and life, the body is always potentially subversive and threatening for any totalitarian system. In those years, the dance survived in the shadows, as a secret ritual shared only by some initiated people.

In the last decade, tango was not only reborn in Buenos Aires but also adopted all around the world. A possible explanation for this revival is that tango seems to be an 'answer' to different individual and social needs.

In today's world of physical loneliness, where people look for friendship, sex and emotion online, dancing tango offers the opportunity of a living encounter, body to body, and it provides, at the same time, the necessary space to live emotional, sensual, social and artistic experiences.

A concentrated sample of life, the *milonga* condenses everything and amplifies the details: characters, social tightness, diversity.

When arriving at the *milonga*, one enters a space both familiar and mysterious. The outside world gets blurry or fades. At this moment, the strong ties with real life and the bonds that connect us with our family, work, commitments, close friends and even with our own identity dissolve. Even worries and affections disappear.

Instead, multiple links are activated, which are random, intense and generally ephemeral, links connecting the circumstantial dance partners, the companions at the table, the dancers that share the same dance floor.

Eventually, some bonds can evolve into friendship, an alliance to perform or teach together, maybe to start a romance.

However, for most people who frequent the *milongas* the sediment, what lasts of these links, ends up constructing a strong weave made of long-lasting friendship and sense of community.

Just like in a large living organism where many processes and experiences occur at the same time, every surviving skill is activated to the fullest at the *milonga*. Intuition, empathy, the peripheral glance and holistic perception are brought to light and trained.

In the *milonga* there is no way to hide: some things are told, others are seen and others again are simply common knowledge. Like in a huge fishbowl, any little wave of emotion or tension will hit the others. Love, anger and revenge are felt in the nuances of gestures, glances and attitudes. Once you are part of the *milonga* there is no escape. If you love or stop loving, if you are happy or sad, everybody will be both your witness and accomplice.

Being a *milonguero* is a cult. With their unique way of seeing life, they are another urban tribe, like the hippies or the rock 'n' rollers.

V From the *Milonga* to the World: Tango Networks, an Urban Tribe

Like in every informal network, tango allows a practice and a belonging that has no boundaries. The same happens with some sports and hobbies: chess, golf, collecting things, which have gained followers around the world. With these activities, people not only can share their passion, but also find the opportunity to socialise.

In tango, moreover, you involve your body and soul. Empathy and wordless communication are developed; you find out about the subtle bonds between a man and a woman, and all this happens in an emotional, artistic and celebratory context.

If tango can be a place to socialise for single people, it is a shared creative activity for couples that helps their communication and keeps their sensuality alive. And in the social fabric, it is a way to reinforce participation and to rebuild broken bonds.

In the history of civilisations, music has always been the emotional and artistic link between communities and cultures. If music means communication, the art of dancing tango is an original way to exchange meaning with other histories and geographies, crossing over the language barriers and favouring the integration of diversity.

And tango also means encounter, the meeting of emotions and experiences. Of people with different identities, backgrounds and languages who can experience, through dance, the contact of bodies, improvisation and the opportunity to show their individual originality.

It is also the appreciation of masculine and feminine roles and of belonging to a human group that is held together by the same passion. Movement and consolidation, fusion and expansion appear both in the couple's dynamics on the dance floor, shared with many other dancing couples, and in the shared universe of tango followers all around the world.

If the world of tango reflects an essential part of the *porteños*' idiosyncrasy, its music, its dance and its poetry contribute with aptitudes and values to today's civilisation. From subculture to transculture, tango is freedom and transgression, and that is why it belongs to the entire world.

Translated by Lorena Kalwill

References

Benedetti, Héctor Ángel, *Las mejores letras de tango*, Buenos Aires 1998.
Del Priore, Oscar, *Yo, Gardel*, Buenos Aires 1999.
Del Priore, Oscar/Amuchástegui, Irene, *Cien tangos fundamentales*, Buenos Aires 1998.
Dinzel, Rodolfo, *Il tango. Quell'ansiosa ricerca della libertà*, Stuttgart 2002.
El Tangauta, column *La vida es una milonga*, no. 62–101.
Ferrer, Horacio, *El libro del tango*, Buenos Aires 1977.
Gache, Rodolfo, *Baile y filosofía*, Buenos Aires 1928.
Gobello, José, *Nuevo diccionario lunfardo*, Buenos Aires 1998.
Gobello, José, *Tangos, letras y letristas*, Buenos Aires 1996.
Ostuni, Ricardo, *Viaje al corazón del tango*, Buenos Aires 2000.
Pujol, Sergio, *Historia del baile*, Buenos Aires 1999.
Richepin, Jean, *A propósito del tango*, Buenos Aires 1988.
Romano, Eduardo, *Las letras del tango*, Rosario (Santa Fe) 1994.

Notes on Contributors

Sonia Abadi is Professor at the Department of Psychoanalysis at the University of Belgrano; founder, coordinator and seminar leader of the module "Pensamiento en Red®" in various institutions and businesses. She received her PhD in Psychoanalysis at the University of Buenos Aires. She is the author of *El Bazar de los Abrazos: Crónicas milongueras* (2001) and of the series *La vida es una milonga*, as part of the magazine *El Tangauta* (1999 – 2003). Sonia Abadi has also created radio programs and produced various theater plays and events about Tango.

Ann Cooper Albright is Professor of Dance and Theater at Oberlin College, performer, choreographer and feminist scholar. She is the author of several books such *as Engaging Bodies: The Politics and Poetics of Corporeality* (2013), *Modern Gestures: Abraham Walkowitz Draws Isadora Duncan Dancing* (2010) and *Choreographing Difference: the Body and Identity in Contemporary Dance* (1997). She is co-editor of *Moving History/Dancing Cultures* (2001) and *Taken By Surprise: Improvisation in Dance and Mind* (2003). Ann Cooper Albright is currently working on an interdisciplinary project entitled *Gravity Matters*, which looks at contemporary embodiment after 9/11.

Marie-Luise Angerer is Professor of Media and Cultural Studies at the Academy of Media Arts Cologne. She was Guest and Visiting Professor at the universities of Bochum, San Diego, Santa Cruz, Ottawa, Sydney, Berlin and Budapest. Among her publications are: *Gender goes Life: Die Lebenswissenschaften als Herausforderung für die Gender Studies* (2008, co-editor Christiane König), *Vom Begehren nach dem Affekt* (2007), *Future Bodies: Visualisierung von Körper in Science und Fiction* (2002, co-editors Kathrin Peters and Zoë Sofoulis), *Body Options: Körper, Spuren, Medien, Bilder* (1999), *The Body of Gender* (ed. 1995).

Gabriele Brandstetter is Professor of Theater and Dance Studies at Freie Universität Berlin. She is co-director of the International Research Centre "Interweaving Performance Cultures." Among her numerous book publications are: *Tanz-Lektüren: Körperbilder und Raumfiguren der Avantgarde* (1995, 2013), *Dance [and] Theory* (2012, co-editor Gabriele Klein), *Notationen und choreographisches Denken* (2010, co-editors Franck Hofmann and Kirsten Maar), *Prognosen über Bewegungen* (2009, co-editors Sibylle Peters and Kai van Eikels), *Tanz als Anthropologie* (2007, co-editor Christoph Wulf), *ReMembering the Body* (2000, co-editor Hortensia Völckers).

Kornilia Chatzimasoura is a social anthropologist. She studied Social Anthropology and History (University of the Aegean, Greece) and continued her studies with a Master degree (DESS) in Research Methods in Social Sciences (Department of Sociology of the University of Poitiers) and a Master degree (DEA) in Biocultural Anthropology (University of Aix-en-Provence). In 2011 she presented her PhD *Become tanguero in the years 2000 in France: a sensitive quest of authenticity* (LASMIC, University of Nice). Her main research interests include: Anthropology of Dance, Anthropology of Sexes, Identity, Authenticity, Performativity, Feelings, and Emotions.

Robin Curtis is Professor of Theory and Practice of Audio-visual Media at the Heinrich-Heine-University in Düsseldorf. Born in Toronto. Filmmaker, curator, and film and media scholar. Feodor-Lynen Fellow, Alexander von Humboldt Foundation 2008–2011 at SUNY Buffalo, NY, USA. Adjunct Professor at New York University, Global Academic Center in Berlin, 2009–2012. Research Fellow at the Collaborative Research Center "Cultures of Performativity" (Freie Universität Berlin), 2002–2010. Her recent publications include: *Conscientious Viscerality: The Autobiographical Stance in German Film and Video* (2006) and *Einfühlung: Zu Geschichte und Gegenwart eines ästhetischen Konzepts* (2008, co-editor Gertrud Koch).

Gerko Egert is a Research Associate at the International Research Center "Interweaving Performance Cultures" at Freie Universität Berlin. He studied Theater Studies and Sociology in Berlin and Potsdam and worked at the cluster of excellence "Languages of Emotion" (Freie Universität Berlin, 2010–2012). In 2012 he was Visiting Fellow at the SenseLab, Concordia University, Montréal. He is currently working on his PhD on touch, movement, and affect in contemporary dance.

Corinne Jola is currently a Postdoctoral Research Fellow at the Neurospin Centre and at Paris 8 after several years of research in the UK (e.g., University of Surrey, University of Glasgow, www.watchingdance.org). She received her MA in Choreography at Laban Trinity College London and her PhD in Cognitive Neuroscience in Zurich. Her research interests are the representation and perception of the human body during complex movement patterns, in particular in relation to the performing arts. She uses contemporary brain imaging techniques (e.g., fMRI, TMS, EEG) in combination with choreographic practices to study how esthetic movements are perceived by a diverse group of spectators.

Krassimira Kruschkova is Head of the Theory Centre at Tanzquartier Wien, Professor of Performance and Theater Studies at the Academy of Fine Arts Vienna. Among her publications are: *SCORES #0 – #4: Uneasy Going* (2013) *What Escapes* (2012) *Touché* (2011), *The Skin of Movement* (2010) (co-editors W. Heun, S. Noeth, L. Mehanovich, and M. Obermayer), *Dies ist kein Spiel* (2009, co-editor A. Böhler), *Ungerufen: Tanz und Performance der Zukunft/ Uncalled. Dance and Performance of the Future* (2009, co-editor S. Gareis), *It Takes Place When it Doesn't: On Dance and Performance Since 1989* (2006, co-editors M. Hochmuth and G. Schöllhammer), *Ob?scene. Zur Präsenz der Absenz im zeitgenössischen Theater, Tanz und Film* (ed. 2005).

Hilge Landweer is Professor of Philosophy at Freie Universität Berlin. Among her numerous book publications are: *Klassische Emotionstheorien: Von Platon bis Wittgenstein* (2008, co-editor Ursula Renz), *Philosophie der Gefühle: Von "Achtung" bis "Zorn"* (2007, co-author Christoph Demmerling), and *Scham und Macht: Phänomenologische Untersuchungen zur Sozialität eines Gefühls* (1999).

Niklaus Largier is Professor of German and Comparative Literature at the University of California in Berkeley. He studied German Language and Literature, Russian Language and Literature, and Philosophy at the University of Zurich. His research deals with the history of medieval and early modern German literature, especially questions of the relations among literature, philosophy, theology, and other fields of knowledge. His most recent books include: *Lob der Peitsche: Eine Kulturgeschichte der Erregung* (2001, translated in 2007: *In Praise of the Whip: A Cultural History of Arousal*), *Die Kunst des Begehrens: Dekadenz, Sinnlichkeit und Askese* (2007).

Erin Manning holds a University Chair in Relational Art and Philosophy at Concordia University in Montreal. She is the director of the SenseLab, a laboratory that explores the intersections between art practice and philosophy through the matrix of the sensing body in movement. Her publications include: *Always More Than One: Individuation's Dance* (2013), *Relationscapes: Movement, Art, Philosophy* (2009), *Politics of Touch: Sense, Movement, Sovereignity* (2007), and *Thought in the Act: Passages in the Ecology of Experience* (together with Brian Massumi, forthcoming).

Brian Massumi is Professor in the Department of Communication Sciences at the University of Montréal. He received his MA and PhD in French Literature from Yale University and completed postdoctoral work at Stanford University. He is the author of several books, among them: *Semblance and Event: Activist Philoso-*

phy and the Occurent Arts (2011), *Parables for the Virtual: Movement, Affect, Sensation* (2002), *A User's Guide to Capitalism and Schizophrenia: Deviations from Deleuze and Guattari* (1992). Together with Erin Manning he founded the journal *Inflexions: A Journal of Research-Creation.*

Katherine Mezur holds a PhD in Theater and Dance from the University of Hawai'i Manoa. She is an independent Scholar and dramaturg. She was a Research Fellow at the International Research Center, "Interweaving Performance Cultures," at Freie Universität Berlin. Her research focuses on Asia-Pacific and transnational performing arts, media, and gender studies. Her books and articles include: *Beautiful Boys Outlaw Bodies, Devising Female-likeness on the Kabuki Stage* (2005), forthcoming: *Kawaii: Cute Girl Cultures in Contemporary Japanese Performance and Media Art*; *Stranger Communities: Art Labor and Dancing In-Transit*, and *Choreographing Zeal in Radical Youth Performance.*

Jean-Luc Nancy is an emeritus Professor of Philosophy at the European Graduate School and at the University of Strasbourg. He had numerous guest professorships in Berkeley, Berlin, Irvine, San Diego etc. He is the author of a great number of internationally renowned books: *God, Justice, Love, Beauty: Four Little Dialogues* (2011), *Corpus* (2008), *Noli Me Tangere: On the Raising of the Body* (2008), *Being Singular Plural* (2000), *The Sense of the World* (1998).

Mark Paterson is in the Department of Communication at the University of Pittsburgh, and Scholar in Residence at the MacAnulty School of Arts, Duquesne University. His books include *Consumption and Everyday Life* (2005), *The Senses of Touch: Haptics, Affects and Technologies* (2007), and is co-editor of *Touching Space, Placing Touch* (2012). He is currently finishing *Seeing with the Hands: Blindness and Philosophy After Descartes and Diderot.* His next book project is *How We Became Sensorymotor: A History of the 'Muscle Sense'*. His 'Sensorymotor' research website is www.sensesoftouch.co.uk.

Dee Reynolds is Professor at the School of Languages, Linguistics, and Cultures at the University of Manchester. She studied French and German at Trinity College and graduated from the Sorbonne-Nouvelle with a Maîtrise ès Lettres Modernes and obtained her PhD from the University of East Anglia in 1987. She is interested in Stéphane Mallarmé's writings on dance, particularly his response to the dancer Loïe Fuller. This led to the discovery of modern dance and the exploration of rhythm and kinesthetic imagination in dance, which were the starting points for her subsequent research on energy and performer-spectator relations in dance.

Sabine Zubarik is currently working on her postdoctoral project on concepts of simultaneity in literature and film. She also is a Tango dancer and teacher with professional international training. She studied Comparative and Hispanistic Literature and Religious Studies at the University of Erfurt (Germany), with exchange semesters in Istanbul, Córdoba, and Santa Barbara. From 2009 until 2011 she was Research Assistant at the cluster of excellence "Languages of Emotion" at Freie Universität Berlin, on the project "Touching and to Be Touched: *movere* in Dance" and received her PhD in Comparative Literature in 2013.

Bei Fragen zur Produktsicherheit wenden Sie sich bitte an:
If you have any questions regarding product safety,
please contact:

Walter de Gruyter GmbH
Genthiner Straße 13
10785 Berlin
productsafety@degruyterbrill.com